Introduction to the Design and Analysis of Experiments

Geoffrey M. Clarke, M.A., Dip.Stats., C.Stat.

Honorary Reader in Applied Statistics, University of Kent at Canterbury
and Consultant to the Applied Statistics Research Unit

Robert E. Kempson, B.Sc., M.Sc., Ph.D., C.Stat.

formerly of the Applied Statistics Research Unit, University of Kent at Canterbury
and of Wye College, University of London

A member of the Hodder Headline Group
LONDON • SYDNEY • AUCKLAND
Copublished in North, Central and South America by
John Wiley & Sons, Inc., New York • Toronto

First published in Great Britain in 1997 by
Arnold, a member of the Hodder Headline Group,
338 Euston Road, London NW1 3BH

Copublished in North, Central and South America by
John Wiley & Sons Inc., 605 Third Avenue,
New York, NY 10158-0012

British Library Cataloguing in Publication Data
A catalogue record for this book is available from the British Library

Library of Congress Cataloging-in-Publication Data
A catalog record for this book is available from the Library of Congress

ISBN 0 340 64555 5

ISBN 0 470 23409 1 (in the Americas only)

Typeset in 10/12pt Times, printed and bound in Great Britain by J. W. Arrowsmith Ltd, Bristol

Contents

Preface

Applications of the basic ideas and methods contained in the *Design and Analysis of Scientific Experiments* have increased greatly over recent years, and it is safe to claim them as one of the most important parts of applied statistics. The original work was done by R. A. Fisher, F. Yates and others working in agricultural research, and was quickly extended to industrial research when people realized that many of their problems were similar in principle to agricultural ones, so the same ideas were relevant for them too. Developments have since spread to many other fields of study, and new applications continue to arise.

Many statisticians, including the authors, have been brought up on the excellent texts written for agricultural and biological work, by Cochran and Cox, Kempthorne, Yates and others, and those for industry by O. L. Davies and the ICI group. Besides their theoretical content, these texts have a practical approach which is lacking in some published more recently. But they do, inevitably, contain topics that are seen nowadays as of specialist, rather than general, interest, as well as lacking a few of the useful later developments in both content and approach. We have been involved in preparing training courses for the Applied Statistics Research Unit of the University of Kent, given to companies and organizations in this country and in various parts of the world, as well as in teaching undergraduate and postgraduate courses. We have tried to use this experience to separate out what seem to be the essential basic contents for everyone, and to combine with these some of the most useful additional topics in biological, industrial, medical and environmental experimentation. Our treatment begins at a level that is suitable for undergraduate courses in mathematics and statistics, contains the topics we have found time to include in a one-term postgraduate course and introduces a few others.

This text is thus offered as an introduction to what is now a wide subject, and once its contents have been grasped, anyone going on to work in a special field of application will have to read books and papers about that field. But unless the basic principles behind good experimental thinking, planning and action really have been grasped, there is a danger of common fallacies growing up in particular limited special fields – we have come across some of these! Even in a text that is much smaller than it could have been, we therefore make no apology for taking up space in the first few chapters to stress the need for proper planning of experimental work and for clear statements of aims and objectives. Without these, efficient design and analysis are not possible. Experimental resources are scarce far more often than they are plentiful, and always costly, so efficiency is essential.

We must assume in these days that most statistical analyses will be done on a computer with a suitable statistical package program. Versions of MINITAB are often available in universities and colleges, SAS, SPSS and several others in industry and business; and GENSTAT is often used by specialist statisticians. Our (insoluble) problem is therefore which to discuss. A good package has a readable, understandable manual which explains how to use each major operation in it. What the user then needs in addition is a clear idea of which statistical method should be applied to the problem under study, exactly what analyses need to be done, and what items of output will be useful (even essential) in writing a report on the experiment. We hope that this text meets these needs, and that our discussion of computer output, for which we have chosen SAS, gives useful illustrations. We will welcome readers' comments on this approach to the problem; what we hope to help readers to avoid is blindly following instructions which may not be clear, or completely understood – or, even if they are, do not really apply to their situation. We have, of course, come across this too!

We acknowledge the help and experience of many colleagues who have taken part in courses with us, and who have helped prepare material upon which some of this text is, inevitably, based – though by now after repeated editing we should be hard pressed to say who first wrote what, and of course the responsibility for what is presented here is ours alone. Particular mention must, however, be made of Professor S. Clifford Pearce, whose ideas and influence have been very much to the fore in the courses the Unit has given, and who has developed statistical computer packages to put them into effect.

We are indebted also to the Literary Executors of the late Sir Ronald A. Fisher, FRS and Dr Frank Yates, FRS and to Oliver and Boyd Ltd, Edinburgh, for permission to reproduce parts of Tables from the sixth edition of *Statistical Tables for Biological, Agricultural and Medical Research*, and to the University of London, the University of Kent at Canterbury, the Institute of Statisticians and the Royal Statistical Society for the use of past examination questions.

G. M. Clarke

March 1996

R. E. Kempson

1
Collecting data by experiments

1.1 Introduction

There are three main ways of collecting data to use in statistical inference: by surveys, through simulation and in controlled experiments. A very widely used method is the *survey*, where people with special training go out and record observations of the number of vehicles travelling along a road, the area of fields that farmers are using to grow a particular food crop, the number of households that own more than one motor vehicle, the number of shoppers using a supermarket, the number of passengers using public transport, and so on. Here the person making the study has no direct control over generating the data that can be recorded, although the recording *methods* certainly need care and control (Barnett (1991), Cochran (1992), Moser and Kalton (1985)).

Then there is *simulation*, where, for example, a computer model for the operation of an industrial system is set up, and in which an important measurement is the percentage purity of a chemical product. A very large number of realizations of the model can be run, in order to look for any pattern in the results. Here the success of the approach depends on how well that measurement can be explained by the model, and this has to be tested by carrying out at least a small amount of work on the actual system in operation; some refinement of the model may be possible as the study proceeds. There can be some overlap with experimental methods such as those described in Chapter 9.

1.2 Experiments

An *experiment* is possible when the background conditions can be controlled, at least to some extent. We may be interested in choosing the best type of grass seed for use on a sports field. The first stage of work will be to grow all the competing varieties of seed at the same place and make suitable records of their growth and development. The competing varieties should be grown in quite small units close together in the field, as in Fig. 1.1; this gives much more control of local environmental conditions than there would have been if one variety had been placed in a strip in the shelter of the trees, another close by the stream and a third out in the more exposed centre of the field as in Fig. 1.2. Experiments in agriculture and horticulture, comparing different varieties of a crop or different ways of treating it during growth, are best carried out using reasonably small units of land, or *plots* as they are often called; we shall also call them *experimental units* or simply *units*. There is no 'right' size for a unit; it depends on the type of crop, the work that is to be done on it and the measurements that are to be taken. The measurements, upon which inferences are eventually going to be based,

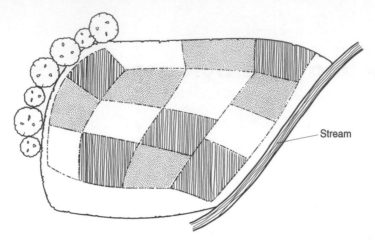

Fig. 1.1 Competing varieties of grass laid out in small unit plots. Key: varieties □ ▦ ▥

need to be taken as accurately as possible. The unit must therefore not be so large as to make recording very tedious because that leads to errors and inaccuracy. On the other hand, if a unit is very small there is the danger that relatively minor physical errors in recording can lead to a large percentage error in the figure recorded. Experimenters, and statisticians who collaborate with them, need to gain a good knowledge of their experimental material, or units, as a research programme proceeds.

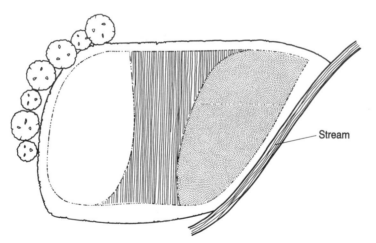

Fig. 1.2 Competing varieties laid out in large strips. Key: varieties □ ▦ ▥

The growth of the seed used for the sports turf will depend on several factors, such as the type of soil – sandy, clay, acid, alkaline, etc. – and its drainage, the climatic conditions during growth and the amount and constitution of fertilizers that may be applied to the land before and during the period of growth. There may be further complicating factors such as the need for a selective weedkiller to remove species that compete with the grass and damage the final quality of the turf. Some factors cannot easily be controlled, and the results of an experiment will, for example, only apply to the particular soil type used: if we carry out an experiment on sandy soil we must not assume that the results will be the same on clay but must repeat the experiment under those conditions. Some factors are not under our control at all, especially climatic factors, but we can at least observe their effects on all the varieties; some varieties may do relatively well in a dry season, compared with the others, but not in a wet season.

However, the fertilizer mixture given to the growing seed *is* under our control; we may vary it deliberately and apply it carefully to the ground so that we know exactly how much each unit receives. The amount of fertilizer is a *treatment* which we have given in the hope of affecting growth. Experiments are carried out to compare different treatments. As we shall see, the word 'treatment' is used in a technical sense to stand for *any* different régimes that are being compared.

1.3 Measurements of yield or response

What exactly should be measured in order to assess the effect of a treatment and to compare different treatments? This requires careful thought at the planning stage of any experiment or programme of experiments, because there can be several different aspects of the effect that are important or interesting. Suppose that we are experimenting with a fruit crop, which is to be sold in shops and markets. The total weight of crop from an experimental treatment is obviously important – the more the weight of the crop, the greater the economic return to the farmer or grower. However, the size of individual fruit is often important also; this can be measured at harvest by weighing random samples of N fruit, where N is a convenient number depending on what fruit we are dealing with. The highest quality of fruit, often called 'Grade 1', will fetch a higher price and so this needs to be recorded separately at harvest, dividing total weight into Grade 1 and Other: the actual definition of Grade 1 will depend on the crop and any local regulations that may be in force for marketing it. Freedom from disease or blemish, due to pests or to some fungus, is likewise important. Therefore samples of fruit have to be assessed for percentage blemish, or disease damage, which can often be done by visual inspection and the use of a somewhat rough-and-ready scoring scale: a large number of fruit scored in this way will give a good average measure of visual appearance. Furthermore, if some of the crop is to be used for making fruit drinks, then a chemical analysis also has to be done on samples from the crop to measure the vitamin content.

If this stage of the planning process is not done well and carefully, some very important measurements might be overlooked until it is too late to obtain them; and so some of the questions which the written report of the experiment should be addressing cannot be answered because an important component in the jigsaw of measurements is not there. Although most of our later examples quote just one measurement for analysis from an experiment, in practice there will usually be several; each may be analysed separately, by the same form of analysis, and the results brought together in the written report.

The measurement taken on a unit plot when given an experimental treatment has often been called the *yield*. Obviously this is often appropriate in agricultural experiments, and it can be so in some industrial trials too. But measurements can be taken on a wide variety of characteristics, as we have just noted, and so we will usually describe a measurement as the *response* of that unit to the treatment it received.

1.4   Natural variation in data

Almost always, the analysis of the data collected in an experiment is – or should be – statistical. Besides the differences we are studying, there will be a random component of variation among the units of material (the plots) used for the experiment. One very good reason why so much of the basic work in designing and analysing experiments took place in biological and agricultural research centres is that random variation is much more obvious in that context. Even in a modern glasshouse, where environmental conditions are controllable much more closely than they once were, it is most unlikely that two plants of the same species will make exactly the same amount of growth in 'identical' conditions. We can take the two plants from the same batch of seedlings raised from the same source of seed, plant them in exactly the same mixture of soil and fertilizers, in two pots of the same size made from the same material and placed side by side on a bench in the centre of the glasshouse, where they are sheltered from draughts and differences in temperature and humidity. We can make sure that we begin with two seedlings that are the same height, as accurately as we can measure them, and we can also make sure that they are treated exactly the same way and given exactly the same amount of water all the time they are growing. Even so, if we measure them again after they have been growing for a month, it is very rare to find that the heights of the two plants are still exactly the same. One will be higher than the other by an amount that is large enough to observe and is not merely measurement error; there *is* a real difference. Taking a whole population of plants, as similar as possible to one another when we first select them, there is always this natural variation. We shall call it *residual variation*, which seems a better phrase than the commonly used *experimental error* since it is not due to error as such.

The presence of such residual natural variation would make the analysis of experiments using these plants very hard if the variation had no pattern and we had no idea how large it was going to be in any particular experiment. But in most populations there is a fairly clear pattern of residual variation, and often it follows – approximately – a *normal distribution*. A good example is the study of wheat and other corn crops, where a field is sown as uniformly as possible with seed and is divided into *plots* of a standard shape and size, each plot containing a large number of individual plants. We could imagine, in theory, that we measure at the end of a season the amount (weight) of grain produced by each plant. In practice this does not really make sense because the whole field will contain some thousands of plants and each individual plot will consist of hundreds. The record of the crop for each unit plot can, however, be taken at harvest, and this is quite often done mechanically. There are at least two advantages in considering only the total crop for each plot: one is that we will be dealing with figures of a moderate size which can be recorded fairly accurately with a recording error that is small relative to the actual size of the measurement. The second advantage is that the *Central Limit Theorem* leads us to expect that the distribution of plot totals will be close to normality because the total record is the sum of contributions from a very large number of individual plants. (See, e.g. Freund and Walpole (1987), Hogg and Craig (1994) for some theory about the Central Limit Theorem.)

When we are taking a record like this, where there is a theoretical reason to expect a normal distribution, we shall carry out the analysis on that basis and use all the valuable properties of the normal distribution to develop the theory that supports the analysis. If we cannot assume normality we must use nonparametric methods that are less powerful. But in fact many responses are at least approximately normally distributed, or can be made so by transforming to another scale of measurement (page 67), so the widespread assumption of normality in experimental analysis is not unjustified.

Example 1.1. The masses of 100 animals from the same breeding population, and of the same sex, were measured to the nearest gram. The results were grouped into intervals of 20 g, as follows:

Mass (g)	70–89	90–109	110–129	130–149	150–169	170–189	190–209
Frequency	3	7	34	43	10	2	1

A histogram of these data (Fig. 1.3) shows a fairly symmetrical picture, with perhaps a slight skewness to the right. The mean of the data is 131.5 and the standard deviation is 20.0, and it is possible to calculate the frequencies that would be expected if a random sample of 100 observations had been drawn from a normal distribution with $\mu = 131.5$ and $\sigma = 20.0$ (see, e.g. Clarke and Cooke (1992)). These frequencies, given to one decimal place, are 1.8; 11.8; 32.4; 35.6; 15.5; 2.7; 0.2. Although the matching is not perfect, we find from an appropriate goodness-of-fit test that the observed frequencies do not differ significantly from these theoretical frequencies calculated from the normal distribution with $\mu = 131.5$ and $\sigma = 20.0$.

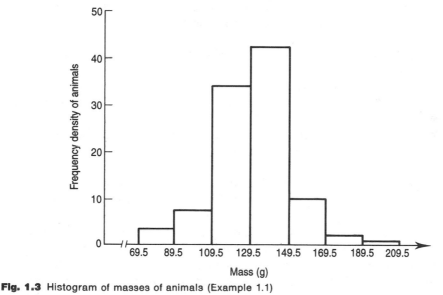

Fig. 1.3 Histogram of masses of animals (Example 1.1)

1.5 Initial data analysis

Often the number of observations available is severely limited, and there is no opportunity to test whether the data collected in one single experiment are approximately normal or not. In fact, tests of normality are not very sensitive and unless a number of routine experiments under well-controlled conditions are available, the assumption of normality cannot easily be checked as a programme of work continues. It is important to have some idea, before we begin the experiments, whether the responses being measured, on the type of units being used in the whole programme, have in the past given data that could reasonably be taken as normally distributed. In looking back at any available *historic* data, as well as making checks during the analysis of new experiments, some of the methods of graphical study that can be found in statistical computer packages are helpful: these are often called *initial data analysis*. The SAS procedure UNIVARIATE gives a test of normality.

Example 1.2. Three different treatments are tested on members of the same population of animals. The treatments are different supplements, A, B or C, to a standard diet and the response that is measured is the increase in weight over a fixed period of time. Ten animals receive each supplement and the weight increases (g) are

A:16, 11, 21, 15, 17, 13, 20, 16, 15, 18;
B:18, 30, 18, 17, 25, 19, 20, 21, 19, 20;
C:14, 5, 12, 16, 13, 18, 16, 20, 18, 22.

Figure 1.4 shows *dot-plots* of these data, drawn on the same scale beneath one another for easy comparison. As we shall see, our standard methods of analysis assume that each set of data is (at least approximately) normally distributed with the *same* variance. We therefore want to look at the following properties of the data.

● Is each set roughly symmetrical about its mean value? If not, the normal assumption may be unsatisfactory.
● Are there any outlying values which do not seem to belong to the same population as the others? If so, we will need to look more closely at those units to see if there is any reason for the response being 'different' from the others.
● Are the levels of variability roughly the same in each set? If not, we should not assume that every set has the same variance.

The methods studied in Chapter 4, based on the *residuals* in the statistical model used for the experiment, address these problems, but we can make a start now using dot-plots (Fig. 1.4).

In treatment A, there is no obvious 'non-normality'. In B, the two largest observations suggest either that these are 'outliers' or that the distribution of the weight increases under this treatment is skew. In C, we would want to look at the animal with the very small increase to see whether it may be suffering disease which is not associated with the treatment; if so, we might exclude it from any analysis, but if not we should be on the lookout for signs that some animals in this population do not do well on treatment C. Whenever possible, observations which seem to depart from the general population pattern should be followed up, by inspecting the unit again when this can be done. Excluding the two possible outliers, B would be giving less variable results than A or C, although if there is *no* genuine reason to exclude them the variance of this set of data is similar to the others.

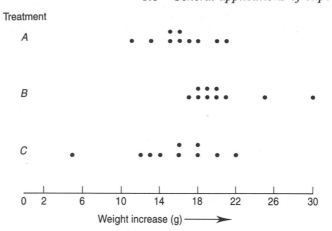

Fig. 1.4 Dot-plots of three sets of data in Example 1.2

1.6 General applications of experimentation

Most of the examples mentioned so far have been biological, involving plants or animals. However, controlled experiments are possible in very many other areas of study. Industrial research and development workers were some of the earliest to discover that the principles that applied in agricultural research very often applied in their work too. No industrial machinery can produce exactly identical results in repeat runs of the same operation, although the variation with modern machinery should be much smaller and more easily controlled than it once was. Even so, the raw materials – chemicals, etc. – used in an industrial process will show variation and this will be enough to affect the responses measured at the end of a process, such as the strength or lifetime of a component being manufactured. Some topics, such as *response surfaces* (Chapter 9), owe much of their development to industrial statisticians and their collaborators.

Human beings vary a lot! We could measure the computer keyboard skills, by some objective test, of a large group of students at the beginning of a course, and choose at random two sets from among those with scores near the average. One set is trained for a particular type of commercial work by means of a standard, well-established training scheme and the other set is trained for the same work in a new way. Provided we choose comparable sets of students at the beginning, a controlled experiment is possible, and the scores of each student at the end of whichever scheme of training they received can properly be compared as a measure of the success of each scheme. The variability is likely to be very much greater here than in an industrial experiment, and higher also than in many biological experiments. We may therefore require a fair number of people to take part in such an experiment, but even so it can be done and very useful results obtained. Many educational and training methods can be studied by controlled experimentation provided that all other factors which may affect a response can be standardized, save only for the methods that are being compared. The students must be from some recognizable 'population', with similar basic knowledge when the experiment begins and preferably with similar IQs or other indicators of their ability to learn when suitably trained. When teaching young children, for example those learning to read, age-groups must be carefully controlled and their teachers must be skilled in all the teaching methods to be used – otherwise a child's measured response may be to the teacher, not the method.

Medical research brings its own real problems in comparing different drugs that may be used for treating a particular disease or condition. It is quite a specialized area, and any reader whose work will mainly be in medical trials should consult references such as Altman (1991), Fleiss (1985), Peace *et al* (1988).

However, large trials of new drugs will necessarily involve several hospitals, several different doctors, surgeons, nurses or consultants who are administering the treatments, and will probably use patients whose condition varies in severity. All these form possible systematic causes of difference among responses, and careful planning is needed to make sure that groups of similar patients, supervised by the same people, are used to make comparisons between the experimental treatments. Different hospitals will almost certainly produce slightly different results! When combining the results from several sites in a large trial, the hope is that the *differences* between treatments will remain fairly constant even if the actual results for one treatment vary from site to site.

Another field of work involving humans is testing reactions to different stimuli, such as light and sound, to compare people of varying age, those doing various kinds of job or those working in more or less favourable environmental conditions. Reaction times can be measured by very accurate recording equipment and useful physiological and neurobiological information can be discovered through experimentation which uses small changes in a person's environmental or working conditions.

Market research can also use experimental methods, and profit greatly from applying close control to its ways of assessing advertising campaigns, display and packaging methods, special promotions, and so on. Sales of the same products in two areas, which have similar sales patterns, may be compared after a local newspaper or television promotion in one area but not in the other; here, past data on sales patterns will be needed to find suitable areas that can sensibly be compared. Financial products – saving schemes, insurance, etc. – may appeal more to one social group than another, to retired people more than others or only to the economically active. Where such groups can be accurately targeted, samples from them can be chosen at random to receive the same information with a view to comparing responses in the different groups. Alternatively, two samples from the same group can receive two different forms of marketing, and the effectiveness of the different forms can be compared. The methods in this book can be applied to a wide variety of fields of study, and the principles originally developed for biological research, together with others discovered from other areas of investigation, keep finding new applications.

1.7   Exercises

1. Draw dot-plots for the data of Example 4.3 (page 49), one for each of the four sets of data, and use them to compare the characteristics of the four sets.
2. Draw dot-plots for the data of Example 4.1 (page 37), one for each set of data. Compare these sets of data and comment on the effects of having unequal numbers of data items in the four sets.
3. Using the data of Example 2.5, draw dot-plots for
   (i) the strengths of items processed for 20 min (L),
   (ii) the strengths of items processed for 15 min (S),
   (iii) the differences ($L - S$) for the 15 mixtures.

   Comment on any similarities and differences among these dot-plots.

2
Basic statistical methods: the normal distribution

The analysis of designed experiments usually assumes that data follow a normal distribution. This is often a satisfactory model, or can be made so by transformation of the data (see Chapter 4). Two or more sets of data, from similar groups of experimental units, will be available. We will consider methods of examining two data sets in this chapter, and extend these to as many sets as necessary in Chapter 4.

2.1 Statistical inference for one sample of normally distributed data

When a random sample of n observations $\{x_i : i = 1 \text{ to } n\}$ is taken from a population in which the measurement x follows the normal distribution with mean μ and variance σ^2 ($\mathcal{N}(\mu, \sigma^2)$ is a useful shorthand for this), we can study the two main characteristics of this distribution by calculating the sample mean

$$\bar{x} = \frac{1}{n} \Sigma x_i$$

and the sample variance

$$s^2 = \frac{1}{n-1} \Sigma (x_i - \bar{x})^2$$

These give *unbiased* estimates of μ and σ^2 (Clarke and Cooke, 1992), and the two parameters *mean* and *variance* give all the information required about a normal distribution. The mean shows its *location*, where on the scale of measurement its 'centre' lies; and the variance shows its *scatter* or *dispersion*, whether it is closely concentrated about its mean or whether it is more spread out or scattered.

Inferences about the value of μ, based on the sample data, are commonly required and there are two useful approaches, the theory of which may be found in introductory textbooks (e.g. Clarke and Cooke, 1992) and is developed further in undergraduate mathematical texts (e.g. Hogg and Craig (1994) or Freund and Walpole (1987)). In a *hypothesis test* we may set up a *Null Hypothesis* about the true value of the mean in the whole population from which the sample was taken; we can specify a particular value for μ and ask whether the data have given us a sample value of $\bar{x}$ which is consistent with this specified μ. If not, then we have 'significant' evidence against this null hypothesis.

Alternatively, a *confidence interval* for the true value of μ may be set up using the sample values $\bar{x}$, s^2; this tells us what range of values of μ in the population could reasonably have led to the sample we have actually observed. We look at these methods in a little more detail here, and refer readers to the textbooks already mentioned for a full treatment.

Distribution of $\bar{x}$ for normally distributed data

It is relatively easy to show for data from any distribution that if

$$\bar{x} = \frac{1}{n}\Sigma x_i$$

and

$$s^2 = \frac{1}{n-1}\Sigma(x_i - \bar{x})^2$$

then the expectations are $E[\bar{x}] = \mu$ and $E[s^2] = \sigma^2$; and also the variance $V[\bar{x}] = \sigma^2/n$. An unbiased estimator of the variance of $\bar{x}$ is therefore s^2/n.

Some more theory (see Hogg and Craig (1994) and Freund and Walpole (1987)) is needed to show that when $x_i : i = 1$ to n are from $\mathcal{N}(\mu, \sigma^2)$ then in repeated sampling $\bar{x}$ will follow the distribution $\mathcal{N}(\mu, \sigma^2/n)$: the result is established most easily using generating functions, e.g. the moment generating function. In other words, provided we deal with normally distributed data, the mean is also normally distributed; this is not the case for other distributions and so the assumption of normality simplifies analyses considerably.

2.2   Hypothesis test

In the analysis of designed experiments when we are studying the values of the mean, we shall be unlikely to know a 'true value' of σ^2. The test of a Null Hypothesis that 'true mean $= \mu$' must therefore be based on the estimated value s^2 and the *t*-test: the statistic

$$t = \frac{\bar{x} - \mu}{s/\sqrt{n}}$$

follows $t_{(n-1)}$.

Example 2.1. The masses (g) of 12 eggs of a standard grade (size) are: 64.1, 60.2, 53.8, 67.2, 56.9, 58.6, 60.0, 66.3, 50.7, 56.0, 63.3, 58.2. Assuming that the distribution of mass in this particular population of eggs is normal, examine the hypothesis that the true population mean weight is 62 g.

The mean and variance of the masses of the 12 eggs are 59.61 and 24.8336, respectively, so that

$$t = \frac{59.61 - 62.00}{\sqrt{24.8336/12}} = -\frac{2.390}{1.439} = -1.66$$

which is not near the 5% critical value of $t_{(11)}$, which from Table I is 2.201, and so does *not* provide significant evidence against the hypothesis. We can therefore accept that this sample could have come from an egg population in which the distribution of mass had a mean value of 62 g.

Confidence interval for the mean

More information can be gained from calculating a confidence interval than from making a hypothesis test; we obtain the range of possible values of μ which, with a probability of 0.95, could have yielded the sample of data that has been observed. Using the observed sample values of $\bar{x}$ and s^2, defined as above, the *95% confidence interval* for the true value of μ is from $\bar{x} - ts/\sqrt{n}$ to $\bar{x} + ts/\sqrt{n}$, where t stands for the critical value of $t_{(n-1)}$ at the 5% (two-tail) point. With probability 0.95, this interval contains the true value of μ.

Example 2.2 Consider the same data as above, on the masses of eggs. We found $\bar{x} = 59.61$ and $s^2 = 24.8336$. We require the 5% point of $t_{(11)}$, which from Table I is 2.201. Therefore

$$ts/\sqrt{n} = 2.201 \times \sqrt{\frac{24.8336}{12}} = 2.201 \times 1.439 = 3.167$$

The interval thus extends from $59.61 - 3.17 = 56.44$ to $59.61 + 3.17 = 62.78$.

The value 62, which we 'tested' above, is included in the interval, and is therefore an acceptable value for μ. A hypothesis ('significance') test therefore will *not* declare the value 62 'significant'. But we have learnt much more than that, for we now know that *no* value between 56.44 and 62.78 would be declared significant at the 5% level.

We can also see roughly how variable our observed data were; since the interval is more than 6 units wide on a measurement of about 60 units, we are not looking at a measurement which was taken very precisely. We will not therefore be able to make highly precise statements about it. Confidence intervals are often more useful than hypothesis tests in extracting information from designed experiments, and in writing reports on them. One reason for this is that they do reflect some of the information contained in the estimate of σ^2; there has often been a tendency to concentrate almost totally on studying means, thereby losing useful additional information given by s^2.

2.3 Comparison of two samples of normally distributed data

Experiments usually involve studying several sets of normally distributed data, but the methods which apply to only two sets can be generalized and so we give these first. As above, we assume that σ^2 is not previously known but has to be estimated from the data. An essential basic assumption will be that *each* set of data comes from a normally distributed population with the *same* value for σ^2; if this is not the case, the t-distribution does not apply to the statistic we need to use in hypothesis testing and in calculating confidence intervals. The problem of 'variance stabilization' in analyses is considered in Chapter 4. In many experimental situations this assumption can be made, since changes in mean due to different treatments often do not lead to changes in variance; a discussion of this also will be found in Chapter 4.

Sums and differences of normally distributed random variables

A further theoretical result, which may be studied in the texts already mentioned in this chapter, is that when x_1 and x_2 are, respectively, $\mathcal{N}(\mu_1, \sigma_1^2)$ and $\mathcal{N}(\mu_2, \sigma_2^2)$ and are independent of one another, then

(a) $x_1 + x_2$ is $\mathcal{N}(\mu_1 + \mu_2, \sigma_1^2 + \sigma_2^2)$ and

(b) $x_1 - x_2$ is $\mathcal{N}(\mu_1 - \mu_2, \sigma_1^2 + \sigma_2^2)$.

Note the result for the variance in (b); variances are always combined by *adding* them, *never* subtracting.

This can of course be extended to the means of random samples from normally distributed variables, since they are also normal; we obtain

(c) $\bar{x}_1 + \bar{x}_2$ is $\mathcal{N}(\mu_1 + \mu_2, \sigma_1^2/n_1 + \sigma_2^2/n_2)$

(d) $\bar{x}_1 - \bar{x}_2$ is $\mathcal{N}(\mu_1 - \mu_2, \sigma_1^2/n_1 + \sigma_2^2/n_2)$

when independent random samples, consisting of n_1 and n_2 observations, respectively, have been taken from the two distributions.

Example 2.3(a). The times taken by a commuter each day to reach the workplace depend on x_1, the time for a rail journey and x_2, the time for a bus journey. The daily values of x_1 follow a normal distribution with mean 38 min and standard deviation 3 min; and those of x_2 follow a normal distribution with mean 12 min and standard deviation 4 min (allowing for the time to change from rail to bus).

The total journey time, $x_1 + x_2$, is the sum of an $\mathcal{N}(38, 9)$ variable and an $\mathcal{N}(12, 16)$ variable. It will therefore follow the distribution $\mathcal{N}(38 + 12, 9 + 16)$, i.e. $\mathcal{N}(50, 25)$. The mean of the total journey time is 50 min and the standard deviation is 5 min.

Example 2.3(d). For our present purposes, the most important of the four results (a)– (d) quoted above is the last one; we often wish to study differences between means.

Returning to the data on egg masses (page 10), let us label that sample number 1, so that $\bar{x}_1 = 59.61$, $n_1 = 12$, $s_1^2 = 24.8336$. A second population of eggs of the same grade, from a different breed of bird, also had a random sample of 12 eggs taken from it, and their masses were 58.7, 62.8, 67.2, 68.0, 68.4, 64.4, 60.9, 63.2, 61.6, 63.5, 74.3, 71.2 g. For these, $\bar{x}_2 = 65.35$, $n_2 = 12$, $s_2^2 = 20.6918$. We can compare the two sets of data in ways that are similar to those we have used for a single sample: either by hypothesis testing or by confidence interval calculations.

Hypothesis test for the difference between two means

Provided the 2 samples of data are independent, the 2-sample *t*-test can be used to compare the means $\bar{x}_1$ and $\bar{x}_2$. But this is only valid when the samples have been taken from populations having the same variance σ^2, as illustrated by Fig. 2.1.

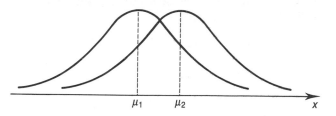

Fig. 2.1 Two normal distributions with the same variance, but different means μ_1, μ_2

There are 2 estimates of variance, one from each sample: s_1^2 and s_2^2. These must be combined to give s^2, a 'pooled' estimate of σ^2, which is then used in the *t*-test. Before doing this, we need to check that s_1^2 and s_2^2 are not significantly different from one another, because it is obviously not reasonable to pool them if they do not seem to be estimating the same thing. The *F*-test, described in the following section, is the appropriate test for two estimates of variance, and the more general problem is discussed in Chapter 4.

Assuming that the *F*-test does not give a significant result, s_1^2 and s_2^2 are combined as a weighted average to form s^2, the weighting coefficients being their degrees of freedom (which are one less than the number of items in each of the samples). The pooled estimate of variance is given by

$$\frac{(n_1 - 1)s_1^2 + (n_2 - 1)s_2^2}{n_1 + n_2 - 2}$$

and will have $n_1 + n_2 - 2$ degrees of freedom. Now we are dealing with 2 independent samples, one of n_1 observations from an $\mathcal{N}(\mu_1, \sigma^2)$ distribution and one of n_2 observations from an $\mathcal{N}(\mu_2, \sigma^2)$ distribution. The estimate of σ^2 based on *both* samples of data is s^2.

A null hypothesis will specify actual values of μ_1 and μ_2, or (more commonly) a value of their difference; we shall most often want to test that this difference is 0 but in general the null hypothesis will be '$\mu_1 - \mu_2 = K$'. In the following example, K will be 0. Since $\bar{x}_1 - \bar{x}_2$ follows $\mathcal{N}(\mu_1 - \mu_2, \sigma^2/n_1 + \sigma^2/n_2)$, the test statistic for the Null Hypothesis '$\mu_1 - \mu_2 = K$' is

$$t = \frac{\bar{x}_1 - \bar{x}_2 - K}{\sqrt{s^2/n_1 + s^2/n_2}}$$

and *t* will follow $t_{(n_1 + n_2 - 2)}$.

Example 2.3(d) continued. Let us test the null hypothesis that the two samples of eggs came from populations with the same mean mass, so that $\mu_1 = \mu_2$. We do not need to specify a value for these means, as it is only the size of their difference that we are interested in. The alternative hypothesis is simply '$\mu_1 \neq \mu_2$'.

The estimates s_1^2, s_2^2 each have 11 degrees of freedom, and so

$$s^2 = \frac{(11 \times 24.8336) + (11 \times 20.6918)}{22} = 22.7627$$

(We shall show below that there is no evidence of any significant difference between s_1^2 and s_2^2, so it is valid to combine them into s^2.)

$$\bar{x}_1 - \bar{x}_2 = 59.61 - 65.35 = -5.74$$

and

$$\sqrt{s^2 \left(\frac{1}{n_1} + \frac{1}{n_2} \right)} = \sqrt{\frac{2}{12} \times 22.7627} = 1.948$$

The test statistic is now

$$t = -\frac{5.74 - 0}{1.948} = -2.95$$

which is significant at the 1% level as $t_{(22)}$ (Table I). We have strong evidence to reject the Null Hypothesis of no difference.

2.4   The *F*-test for comparing two estimated variances

Given 2 samples of data from normal distributions, there are at least two reasons why we may wish to test whether they could have come from populations with the same variance (but not necessarily the same mean). We have just referred to one of them: when pooling estimates of variance we wish to be reasonably sure that they are estimates of the same σ^2. Another useful item of information is whether an experimental treatment may have altered the variance of some response that we are measuring, as well as whether it has altered the mean.

The distribution needed for this test (Table II) is the *F*-distribution, and it examines the *ratio* (not the difference) of the two estimates of variance. As usual, label the 2 estimates s_1^2 and s_2^2, and their degrees of freedom v_1 and v_2. The actual critical points in the distribution of F for any particular problem depend on both degrees of freedom (d.f.), and because there are two degree-of-freedom parameters involved (not merely one, as for t) F tables take a lot of space to print. By far the most common use of F is in the Analysis of Variance (Chapter 4), and for this reason the distribution is tabulated in *one-tail* form: Table II shows only the values that are 'significantly large' for each pair of degrees of freedom. These critical values are presented for the three conventional levels 5%, 1% and 0.1%, and also for the 2.5% level.

Because we have a one-tail F table, the test of equality of 2 variances has to be made by writing their ratio with the larger numerical value on top. But for this use of the *F*-test we really require a two-tail version, since the null hypothesis will be '$\sigma_1^2 = \sigma_2^2$' and the alternative hypothesis '$\sigma_1^2 \neq \sigma_2^2$'. We can do this at a nominal 5% level by using the 2.5% upper critical points from Table II (but not at 1% or 0.1%, for which more extensive tables should be consulted).

Example 2.3(d) concluded. Before using s^2, we check that there is no significant difference between the two estimates s_1^2, s_2^2. Each has 11 d.f., and the larger is s_1^2, so we write

$$F_{(11,11)} = s_1^2/s_2^2 = \frac{24.8336}{20.6918} = 1.20$$

which does not approach significance since the upper 2.5% point of $F_{(11,11)}$ is approximately 3.48. The upper d.f., shown along the top row of Table II, are those for the estimate in the numerator, s_1^2, and the d.f. shown in the first column of the table are for the denominator, s_2^2. At the 2.5% level, $F_{(10,11)}$ is 3.53 and $F_{(12,11)}$ is 3.43, and we have to interpolate between these two values to obtain $F_{(11,11)}$. Since our observed value 1.20 is very much smaller than this critical point, there can be no objection to pooling these two estimates to produce s^2 for use in the two-sample t-test.

2.5 Confidence interval for the difference between two means

When studying a single mean, a confidence interval for its true value in the population, based on sample estimates of the mean and standard deviation, has limits which can be stated as *'the estimated mean plus or minus t times its standard error'* (the phrase 'standard error' is often used instead of 'standard deviation' when speaking of a mean rather than a single observation).

In the same way, the limits of an interval for a difference between means are *'the estimated difference plus or minus t times the standard error of the difference'*. The standard error (or standard deviation) of a difference between two means is $\sqrt{s^2/n_1 + s^2/n_2}$.

Example 2.4. For the two samples of eggs, we found $s^2 = 22.7627$, with 22 degrees of freedom; n_1 and n_2 are both 12; and the observed means are $x_1 = 59.61$ and $x_2 = 65.35$. Let us find a 95% confidence interval for the true difference ($\mu_2 - \mu_1$).

The estimated (observed) difference is $65.35 - 59.61 = 5.74$, and its standard error (standard deviation) is $\sqrt{(2/12) \times 22.7627} = 1.948$ as we have already found. Also the 5% two-tail critical value of $t_{(22)}$ is 2.074. Therefore 'the estimated difference plus or minus t times its standard error' is $5.74 \pm 2.074 \times 1.948$, which is 5.74 ± 4.04, giving the interval 1.70 to 9.78.

We saw above that these means were 'significantly different', which is confirmed by finding that 0 is not contained within the interval. Again the interval is rather wide relative to the size of $\bar{x}$, which simply reflects the inherent variability of the egg masses.

In later chapters we shall sometimes want to examine differences between pairs of means even when the complete experiment contains many more than two treatments.

2.6 'Paired data' t-test when samples are not independent

Some experiments must be carried out on units (plots) where we expect considerable unit-to-unit difference. An example is when plants are grown in a glasshouse and their leaves are treated with a chemical fungicide to prevent a fungus disease from attacking them. Their leaves vary sufficiently in age, size and smoothness of surface to alter the effect of the treatment quite substantially. The records from an experiment will appear to be very variable, and may not give results of any useful degree of precision, unless we can remove the leaf-to-leaf component of the variation from the analysis.

Suppose that a new fungicide is compared with a standard one. Since leaves often grow in pairs, we could take a pair and allocate the new fungicide to one leaf, chosen at random, and the standard to the other (e.g. the allocation could be determined by tossing a coin; see Chapter 3). It might be better, if possible, to use individual leaves, and paint the new fungicide carefully on to one half of a leaf and the standard on to the other – allocation of halves would be at random. Comparison of the 2 fungicides can thus be made within, rather than between, leaves (or pairs of leaves), and this considerably reduces the variability of the results because that part of the variation which is really due to the leaves (not to the fungicides) has been removed.

However, the 2 halves of a leaf, or the members of a pair of leaves, are certainly not independent of one another, and the theory for the two-sample *t*-test no longer applies. We must reduce the data to a *single* sample of *differences* between the responses to new and standard fungicides, a difference being calculated for each pair of leaves or half-leaves. After doing this, it is usually reasonable to assume that these differences follow a normal distribution, and to use the single-sample *t*-test for making inferences about the difference between new and standard fungicides. As we shall see in Chapter 4, removing the inherent differences between leaves, so that the comparison of treatments is made within leaves, is an example of the use of *blocking* in experimental design.

Example 2.5. An industrial process uses a mixture of materials to make a small component; these materials can be obtained from different sources of supply, so that each mixture may be slightly different from the others. After manufacture, the strength of components may be measured in a standard test.

The length of time for which the process is allowed to run may also affect the strength, and two times are being compared in an experiment: the longer time (*L*) is 20 min and the shorter time (*S*) is 15 min. So as to eliminate possible differences due to the sources of supply of the materials, two units are taken from each mixture; one of these is chosen at random to receive *L* and the other receives *S*. The strengths of components made from 15 pairs of units are measured, and the data are shown in Table 2.1.

Table 2.1   Data for Example 2.5

Mixture	1	2	3	4	5	6	7	8	9	10	11	12	13	14	15
L	34.2	28.5	26.4	22.8	24.1	29.4	22.8	27.6	28.9	23.6	30.5	28.3	30.9	24.5	31.3
S	28.2	26.2	25.1	19.7	22.7	23.6	24.3	23.4	27.0	26.9	27.8	28.3	26.7	24.3	31.7
Difference (*L−S*)	6.0	2.3	1.3	3.1	1.4	5.8	−1.5	4.2	1.9	−3.3	2.7	0.0	4.2	0.2	−0.4

Hypothesis test

We shall test the null hypothesis that there is no difference in mean strength between components manufactured with the process running for 20 min and those when it has run for 15 min.

The differences, d, between L and S for the same mixture are shown in the last row of the table. Their mean $\bar{d}=1.86$, $n=15$ and $s_D^2=6.8440$ with 14 degrees of freedom. The null hypothesis in terms of the differences will be '$\mu_D=0$' and the alternative hypothesis '$\mu_D \neq 0$'. Therefore the statistic

$$t=\frac{\bar{d}-0}{s_D/\sqrt{n}}=\frac{1.86-0}{\sqrt{6.8440/15}}=\frac{1.86}{0.675}=2.75$$

is distributed as $t_{(14)}$ on the null hypothesis; this value is significant at the 5% level, though not quite at 1%. So we have evidence against the null hypothesis: it does *not* appear that L and S give components of the same strength.

Confidence interval

A confidence interval for the true difference between the mean strength of components made under condition L and those under S must be found from the differences d. Its limits will be, in the usual way, 'observed mean $\bar{d} \pm t \times$ standard error of $\bar{d}$', and t will have 14 d.f. For these data, we already have $\bar{d}=1.86$ and the standard error of $\bar{d}=s_D/\sqrt{15}=0.675$. Also $t_{(14)}=2.145$ at the 5% critical level, so the limits of the 95% confidence interval for the difference between the means of L and S are $1.86 \pm 2.145 \times 0.675$, namely 0.41 and 3.31.

The mean strength of components made using time L is, with probability 0.95, between 0.41 and 3.31 units greater than for components made using time S.

2.7   Linear functions of normally distributed variables

We have already noted that sums and differences of normal variables also follow normal distributions, and so do sums and differences of means of normal variables. There is one more result of this type which we shall use in Chapter 4 when we consider more general *contrasts*.

If x, y and z are all normally distributed, independently of one another, and a, b and c are numbers (constants), it can be shown (e.g. by using moment generating functions) that $ax+by+cz$ is also normally distributed. A combination of x, y and z formed in this way is a *linear function* of them; the numbers a, b and c can take any values, positive or negative.

Suppose that x is $\mathcal{N}(\mu_x, \sigma_x^2)$, with a corresponding notation for y and z. The mean of the function $ax+by+cz$ is $a\mu_x+b\mu_y+c\mu_z$ and its variance is $a^2\sigma_x^2+b^2\sigma_y^2+c^2\sigma_z^2$. Also, as we have just remarked, it follows a normal distribution.

So, for example, when $a=1$, $b=-2$ and $c=1$ we can say that $x-2y+z$ will be $\mathcal{N}(\mu_x-2\mu_y+\mu_z, \sigma_x^2+4\sigma_y^2+\sigma_z^2)$, noting that as usual the variances must be combined by adding but that the means carry the plus and minus signs as in the linear function.

This result can be extended to any number of independent random variables x_1, x_2, x_3, $x_4, \ldots$ in the same way. In fact, the result for the means of normal variables is a special case of this where we add up the n x's, all with the same distribution and all multiplied by the constant $1/n$.

2.8   Linear models including normal random variation

The methods we have been discussing can be expressed in a somewhat different way which is particularly useful for handling data from designed experiments. In Section 2.1, we were dealing with n observations all independently taken from $\mathcal{N}(\mu, \sigma^2)$. We may describe the set of observations $\{x_i : i = 1 \text{ to } n\}$ by writing

$$x_i = \mu + \varepsilon_i, \qquad i = 1, 2, \dots, n$$

where the set $\{\varepsilon_i\}$ are all drawn independently from $\mathcal{N}(0, \sigma^2)$. In our earlier discussion we wished to estimate μ and/or study a hypothesis about it.

In Section 2.3 we dealt with two sets of data, which in the same way could be specified as

$$x_{1i} = \mu_1 + \varepsilon_{1i}, \qquad i = 1, 2, \dots, n_1$$

and

$$x_{2j} = \mu_2 + \varepsilon_{2j}, \qquad j = 1, 2, \dots, n_2$$

where both $\{\varepsilon_{1i}\}$ and $\{\varepsilon_{2j}\}$ are drawn independently from $\mathcal{N}(0, \sigma^2)$. Interest in this case generally involves differences between μ_1 and μ_2.

The paired-data t-test of Section 2.6 introduces a further source of variation; in Example 2.4 the mixtures are likely to differ from one another systematically, not purely at random, and so we should not be including that element of variability in ε_{ij}. In that situation we could model the two sets of data as

$$x_{1i} = \mu_L + m_i + \varepsilon_{1i}, \qquad i = 1, 2, \dots, 15$$

and

$$x_{2i} = \mu_S + m_i + \varepsilon_{2i}, \qquad i = 1, 2, \dots, 15.$$

Here $\{\varepsilon_{1i}\}$ and $\{\varepsilon_{2i}\}$ are both independently distributed as $\mathcal{N}(0, \sigma^2)$, and the set of values $\{m_i\}$ allows for the systematic differences from one mixture to another. In the way the data were analysed in Section 2.6, by considering differences for each value of i, we generated a set of data whose mean is $(\mu_L - \mu_S)$, because the m_i term cancels out. Also the variance in this set of data is constant and the differences $(x_{1i} - x_{2i})$ follow a normal distribution.

We shall generalize all these ideas about 'linear models' in Chapter 4 and the remainder of this book.

2.9   Exercises

1. Fifteen samples of sand to be used for growing seedlings in pots are sterilized by 3 different methods, 5 by each method A, B, C. Method A requires a higher temperature than the others, and so the mean amount of a chemical present after using method A is to be compared with the mean for the other 10 samples from methods B and C. Find an expression for the variance of this comparison, in terms of σ^2, the variance of a single measurement (which is assumed to be the same for all three methods).

If the means for *A*, *B*, *C*, respectively, are 0.106, 0.112, 0.114 (in suitably coded units), an estimate of σ^2 from all the data is 2.6133×10^{-5}, and the observations are normally distributed,

(i) test whether the mean of *A* differs from that of *B* and *C*;
(ii) construct a 95% confidence interval for the true value of the difference between these two means.

2. An industrial process consists of three sections *P*, *Q*, *P*, the first and last containing the same sequence of operations. Find the distribution of the total time for the process if *P* is distributed normally with $\mu = 25$ min and $\sigma^2 = 15$ min^2 and *Q* is distributed normally with $\mu = 45$ min and $\sigma^2 = 20$ min^2.
3. If the process described in Question 2 needs a 5-min break period before and after *Q*, recalculate the distribution of the total process time.
   One run of the process took 2 h. Was that significantly longer than the expected time?
4. A motorist makes the same journey each morning, and returns by the same route each evening. Besides random variations in the time taken, there may be more systematic day-to-day effects due to weather, road conditions, road works, local events, etc. As a simple approximation, it may be assumed that any such effects present in a morning are also present the same evening. Eighteen days' records are given below; the journey never takes less than 30 min and the data have therefore been coded by subtracting 30 from them.

Day	1	2	3	4	5	6	7	8	9	10	11	12	13	14	15	16	17	18
Out	6.5	8.1	2.4	2.7	5.3	4.2	3.2	6.1	2.9	7.8	5.5	8.2	3.3	5.4	4.7	2.8	6.0	4.5
Return	7.7	9.0	3.3	4.5	5.7	5.5	4.3	7.1	3.5	9.0	6.0	8.3	5.0	5.8	5.5	4.1	7.1	6.1

(i) Set 95% confidence limits for the difference between the return and out times of the daily journey. What further assumptions have been made in calculating these?
(ii) If on one day the 'out' journey takes 4.9 min, what is the expected 'return' journey time?

3
Principles of experimental design

3.1 Introduction

Designing an experiment involves two major decisions. First, what is the detailed purpose of the experiment and what questions should the results help us to answer? Specifying this carefully determines exactly what treatments should be included in the experiment: we will call this the *treatment structure*. Often there is not enough thought given to this at the planning stage, when the experiment is being designed and suitable material is being obtained to form the experimental units or plots. We return to this in Section 3.2, and readers will gain a great deal by asking themselves whether each of the examples, analysed in Chapter 4 and later, could have given different information if changes – often quite minor – had been made to the treatment structure.

Second, what are the characteristics of the experimental units that will be used; in particular are they homogeneous or do they fall into a number of distinct groups? This determines the actual *layout*, or *design*, which is required. There is a temptation to concentrate more on this aspect of the planning because, as we shall see, there are sometimes interesting mathematical problems in finding suitable designs. We should, however, avoid making the layout any more complicated than is necessary because the less straightforward an experiment is to perform, the more likely it is to suffer some mishap in carrying it out. This in turn will make the analysis less straightforward. Although a harder analysis may cause no problems when a computer is available to do the arithmetic, it is still true that designs which have not been carried out as originally planned almost always have lower precision than is possible: that is to say, differences among treatment means have larger variances than the designer intended.

3.2 Treatment structure

There is a wide variety of possibilities. We give several here, and some other special cases are dealt with as they arise in later chapters.

(1) A *standard* treatment (often called a *control*) may be included as a base against which to compare all the others. For example, in studying the growth and cropping of new varieties of corn, vegetables or fruit we shall want to find out how each behaves relative to something that we already know about; that baseline will be a standard variety already grown in the region where we are hoping to recommend some of the new ones. The farmers or the growers are familiar with the standard; they will certainly want to know how any new varieties compare with it, and they may often need quite a bit of persuading to change to something new.

(2) As a minor, but sometimes very important, variation on (1), there may be a standard treatment against which others are to be compared, but in addition a *zero* treatment which is in fact no treatment at all. Trials of new drugs for recurring conditions (such as 'cures' for headaches) are sometimes confused by patients saying they have recovered when in fact they have been given an inert substance, a pill made up to look just like one of the experimental treatments but containing no active material. In clinical trials this is called a *placebo*, something that 'pleases' us (and that is all it does!).

It follows that designs sometimes need to incorporate *both* a *placebo* and a *standard*, the first to check if patients recover without any treatment and the second as a base for comparing new treatments. Unfortunately, the placebo is sometimes called the 'control'. This is confusing, so we shall prefer to use the word 'standard' for the basis as described in (1); what is often referred to in publications as an *untreated control* we shall call a *placebo* in the medical sense and *zero* in other types of experiment.

(3) The treatments to be used may represent various combinations of levels (amounts) of different factors, and these factors may or may not act independently of one another. This is a very important class of experimental situation which we begin to consider in Chapter 5 and return to several times after that.

(4) Increasing amounts of one factor (or of several factors) may form the treatments to be compared, and a graph may be plotted of the responses measured for the different treatment levels. A curve (or surface) may be fitted to show any pattern in these responses. Again we mention this in Chapter 5, but there has been much useful development of methods for these situations, which we describe in Chapter 9.

(5) Treatments may fall into groups, and we may want to make comparisons between the different groups as well as looking at differences among the treatments within groups. The method of *contrasts* (Section 4.8) is very well applied to this sort of problem.

For example, we may supply the essential element nitrogen in an agricultural fertilizer through two different chemicals, which we will call N and S. The manufacturers have two different physical forms of N (call these G and H), and also of S (call these P and Q). An experiment is therefore going to contain *four* treatments G, H, P and Q, each of which is applied to several different plots, all of the same size, on an experimental site. These treatments will each be applied at the same time in the season, at levels that provide the same amount of nitrogen per unit area.

One comparison that suggests itself is N against S: which chemical (if either) leads to higher crop yield and/or quality. In order to examine this, we want to take the mean yield (or measure of quality, e.g. top grade crop) of all those plots receiving N and the mean of all those receiving S, and compare these two means. This is in fact the comparison of the two *groups*, (G, H) against (P, Q). Having made this comparison we can further ask whether G is different from H, or whether they are equally effective forms of applying N; and similarly we ask whether P is different from Q as a method of applying S.

We shall see in Section 4.8 that these are three *independent*, or *orthogonal*, comparisons among the four treatments G, H, P and Q. There can only be three independent comparisons among four treatments (though there is more than one way of choosing them), and any further comparisons that we may suggest will give no further information, mathematically speaking. If there are v treatments in an experiment, there are only $(v-1)$ independent comparisons that can be extracted. The skill lies in choosing useful ones. Assuming that *G* and *H* were not the same two physical formulations as *P* and *Q*, there would be no point in making cross-group comparisons (like *G* against *Q*) because we should not be able to say anything useful about them in a report.

This experiment can *not* tell us anything about the right level of nitrogen for the plants, because only one level has been used. Nor can it tell us anything about whether any of *G*, *H*, *P* or *Q* do better when applied at different times of the season, because only one time has been used. Usually there are far more possible and interesting questions that could be asked than can be answered in one experiment; otherwise the experiment becomes over-large, in terms both of the amount of material available for units (plots) and the time and resources available to do the necessary work.

3.3   Changing background conditions – the need for comparison

In this book we shall always be proposing methods that involve *comparing* two or more treatments, rather than estimating absolute values of individual treatment means. This is because in biological, agricultural and even in highly controlled industrial work the same treatment applied to the same type of experimental unit at different times will almost always give different mean responses, sometimes very different means. We should expect this if we are conducting a series of agricultural experiments in different regions, or medical trials using several different hospitals, or industrial experiments where several different firms take part. But it can also happen in research institutes, in the same hospital at different times, and in the same firm when an experiment is repeated, or when the same treatment is included in more than one experiment. One reason for this is that it is impossible to find *exactly* the same type of experimental unit for different experiments, especially at different times. Another is that even in very well-controlled situations there may be changes – often small but still enough to have an effect – in the background conditions. Such things as temperature, humidity, amount of light or speed of a machine can actually be very hard to reproduce exactly.

For all these reasons there is not usually much value in doing experiments that provide estimates only of treatment means. It is *differences* between treatments that are most likely to be reproducible in different experiments at different times and places, and give a sound basis for choosing between better and poorer treatments for future use. In Example (5) of Section 3.2 we would only be justified in restricting ourselves to these four treatments if one (at least) of them was a standard, in the sense that enough was already known about its behaviour for us to base comparisons on it. If all the four physical formulations had been new ones, or had not previously been used on this crop, or in this region, we would have been very wise to include in the list of treatments a standard which was the form in which the desired amount of nitrogen had been applied in the past. Before other comparisons were made, we might examine whether on average the new treatments appeared to differ from this known standard. At the very least it would provide a sort of 'calibration' of this set of experimental results against those obtained earlier or elsewhere.

3.4   Replication

In the *t*-tests of Chapter 2, we used *estimates* of the value of σ^2, the variance of each observation. Usually, even in closely controlled conditions such as an industrial experiment, it is better to estimate σ^2 afresh for each experiment, because there is the possibility that the units of material used for the present experiment may be slightly different from those used previously. Sometimes, however, in a routine series of experiments in industry, variability is so similar from one experiment to another that we can use 'historical', i.e. previous, data to give the value of σ^2 based on a very large number of earlier experiments. In any other field of work we shall want to plan an experiment so as to give us an estimate of σ^2.

This estimate comes from having measurements on more than one plot (unit) receiving each treatment, i.e. from *replication* of the experimental treatments. In the standard methods of analysis of data, based on linear models, we use a 'pooled' estimate of variance calculated from all the observations on all the treatments, and we aim to make this estimate have a minimum of about 10 degrees of freedom. The amount of replication of a complete experiment needed to produce this will depend on the number of treatments and on the design, as we shall see in the following chapters. There is no hard and fast rule about the 'right' number of replicates for every experiment – only these guidelines about degrees of freedom.

Another consideration is to avoid the danger of basing inferences on data that contain 'rogue' observations or *outliers* that are not typical of the effect of a treatment. There needs to be sufficient replication so that outliers can be noticed if they occur, and either we may treat them as *missing values* (pages 69–71, 207–8) or if the experimental material is still available we can go back and look again at any unit that seems to have given an odd result. In theory, some large experiments need only a single replicate (see Chapters 5, 6 and later) but many scientists would be unhappy to base inferences on an unreplicated experiment unless these could be checked at some later stage in the research programme.

3.5 Randomization

Suppose that two treatments A and B are to be compared, and that we have 20 units altogether which can be used for the experiment. Suppose further that we plan to make the comparison by taking a single measurement x on each unit and comparing the mean $\bar{x}_A$ for those units that receive A with $\bar{x}_B$, the mean for those receiving B. Is there any need to select, in any special way, which units are to receive each treatment? First, it is *not* necessary always to have the same number of As as Bs: we need not arrange to put exactly 10 on each treatment. But against this we may well argue that there is no reason to have more replication of one than of the other because that would cause one mean to be estimated more precisely than the other, since the variance of a mean is σ^2/n when the mean is based on n observations. There are actually some occasions when different replication is useful (pages 37, 47), but for the present we shall assume that we want to select 10 of the 20 units to receive A.

Do we take the first 10 that come to hand – the first 10 in a row of plants, the first 10 animals we can catch for a nutrition trial, the first 10 patients being treated for an illness, the first 10 runs of a particular machine in a day's work at an industrial bench, the first 10 children in a list to be given Method A of teaching a particular topic? If we do this, we can be severely criticized for 'fixing' the results by choosing the 10 best-sheltered plants for one treatment and the 10 worst-sheltered for the other; or the 10 least-active animals for one treatment because they were easy to catch; or the 10 patients who were diagnosed first, or who have the disease more seriously and are in most urgent need of treatment; or using the running-in time of the machine before it was properly warmed up, or taking 10 children all from the same small group out of the whole age group in their school. There are often critics looking for flaws in experimental results, especially if results are different from any they may have found themselves or if they do not want to believe the results for reasons which may or may not be scientific. Too often their task is made easy by poor planning which lays itself open to obvious criticisms like those mentioned above.

We need to choose the units by an entirely non-subjective process that gives us no say in which unit shall receive which treatment. *Randomization* achieves this, and as a bonus it helps to validate the statistical tests for orthogonal designs made in Chapter 4 even if the assumption of a normal distribution for the residual terms, ε_{ij}, in a linear model may not be satisfied. In fact, much discussion about the process of randomization in more complex designs is still going on (Youden (1972), Kempthorne (1977), Preece, Bailey and Patterson (1978)).

A *table of random digits* (Table IV), in which every position is equally likely to be occupied by any of the digits 0, 1, 2, . . . , 9 can be used to make the choice of which units receive which treatment. First we need to identify *all* the units that are going to be used for the experiment – all the plots in a field, all the children in an educational experiment, all the animals that can be used in a nutritional trial. This in itself is a useful exercise: do we have enough of them, and are they homogeneous as a group, without any non-random sources of variation among them? If we really can find 20 fully comparable units we must then number them 01–20. An extract from a table of random digits is:

    78952 52966 72757 52481 96277 55914 09539 89401 17841 95417 67633 74283
    26514 05525 21285 76710 36950 49940 67995 66186 16757 23751 11799 74392
    52205 63948 00221 . . .

Since our units have two-figure numbers, we must read *pairs* of digits from the table; we shall identify these pairs of digits with individual units. There are no units numbered 78, 95, 25 or indeed anything above 20, so we shall obviously waste a great quantity of digits if we only use those that directly match one of the available units.

First we reduce the pairs of digits read from the table *modulo 20*, i.e. we take away as many multiples of 20 as possible from each pair. This will give numbers in the range 01–20: 18, 15, 05, 09, 06, 12, 15, 15, 04, 01, 16, 07, 15, 19, 14, 09, ... If we encounter 00, we call it 20. In a simple example like this one, every unit 01–20 has exactly the same number of pairs attached to it: the random digits 01 give unit 1, and so do 21, 41, 61, 81. We can find exactly five pairs of random digits that correspond to any of the units; for example unit 19 is used whenever we meet the digits 19, 39, 59, 79, 99, and unit 20 is used for any of the digits 20, 40, 60, 80, 00.

The first ten units identified by this process of random selection will receive A, and these are 18, 15, 05, 09, 06, 12, 04, 01, 16, 07; or, in order, 01, 04, 05, 06, 07, 09, 12, 15, 16 and 18. The other ten units receive B. There is still, of course, a small amount of wastage when unit numbers repeat themselves during selection; we obtained 15 several times in the selection we have just made and all except the first appearance must be rejected.

Most problems of selection are less straightforward than this one. Suppose that there are six treatments, which we refer to as A, B, C, D, E, F, to be compared in the same experiment. There are 4 replications of A and B, 6 replications of C and D and 5 replications of E and F. Therefore we need 30 experimental units altogether, and we must begin by numbering them 01–30.

Again we need to read digits in pairs from a table; we will use the same extract as above although in practice we should take a fresh run of digits for every new problem. We can read from Table IV starting at any position and moving upwards or downwards, left or right, or diagonally, to obtain new randomizations as required. This time we read *modulo 30*, reducing the pairs of digits by as many multiples of 30 as possible. The pairs 01–30 correspond directly to the units, numbers 01 to 30; 31 to 60 and 61 to 90 give two more complete sets which reduce, modulo 30, to 01–30. But we cannot use the pairs 91–99 and 00 because they do not cover the whole 30 available units: if we did use these pairs some units would have four pairs of digits associated with them and the others only three, so that those with four pairs would be more likely to be selected early in the process, contrary to the equal chances of selection required by randomization.

After this adjustment modulo 30, and after rejecting any units that appear more than once, the first four units selected will receive treatment A, the next four B, the next six C, and so on. Using the above run of random digits, the pair 78 has 60 subtracted from it to give 18; 95 cannot be used; 25 gives unit number 25 directly; 29 gives 29; 66 gives 06; 72 gives 12; 75 gives 15; 75 occurs again and is discarded; 24 gives 24; 81 gives 21; and so on. The first four units, numbers 18, 25, 29, 06, will receive treatment A; units 12, 15, 24, 21 will receive B; 27, 14, 09, 23, 01, 17 will receive C; 19, 07, 03, 26, 10, 28 will receive D; and 04, 16, 11, 13, 22 will receive E. Now all the remaining units must be given F; they are 02, 05, 08, 20, 30. This process requires a considerable number of random digits because of the number of pairs that have to be rejected as repeats. (Readers may think of various ways of reducing this wastage, but should spend a moment convincing themselves that 'better' ways really are random throughout the selection process.)

If this was a field experiment, with land split up into square plots, the final layout might look like Fig. 3.1.

1 C	2 F	3 D	4 E	5 F	6 A
7 D	8 F	9 C	10 D	11 E	12 B
13 E	14 C	15 B	16 E	17 C	18 A
19 D	20 F	21 B	22 E	23 C	24 B
25 A	26 D	27 C	28 D	29 A	30 F

Fig. 3.1 Completely randomized allocation of treatments to units

3.6 Blocking

In Section 2.8 and in Example 2.4, we saw that a systematic source of variation can exist among the units of material available for an experiment. In that case it was the mixtures used in the process, and we wrote down a linear model which included a term m_i for that source of variation. Very often the units available for an experiment are not completely homogeneous; there are some systematic variations among them but they can be split up into groups each of which is reasonably homogeneous within itself. Example 2.4 had 30 units available, which fell naturally into 15 groups – the mixtures – each containing two units.

The design of the experiment must then include *blocking* to remove that source of systematic variation from the estimate of residual natural variation. Otherwise our statistical analyses will not be valid, because they assume that the natural variation is purely random. In Example 2.5, a paired-sample *t*-test could be used because the two treatments were each applied to units from every mixture. Having two units of material from each mixture, we could toss a coin to see which should receive treatment L; the other then receives S. The mixtures are the *blocks*: each block contains each treatment once. The differences between the blocks are the differences between the mixtures, i.e. the systematic element of the variation present among the units: these differences are accounted for by $\{m_i\}$. The comparison between treatments, which is really what we want to make, has been freed from this systematic element by the blocking (i.e. the pairing).

In the original use of these methods in agriculture, where unit plots are often quite large, blocking was used to take out fertility or climatic trends over the whole area of the experiment. Figure 3.2 facing indicates how this may be done. Suppose there is a fertility trend from top to bottom of the diagram, perhaps because the land slopes downwards so that all the plots at the top have rather rocky soil, and the soil depth and quality improve as we go down into a valley. If we have seven treatments to be compared, labelled A–G, and we randomize them over the whole 28 plots by a similar method to that used in constructing Fig. 3.1, there is a real danger that some treatments will be allocated mostly to the better plots further down the slope and others mostly to the poorer plots higher up. Comparisons among the treatment means will not therefore be made independently of the fertility trend. So as to achieve this independent comparison, as far as possible with the available experimental units, we introduce blocks.

Blocks must be homogeneous within themselves, and the systematic differences among units must be taken out between blocks. In this example we achieve this by laying out blocks at right angles to the trend, as shown in Fig. 3.2 (blocks are usually labelled with roman numerals). Each block is the right size to contain each treament once (in the simplest case: there is further discussion in Chapter 4 and later). The response measured on each unit can be split into components due to which treatment it receives, and to which block it lies in, together with the usual random natural variation. This is the *randomized complete block* (Chapter 4), which is the most common experimental design in general use.

A randomized complete block design can be constructed using a run of random digits. First the units are divided into blocks so as to remove the trend, and then the order of treatments is determined at random independently for each block. Given seven

BLOCK

1 G	2 E	3 B	4 F	5 D	6 A	7 C	I
8 F	9 B	10 G	11 E	12 A	13 D	14 C	II
15 E	16 C	17 D	18 A	19 G	20 F	21 B	III
22 G	23 F	24 C	25 D	26 B	27 E	28 A	IV

Direction of trend ↓

Fig. 3.2 Allocation of treatments in random order to each of four blocks

treatments to be set out in four blocks (28 units in all) we need four separate *random permutations* of the numbers 1–7. As shown in Fig. 3.2, we shall associate the letters A–G with 1–7 in order, and so obtain the random layout of the treatments A–G. This does not need a very long run of digits, being usually less wasteful than the procedure we adopted above for complete randomization in Fig. 3.1. We use the run

78952 52966 72757 52481 96277 55914 09539 89401 17841 95417 67633 74283 26514

which will be sufficient for the whole exercise. We must discard 8, 9 and 0 since there are no treatments corresponding to them. The first random permutation, read from the remaining digits, is 7 5 2 6 4 1 3 (the 3 following at once as soon as the first six numbers have been found). The corresponding randomization of treatments in block I is G E B F D A C. Continuing through the run of random digits, we obtain for block II the permutation 6 2 7 5 1 4 3 and hence F B G E A D C. Then in block III we find 5 3 4 1 7 6 2 which gives E C D A G F B and finally for block IV the permutation is 7 6 3 4 2 5 1 leading to G F C D B E A. Laying the permutations out in a standard order, in this case from left to right, in each block, we obtain the final design of Fig. 3.2.

3.7 Sources of variation

The example of a fertility trend in Section 3.6 is a common one in field work. Also there may be a climatic trend, perhaps due to temperature changes down the slope, more likelihood of frost at the top or more likelihood of humidity through mist lying at the bottom of the slope. There may be more shelter on one side of an experiment than on another and this type of trend often arises through a valley, at right angles to that shown in Fig. 3.2; this leads to the need for designs that can take out two sources of systematic variation, such as Latin square designs (page 52).

When there are blocks in an experimental design they can be used to remove any known or suspected source of variation. Field experiments often need two or more days at harvesting time to record the crop yields; bad weather can interrupt this. It is very good practice to record one block completely before beginning the next, and if it is necessary to suspend operations and come back later, any systematic differences caused by the delay will be removed in the blocks term in the analysis. If block I is recorded before a rainstorm, which causes a day's delay before recording can be resumed, the remaining blocks may differ systematically from block I because the rain has damaged the crop, or because the crop is now slightly more mature, perhaps over-mature, or because the crop was easier to pick when dry than while still wet on day 2. Another use of blocks is to cut down the whole experimental area into manageable parts for cultivation operations such as ground cultivation done partly by hand and therefore not very quickly; one block is done at a time, perhaps on successive days, so that any differences due to the time delay are taken out in blocks. Often when there is no obvious trend that should be removed, blocks will still be included to make up convenient administrative units; they would then consist of compact groups of plots such as those shown in Fig. 3.3. Blocks may be *any* shape that will best achieve the aims of having units homogeneous within each block, with the systematic differences taken out between blocks.

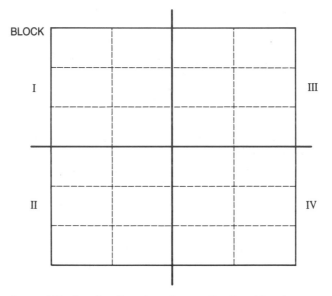

Fig. 3.3 Possible shape of blocks when there is no known directional trend

In a field trial, the plants near the edges of the experiment may be less sheltered than those in the middle. Blocks can be used to remove any possible systematic shelter effect. In a glasshouse, or other protected cropping, plots near the windows of the glasshouse, or at the edge of the protection, may be more exposed, so that an effect from the window or from the edge inwards needs removing. If the land has been used previously, as it may be on a research institute, blocking should whenever possible take out likely differences due to past use. Blocks can be different shapes – some square, some long and thin, and so on – so as to remove differences most effectively.

Animals are notoriously variable, and there are often systematic differences between litters of animals from different parents. Almost always the two sexes differ. So blocks will be made up of litter-mates of the same sex, for use in comparing treatments such as different diets. This can severely limit available block sizes, and the methods of Chapters 6 and 11 may be needed. In human experiments such as comparisons of old and new drug treatments hospitals taking part in a trial may differ systematically and so should be the basis of blocks; individual doctors assessing patients may also show systematic differences (for example in their use of a scoring scale) which will need to be removed before treatments are compared.

In industrial experiments, different machines used for the same process may give systematically different results, so that machines have to form blocks; different people operating the same machine can also give different results for the same operation; and the time of day or of the week may also need to be the basis for blocking. When raw materials are bought in from different suppliers, it is wise to use those from only one supplier in each block of experimental units.

Experiments in market research may try to find, for example, the best advertizing approach for selling some product. Results may differ in different parts of the country, so blocking by areas is useful; it could be even better to run a series of experiments, one in each main area. Socio-economic groups in the population can be a very useful basis for blocking, since they may react at different levels to the product; perhaps one group buys systematically more of the product than do others.

Other instances of the need for blocking can be found in the examples and exercises throughout the following chapters.

3.8 Planning the size of an experiment

A question very often put to a statistician is 'how many plots?' (or units) an experiment will need; and of course usually the enquirer means 'how few?', i.e. the minimum that would make the experiment worth doing at all. The question cannot be answered until the statistician is given more information.

In this book we shall aim to use confidence intervals whenever possible, and not put too much emphasis on significance tests. However, when deciding the minimum size for an experiment it is useful to consider the difference between two treatment means: if we replicate each treatment r times, what is the least difference between the means that we can claim is 'significant at the 5% level'? The two-sample t-test (Section 2.3), using the common 'pooled' estimate of variance from all the data (s^2), tells us that $(\bar{x}_1 - \bar{x}_2)/\sqrt{2s^2/r}$ is distributed as t with the same degrees of freedom as the estimate s^2: say f degrees of freedom. Of course we will not know exactly what f is until we have decided on r; nor will we know s^2 until we have carried out the experiment. But we can use any knowledge that is available on past work with the same type of experimental material, the same sort of units, to make a guess of s^2. Without this informed guess we cannot give a reasonable answer to the original question.

The experimenter is then asked what is the minimum difference $(\bar{x}_1 - \bar{x}_2)$ that it is important to establish as being 'significant'. This difference, when divided by $\sqrt{2s^2/r}$, must be at least as great as the 5% value of $t_{(f)}$; if we knew f we could read this 5% value from Table I. Since we do not know f, we may take an approximate starting guess of 2 for this value, and hence find an approximate value for r. Then we can find f, which will depend in general on the type of experimental design used as well as on the actual value of r. If we are only studying two treatments, f will be $(r-2)$ in a completely randomized layout, though considerably less when other designs need to be used. This leads to a more accurate value for the 5% point of $t_{(f)}$ which allows us to make a second iteration of the same calculations to obtain a better estimate of r.

Example 3.1. In an experiment where it is expected that s^2 will be approximately 25, a difference in means of 6 units is thought important. As a first approximation to find r, the replication needed to show this difference at the 5% level of significance, set $(\bar{x}_1 - \bar{x}_2)/\sqrt{2s^2/r}$ equal to 2, so that $6/\sqrt{2 \times 25/r} = 2$, i.e. $6 = 2\sqrt{50/r}$, or $3 = \sqrt{50/r}$, which gives $9 = 50/r$ and therefore $r = 6$ to the nearest integer above.

If only two treatments are being compared, each replicated 6 times, the pooled estimate s^2 will have 10 degrees of freedom (assuming that no blocking has been included in the experiment); so its critical 5% point will be 2.228. The calculation now becomes $6/\sqrt{2 \times 25/r} = 2.228$, and so

$$\left(\frac{6}{2.228}\right)^2 = \frac{50}{r}$$

which requires $r = 7$. There is no need to refine this any further.

Seven replicates of each of two treatments is a very convenient size of experiment; if an experiment becomes too large the necessary amount of work is likely to lead to inaccuracy (so increasing s^2), and if a large number of units is needed it is hard to find enough homogeneous ones. In the present example, we could most likely increase r to 8 or 9, to allow for the possibility that s^2 when we actually do the experiment is larger than we have guessed; the experiment would still be a very reasonable size.

If the experimenter had aimed to detect a difference of only 3 units between the two means, we would find that we needed $r = 23$. This might just be possible if there are only two treatments in an experiment, but there is likely to be too much replication if three or more treatments are included in the same experiment. We would have to be less ambitious about the size of difference to be detected – it could not be as small as 3 units.

Besides these rough calculations for finding r, we should always look carefully at the practical technique of carrying out the experiment and taking the measurements, because the value of s^2 can be kept small by good technique and careful recording. These practical ways of controlling variability, and so reducing the size of difference that the experiment can detect (i.e. increasing its *sensitivity*), are a great deal better than simply trying to reduce $\sqrt{2s^2/r}$ by increasing r. Dyke (1988) discusses field experiments very thoroughly, and the principles behind good field experimentation apply in many other areas of work.

3.9 Exercises

1. Using the Table of Random Digits (Table IV), construct examples of the following types of design:

   (i) a completely randomized layout with each of nine treatments replicated three times;
   (ii) a randomized block layout where there are to be three replicates of each of eight treatments, but the land available consists of six columns each containing four plots and it can be assumed that there is only a gentle trend across the columns;
   (iii) a randomized block layout in which each block is to consist of two replicates of a standard treatment S, two of a 'control' O, and five new treatments A–E once

each; 36 plots are available, of which 18 are on one side of a track through the field and 18 on the other side;

(iv) a completely randomized layout with twice as many replicates of treatment A as of all the others B, C, D, E, F and the maximum number of units available is 40.

2. Suggest economical ways of using digits to make the randomizations discussed in Section 3.5.

3. The mean height to which wheat plants grow in a fixed period of time is 36 cm, and the standard deviation of height is 5·6 cm, based on an estimate having 21 degrees of freedom from four complete replicates. What is the least difference in mean height, between plants from two different treatments, that can be demonstrated as 'significant' (at the 5% level) in this experiment?

4. In a field experiment using plots of $4 \, m^2$, it is expected from past experience that the coefficient of variation $(s/\bar{y})$ of crop yield will be about 0.08 (i.e. 8%). What is the minimum replication that will enable a 10% difference in treatment mean yields to be detected, at the 5% level, in a significance test? A completely randomized layout is to be used. If plots are reduced to $2 \, m^2$, so that this replication can be doubled, the coefficient of variation will increase to 0.12 (i.e. 12%). What difference can now be detected as significant?

4
The analysis of data from orthogonal designs

4.1 Introduction

The analysis of the observations from designed experiments is based on linear model theory with normally distributed random natural variation; we shall call this *residual variation*, avoiding the word 'error' (page 4). In Chapter 11 we develop the matrix-based theory for general block designs, which is the background to statistical computing packages. However, in order to understand the analysis of the most common simple designs and the output which may be obtained from using packages, we now give direct algebraic derivations of the terms that appear in an Analysis of Variance (often called ANOVA for short) and justify the statistical tests used.

In a *completely randomized* design, the linear model is

$$y_{ij} = \mu + \tau_i + \varepsilon_{ij} \qquad (i = 1 \text{ to } v; j = 1 \text{ to } r_i)$$

where treatment i is replicated r_i times and $\Sigma r_i = N$, the total number of units (plots) used in the experiment. We assume that the $\{\varepsilon_{ij}\}$ are independent of each other, and all follow a normal distribution with the same variance σ^2. Since the mean μ is already in the model, the $\{\varepsilon_{ij}\}$ will thus be $\mathcal{N}(0, \sigma^2)$, and $\text{cov}(\varepsilon_{ij}, \varepsilon_{i'j'}) = 0$.

Also the presence of μ implies that the treatment effect τ_i is a *deviation* from the overall mean of the N observations, and so $\Sigma r_i \tau_i = 0$.

The method of *least squares* is used to estimate the parameters in the model, which are μ, $\{\tau_i\}$ and σ^2. It is not hard to show that, assuming normally distributed $\{\varepsilon_{ij}\}$, the least-squares equations that must be solved are also those that would arise from maximum likelihood estimation, so that the estimates of the parameters will have all the properties of maximum likelihood estimators.

The full linear model is in a form suitable for an Alternative Hypothesis. Since we will be interested in differences among treatments, a suitable Null Hypothesis is '*all τ_i are zero*', or in symbols $y_{ij} = \mu + \varepsilon_{ij}$. In practical uses of designed experiments, this Null Hypothesis is not often of much interest, but it provides a basis for calculations that give estimates of all the parameters needed to compare any pairs or groups of treatments.

We must minimize

$$\Sigma = \sum_i \sum_j (y_{ij} - \mu - \tau_i)^2 \qquad (i = 1 \text{ to } v; j = 1 \text{ to } r_i)$$

Differentiating in turn with respect to μ and $\{\tau_i\}$, and setting the derivatives equal to 0, we have what are always called the *normal equations*:

$$-2 \sum_i \sum_j (y_{ij} - \hat{\mu} - \hat{\tau}_i) = 0 \qquad \text{(i)}$$

and

$$-2 \sum_{j=1}^{r_i} (y_{ij} - \hat{\mu} - \hat{\tau}_i) = 0, \qquad \text{for each } i \qquad \text{(ii)}$$

We need some standard notation for sums of data.

$$\text{The total of all the observations} = \sum_{i=1}^{v} \sum_{j=1}^{r_i} y_{ij} = G$$

$$\text{the total number of observations} = \sum_{i=1}^{v} r_i = N$$

and the total of all the observations receiving treatment i

$$= \sum_{j=1}^{r_i} y_{ij} = T_i, \qquad \text{for each } i = 1 \text{ to } v$$

The normal equations become:

$$G = N\hat{\mu} + \sum_i r_i \hat{\tau}_i$$

and since $\sum_i r_i \hat{\tau}_i = 0$ this is

$$G = N\hat{\mu} \qquad \text{(1)}$$

together with

$$T_i = r_i \hat{\mu} + r_i \hat{\tau}_i \qquad \text{for each } i \qquad \text{(2)}$$

Therefore

$$\hat{\mu} = \frac{G}{N} \quad \text{and} \quad \frac{T_i}{r_i} = \hat{\mu} + \hat{\tau}_i$$

or

$$\hat{\tau}_i = \frac{T_i}{r_i} - \frac{G}{N}$$

The estimate of a treatment mean is $\hat{\mu} + \hat{\tau}_i$, and is provided by the mean T_i/r_i of the observed data from units receiving that treatment.

Write $\bar{y}_i$ for T_i/r_i, and also $\bar{y} = \sum_i \sum_j y_{ij}/N$; then $\bar{y}$ is the estimate of μ.

The residual sum of squares is the minimum value of $\sum$, when $\hat{\mu}$ and $\{\hat{\tau}_i\}$ are substituted into it:

$$\sum_{\min} = \sum_i \sum_j \left(y_{ij} - \frac{T_i}{r_i} \right)^2 = \sum_i \sum_j (y_{ij} - \bar{y}_i)^2$$

which is the sum of squares *within* treatments, pooled for all the treatments.

Consider the *total sum of squares* $\sum_i \sum_j (y_{ij} - \bar{y})^2$. Writing

$$y_{ij} - \bar{y} = (y_{ij} - \bar{y}_i) + (\bar{y}_i - \bar{y})$$

we have

$$(y_{ij} - \bar{y})^2 = (y_{ij} - \bar{y}_i)^2 + 2(y_{ij} - \bar{y}_i)(\bar{y}_i - \bar{y}) + (\bar{y}_i - \bar{y})^2$$

Summing over j first and then over i, the left-hand side is the total sum of squares, and the first term on the right-hand side is $\sum_{\min}$, the sum of squares within treatments or the residual sum of squares. The last term on the right-hand side is the sum of squares between the treatment means. The middle term is 0, because summing over j gives

$$+2 \sum_i (\bar{y}_i - \bar{y}) \left[\sum_j (y_{ij} - \bar{y}_i) \right]$$

and the sum in square brackets is 0, for each i, being the sum of the deviations of a set of values from their mean.

On the Null Hypothesis, all $\{y_{ij}\}$ are independent normal random variables, $\mathcal{N}(\mu, \sigma^2)$, so the total sum of squares is distributed as σ^2 times $\chi^2_{(N-1)}$. Also $\sum_i \sum_j (y_{ij} - \bar{y}_i)^2$ is the sum of v sums of squares (over j), each of these being distributed as $\sigma^2 \chi^2_{(r_i-1)}$; it is therefore itself σ^2 times a χ^2 whose degrees of freedom are $\sum_{i=1}^{v} (r_i - 1) = N - v$. By Cochran's Theorem (Appendix 4A) it follows that $\sum_i \sum_j (\bar{y}_i - \bar{y})^2$ is also σ^2 times a χ^2, which will have $(N-1) - (N-v) = v - 1$ d.f. We display all this information in the Analysis of Variance table, whose theoretical form is shown in the table below.

Analysis of Variance for a completely randomized design			
Source of variation	d.f.	Sum of squares	Distribution on Null Hypothesis
Between treatments	$v-1$	$\sum_i \sum_j (\bar{y}_i - \bar{y})^2$	$\sigma^2 \chi^2_{(v-1)}$
Residual	$N-v$	$\sum_i \sum_j (y_{ij} - \bar{y}_i)^2$	$\sigma^2 \chi^2_{(N-v)}$
Total	$N-1$	$\sum_{i=1}^{v} \sum_{j=1}^{r_i} (y_{ij} - \bar{y})^2$	$\sigma^2 \chi^2_{(N-1)}$

In order to compute sums of squares, we define *summation terms*:

$$S = \sum_i \sum_j y_{ij}^2$$

$$S_0 = G^2 / N$$

$$S_T = \sum_i T_i^2 / r_i$$

The total sum of squares

$$\sum_i \sum_j (y_{ij} - \bar{y})^2 = \sum_i \sum_j (y_{ij}^2 - 2\bar{y}y_{ij} + \bar{y}^2)$$

$$= \sum_i \sum_j y_{ij}^2 - 2\bar{y}\left(\sum_i \sum_j y_{ij}\right) + N\bar{y}^2$$

$$= S - N\bar{y}^2 \qquad \text{since } \sum_i \sum_j y_{ij} = N\bar{y}$$

$$= S - G^2/N \qquad \text{since } \bar{y} = G/N$$

$$= S - S_0$$

S is the 'raw' sum of squares and S_0 is the correction term, which adjusts S so that the sum of squares is taken about the mean: $S - S_0$ is the *corrected total sum of squares*.

(S_0 has often been called the 'correction factor' (CF), but since we do not multiply anything by it, 'factor' is clearly not the appropriate word! Something that is added or subtracted is a *term*.)

Likewise

$$\sum_i \sum_j (\bar{y}_i - \bar{y})^2 = \sum_i \sum_j (\bar{y}_i^2 - 2\bar{y}_i\bar{y} + \bar{y}^2)$$

$$= \sum_i \sum_j \bar{y}_i^2 - 2\bar{y} \sum_i \sum_j \bar{y}_i + N\bar{y}^2$$

$$= \sum_i \sum_j (T_i/r_i)^2 - 2\bar{y} \cdot N\bar{y} + N\bar{y}^2$$

$$= \sum_{i=1}^{v} r_i(T_i/r_i)^2 - N\bar{y}^2$$

$$= \sum_i T_i^2/r_i - G^2/N$$

$$= S_T - S_0$$

Clearly S_T will be small if the treatment means are all close to the grand mean $\bar{y}$, i.e. if the treatments are not very different in their effects; and S_T will be large when there are large differences between treatments.

These terms are computed and put into the Analysis of Variance as shown in the next table. The residual sum of squares (RSS) is best found by subtracting the treatments sum of squares from the total sum of squares.

Form of computing for Analysis of Variance

Source of variation	d.f.	Sum of squares	Mean square
Treatments	$v-1$	$\sum_{i=1}^{v} T_i^2/r_i - G^2/N$	$(S_T - S_0)/(v-1)$
Residual	$N-v$	S.S. Resid. (found by subtraction)	S.S. Resid./$(N-v)$
Total	$N-1$	$\sum_{i=1}^{v} \sum_{j=1}^{r_i} y_{ij}^2 - G^2/N$	

In order to carry out some statistical inference, we must first reduce the corrected sums of squares to comparable units. This we do by dividing each sum of squares by its degrees of freedom, obtaining what are always called *mean squares*. Remembering that the expected (mean) value of a χ^2 random variable is equal to σ^2 times its degrees of freedom, and denoting mean squares by M with appropriate suffixes, we have $E[S-S_0]=(v-1)\sigma^2$ and so $E[M_{\text{TRTS}}]=\sigma^2$. Similarly, the residual sum of squares has expected value $(N-v)\sigma^2$, so that $E[M_{\text{RESID}}]=\sigma^2$. Cochran's Theorem tells us that these two estimates come from independent χ^2 distributions, and so the ratio of the two estimates of σ^2 is distributed as $F_{[(v-1),(N-v)]}$ under the Null Hypothesis.

The result of carrying out this *F*-test tells us whether or not it is reasonable to reject the Null Hypothesis '*All $\{\tau_i\}$ are zero*', or, equivalently, '*All treatments have the same effect*' – the *Null* Hypothesis of *no* difference. In practice, as we shall see shortly, this by itself is often the least important hypothesis we might wish to test about treatment differences, although in the early stages of a research programme we may simply want to find out which of the treatments may be worth following up in later experiments with no more specific comparisons to be made.

On the Null Hypothesis, both mean squares, treatments and residual give estimates of σ^2 (and we never compute a 'total mean square' which has no useful purpose). However, we need an estimate of σ^2 in general which can also be used when the Alternative Hypothesis is accepted, in comparisons between the means under different treatments.

When the Alternative Hypothesis is rejected, the linear model is

$$y_{ij}=\mu+\tau_i+\varepsilon_{ij}$$

and so

$$\bar{y}_i=\mu+\tau_i+\bar{\varepsilon}_i \qquad \text{for each } i=1 \text{ to } v$$

The residual sum of squares now is

$$\sum_{i=1}^{v}\sum_{j=1}^{r_i}(y_{ij}-\bar{y}_i)^2=\sum_i\sum_j(\varepsilon_{ij}-\bar{\varepsilon}_i)^2$$

Fixing i, summation over j gives a sum of squares of r_i deviations of $\mathcal{N}(0,\sigma^2)$ variates from their mean $\bar{\varepsilon}_i$, and so has the expected value $\sigma^2(r_i-1)$. Summing over i gives $E[\text{RSS}]=\sigma^2\sum_{i=1}^{v}(v_i-1)=\sigma^2(N-v)$, exactly as on the Null Hypothesis.

Therefore on either hypothesis the residual sum of squares has the same expectation and the residual mean square has the expected value σ^2. The residual mean square gives the unbiased estimate of σ^2 which can always be used as the estimated variance of each observation, and is denoted by $\hat{\sigma}^2$.

The table (Table II) of the *F*-statistic is always printed in one-tail form because this is how we use it in Analysis of Variance. To justify this, let us examine the expected value of the mean square for treatments on the Alternative Hypothesis. Now we shall have

$$(\bar{y}_i-\bar{y})=(\mu+\tau_i+\bar{\varepsilon}_i)-(\mu+\bar{\tau}+\bar{\varepsilon})=(\tau_i-\bar{\tau})+(\varepsilon_i-\bar{\varepsilon})$$

in an obvious extension of our notation for means. But since the constraint $\sum_{i=1}^{v} r_i \tau_i = 0$ is applied (page 32), it follows that

$$\bar{\tau} = \frac{\sum_i r_i \tau_i}{\sum_i r_i} = 0$$

So

$$(\bar{y}_i - \bar{y}) = \tau_i + (\bar{\varepsilon}_i - \bar{\varepsilon})$$

Hence

$$\sum_i \sum_j (\bar{y}_i - \bar{y})^2 = \sum_i r_i \tau_i^2 + 2 \sum_i r_i \tau_i (\bar{\varepsilon}_i - \bar{\varepsilon}) + \sum_i r_i (\bar{\varepsilon}_i - \bar{\varepsilon})^2$$

and the constraint $\sum_i r_i \tau_i = 0$, together with independence of treatment effects and residuals, will ensure that the middle term on the right-hand side is zero when expectations are taken. So

$$E\left[\sum_i \sum_j (\bar{y}_i - \bar{y})^2 \right] = E\left[\sum_i r_i \tau_i^2 \right] + E\left[\sum_i r_i (\bar{\varepsilon}_i - \bar{\varepsilon})^2 \right]$$

the final term being exactly the one that arises on the Null Hypothesis. We therefore add to this a term that cannot be negative, $\sum_i r_i \tau_i^2$, and that is only zero when all $\{\tau_i\}$ are zero, on the Null Hypothesis. On the Alternative Hypothesis therefore, the expected value of the treatments sum of squares, and of the treatments mean square, definitely exceeds the expected value on the Null Hypothesis; so a one-tail F-test is appropriate.

A summary of the analysis so far is:

- if $F_{[(v-1),(N-v)]}$ is significant in a one-tail test, reject the Null Hypothesis of no treatment differences in favour of the Alternative Hypothesis that there are some differences;
- on either hypothesis, the residual mean square gives an unbiased estimate of the variance of each observation, σ^2, which is called $\hat{\sigma}^2$ and can be used in further examination of the treatments.

Example 4.1. The fuel consumption of a motor vehicle is measured under controlled conditions with four different mixtures A, B, C, D. The results (km/L) are:

A:	13.31	14.04	13.68	13.75	13.12	14.11	13.96
B:	14.28	14.47	14.03	15.62	15.10		
C:	15.04	14.77	15.13	15.45	14.98	15.51	
D:	14.66	13.93	15.05	14.21	14.42	14.30	14.25

From this we can get summary statistics as shown in the table below.

(1) Treatment	(2) Replicates	(3) Total	(4) Σy^2	(5) T_i^2/r_i
A	7	95.97	1316.5907	1315.7487
B	5	73.50	1082.1346	1080.4500
C	6	90.88	1376.9344	1376.5291
D	7	100.82	1452.8760	1452.0961
	$N=25$	$G=361.17$	$S=5228.5357$	$S_T=5224.8239$

Within A, the variance is $(1316.5907-1315.7487)/6=0.140333$; within B it is $(1082.1346-1080.4500)/4=0.421150$; within C it is $(1376.9344-1376.5291)/5=0.081067$; and within D it is $(1452.8760-1452.0961)/6=0.129983$.

The total sum of squares is $S-S_0$. S is the sum of column (4)$=5228.5357$. $S_0=G^2/N=(361.17)^2/25=5217.7508$. Hence $S-S_0=10.7849$.

The sum of squares for treatments (which are the four mixtures) is S_T-S_0. S_T is the sum of column (5)$=5224.8239$. Hence $S_T-S_0=7.0731$.

Analysis of Variance

Source	d.f.	Sum of squares	Mean square
Treatments	3	7.0731	
Residual	21	3.7118	$0.1767=s^2$
Total	24	10.7849	

The estimate s^2 is the weighted average, using their degrees of freedom as weights, of the variances within A, B, C and D:

$$s^2=[(6\times0.140333)+(4\times0.421150)+(5\times0.081067)+(6\times0.129983)]/21=0.17675$$

4.2   Comparing treatments

The means are A, 13.71; B, 14.70; C, 15.15; D, 14.40. As we shall see shortly, and as we discussed in Chapter 3, a properly planned experiment will have specified comparisons that are important, and these are the only ones that are examined. But early in a research programme it may not be possible to be specific, and we need methods of extracting some information from the data about possible differences that would be worth examining in future experiments.

A simple way to do this is to compare means in t-tests. In this set of data, the means for A and B differ by 0.99, and the variance of their difference is estimated by

$$\frac{s^2}{7}+\frac{s^2}{5}=0.060583$$

the standard error of the difference is 0.246. Now the ratio 0.99/0.246 is distributed as

$t_{(21)}$, and so its value, which is 4.02, is highly significant (Table I). It seems very likely that there is a difference between these two mixtures. Furthermore, for A and C,

$$\frac{15.15 - 13.71}{\sqrt{(\frac{1}{7} + \frac{1}{6})(0.1767)}} = \frac{1.44}{0.234} = 6.15$$

which points even more strongly to a difference. Continuing these comparisons, the $t_{(21)}$ values for $A - D$, $B - C$, $B - D$ and $C - D$ are -3.11, -1.79, 1.22 and 3.25, which suggest that A and D differ, C and D differ, but B does not differ from C or from D.

However, only three comparisons can give independent information, because there are only three degrees of freedom among four treatments: the difference $(B - C)$ is simply $(A - C)$ *minus* $(B - C)$, and so on. We have made six t-tests; by doing this we violate the stated probability level (5%, 1% or 0.1%) at which we claim to be working. Various methods of adjusting for this, to allow all possible comparisons to be made, have been proposed; they are *multiple range* or *multiple comparisons* tests and the one due to Duncan (1955) is most commonly found in computer programs. We shall consider this problem further (page 46) after introducing randomized complete block designs, because it is easier to see how they work when all treatments have equal replication.

In order to help the non-statistical reader understand a report of an experiment, it is often useful to set out the treatment means in order of size and to mark with brackets those that do not differ significantly at each of the three standard significance levels. Of course we can only do this by effectively making all possible t-tests, so we are in great danger of giving false ideas of the extent of significance among the treatments unless we use the results with caution and intelligence. A written report should always make this clear, and not merely give the numbers that came out of the calculations. There is no alternative, early in a research programme, to making a large number of comparisons, though there is more than one way of doing that (see pages 51–2); but later in a programme *contrasts* should always be specified (page 47).

A summary, to be used with care and intelligence, of the results of this experiment, may be presented as shown in the table below. Means bracketed together do *not* differ significantly.

Treatment	Mean	5%	1%	0.1%
A	13.71			⎫
D	14.40	⎫	⎫	⎬ ⎫
B	14.70	⎬ ⎫	⎬ ⎫	⎭ ⎬
C	15.15	⎭ ⎭	⎭ ⎭	⎭

Without any further information about the mixtures, we cannot really say very much with confidence about their differences; but if we knew, for example, that A was a standard basic mixture to which other different ingredients had been added to form B, C and D we should be in a position to test useful *contrasts*. For example, we might limit our enquiry to whether any of B, C or D differs from A. The table above of course allows us to do this; with this aim in mind the report would state that there was evidence that all the others did differ from A, especially B and C which achieved a high level of significance. We would not say anything about differences among B, C and D in this situation.

4.3   Confidence intervals

Using some of the calculations we have already made for significance tests, we can gain valuable information by finding confidence intervals. For the difference between the means under mixture B and mixture A, we actually observed $14.70 - 13.71 = 0.99$; the standard error (standard deviation) of this difference is 0.246. The estimate s^2 has 21 d.f. and so $t_{(21)}$ has to be used in all calculations: $t_{(21)} = 2.080$, 2.831 and 3.527 at the 5%, 1% and 0.1% levels, respectively (Table I).

A 95% confidence interval for the true value of the difference between the means for mixtures B and A is therefore $0.99 \pm (2.080 \times 0.246)$, i.e. 0.99 ± 0.51, or (0.48 to 1.50). This is not a wide range for measurements that have mean values of round about 14, and we may feel that this experiment has given reasonably precise results.

If we want a higher probability that our interval does contain the true value of the difference between means, we may compute a 99% confidence interval using the 1% point of $t_{(21)}$ from Table I, which is 2.831. This interval is $0.99 \pm 2.831 \times 0.246$, i.e. 0.99 ± 0.70, which is (0.29 to 1.69). Finally the 99.9% confidence interval, using the 0.1% point of t, is $0.99 \pm 3.527 \times 0.246$, i.e. 0.99 ± 0.87 which is (0.12 to 1.86). Even this last interval does not include 0 within it, and so the data indicate that A and B are significantly different at the 0.1% level as we have already found.

Confidence intervals for the difference between means for mixtures D and A use the observed difference of 0.69 and the standard error (standard deviation) of the difference:

$$\sqrt{\frac{2 \times 0.1767}{7}} = 0.225$$

so the 95% confidence interval for the true value of this difference is $0.69 \pm (2.080 \times 0.225)$, i.e. 0.69 ± 0.47, which is (0.22 to 1.16). For a 99% interval, we substitute 2.831 for 2.080 and find 0.69 ± 0.64, which is (0.05 to 1.33); and for 99.9% confidence we use 3.527 for t to obtain 0.69 ± 0.79, which is (-0.10 to $+1.44$). The results are of course consistent with what we found above from significance tests, since the only level at which 0 is included in the confidence interval is 0.1%.

4.4   Homogeneity of variance

We assume in setting up the Analysis of Variance that each set of data, for the different mixtures, has the same variance; s^2 is a pooled estimate of this population value σ^2. For these data this may be in some doubt because the variance of the data for B is five times that for C. If we were studying just two variances we would be able to use an F-test, but for more than two we need another test. In fact, tests of the Null Hypothesis 'all variances for different mixtures are the same' would require several more degrees of freedom than there are in this example to reach any real sensitivity; and *Bartlett's Test*, which we shall now describe, although it is the likelihood ratio test for that Null Hypothesis against the Alternative that there are some differences, is almost as sensitive to non-normality of data as it is to differences in variance.

For the theory underlying Bartlett's test, see for example Kendall *et al.* (1991). We assume that there are k sets of data, for k different treatments, to be compared. We wish, before doing an Analysis of Variance, to test the Null Hypothesis that all the estimated variances within each treatment are in fact estimating the same population value σ^2. Let the variances be $\{s_i^2\}$ with degrees of freedom $\{f_i\}$, for $i = 1$ to k.

Define

$$s^2 = \sum_{i=1}^{k} f_i s_i^2 \Big/ \sum_{i=1}^{k} f_i$$

and

$$M = \left(\sum_i f_i \right) \log_e s^2 - \sum_i (f_i \log_e s_i^2)$$

Also

$$C = 1 + \frac{1}{3(k-1)} \left(\sum_i (1/f_i) - 1 \Big/ \sum_i f_i \right)$$

Then M/C follows the distribution $\chi^2_{(k-1)}$ (Table III).

	f_i	s_i^2	$f_i \log_e s_i^2$
A	6	0.140333	-11.78242
B	4	0.421150	-3.45906
C	5	0.081067	-12.56240
D	6	0.129983	-12.24211
	21		-40.04599

s^2 is the pooled estimate already calculated, i.e. 0.17675.

$$M = 21 \log_e s^2 + 40.04599 = 3.6526$$

$$C = 1 + \tfrac{1}{9} \left\{ \left(\tfrac{1}{6} + \tfrac{1}{4} + \tfrac{1}{5} + \tfrac{1}{6} \right) - \tfrac{1}{21} \right\} = 1.0817$$

so the test statistic

$$\chi^2_{(3)} = \frac{3.6526}{1.0817} = 3.377$$

which is not significant at 5% and so gives no evidence of real differences among the four variances.

Because of the lack of sensitivity, it is often more helpful to look at graphical displays of residuals (page 66) than to carry out tests. The graphical approach sometimes sheds more light on the reasons for variability, such as having one or two outlying values in some treatments, and if the experimental units are still available, as in a long-term field experiment, it is wise to go and look at the units that have given unusual values and even to re-measure them. Treatments that seem to lead to greater variability in response among units may not be useful for general practical applications, and in Chapter 8 some of the recent ideas about studying variability in industrial processes are discussed.

4.5   The randomized complete block

The *randomized complete block* (often called simply the *randomized block*) design is very useful when the units available for an experiment are not all entirely homogeneous. As we saw in Chapter 3, a completely randomized design fails to control any systematic differences that may be present among the units; some systematic variation therefore remains in our estimate of σ^2, making it no longer a valid estimate of random natural variation. When there is only one source of systematic variation present, that is made the basis of blocking, and the simplest form of block design is the randomized complete block in which each treatment appears just once in each block. As we shall see (page 47) there is no difficulty in giving some especially important treatments extra replication, provided that every block contains exactly the same set of treatments; if different blocks contain different selections of treatments, the methods of Chapter 11 are necessary.

The simplest form of model for the Alternative Hypothesis in a randomized complete block is

$$y_{ij} = \mu + \tau_i + \beta_j + \varepsilon_{ij} \qquad i = 1 \text{ to } v, j = 1 \text{ to } r$$

where y_{ij} is the observation recorded on the unit receiving treatment i and lying in block j. All the other terms in the model have the same properties as for a completely randomized design, and β_j is a 'block effect' similar to the 'treatment effect' τ_i. As long as every treatment appears just once in every block, the two constraints on $\{\tau_i\}$ and $\{\beta_j\}$ will be

$$\sum_{i=1}^{v} \tau_i = 0, \qquad \sum_{j=1}^{r} \beta_j = 0$$

The least-squares solution for μ, $\{\tau_i\}$, $\{\beta_j\}$ comes from minimizing $\sum = \sum_{i=1}^{v} \sum_{j=1}^{r} (y_{ij} - \mu - \tau_i - \beta_j)^2$, using the two constraints just mentioned. The normal equations are:

for $\hat{\mu}$: $\quad -2 \sum_i \sum_j (y_{ij} - \hat{\mu} - \hat{\tau}_i - \hat{\beta}_j) = 0$

for $\hat{\tau}_i$: $\quad -2 \sum_j (y_{ij} - \hat{\mu} - \hat{\tau}_i - \hat{\beta}_j) = 0 \qquad$ for each $i = 1$ to v

for $\hat{\beta}_j$: $\quad -2 \sum_i (y_{ij} - \hat{\mu} - \hat{\tau}_i - \hat{\beta}_j) = 0 \qquad$ for each $j = 1$ to r

leading to

$$G = N\hat{\mu} + r \sum_i \tau_i + v \sum_j \beta_j$$

$$= N\hat{\mu} \quad \text{using the constraints above}$$

$$T_i = r\hat{\mu} + r\hat{t}_i + \sum_j \hat{\beta}_j$$

$$= r(\hat{\mu} + \hat{t}_i) \qquad \text{for each } i = 1 \text{ to } v$$

$$B_j = v\hat{\mu} + \sum_i \hat{t}_i + v\hat{\beta}_j$$

$$= v(\hat{\mu} + \hat{\beta}_j) \qquad \text{for each } j = 1 \text{ to } r$$

G and T_i are defined as on page 33, and B_j is the total of all the observations from units in block j.

Hence

$$\hat{\mu} = G/N; \qquad \hat{t}_i = \frac{T_i}{r} - \frac{G}{N}; \qquad \hat{\beta}_j = \frac{B_j}{v} - \frac{G}{N}$$

The residual sum of squares $\sum_{\min}$ is the value of $\sum$ when $\hat{\mu}$, $\{\hat{t}_i\}$, $\{\hat{\beta}_j\}$ are substituted:

$$\sum_{\min} = \sum_{i=1}^{v} \sum_{j=1}^{r} (y_{ij} - \bar{y}_i - \bar{y}_j + \bar{y})^2$$

in which $\bar{y}_i = T_i/r$, $\bar{y}_j = B_j/v$ and $\bar{y} = G/N$, leading to the residual sum of squares

$$\sum_{i=1}^{v} \sum_{j=1}^{r} \{(y_{ij} - \bar{y}) - (y_i - \bar{y}) - (y_j - \bar{y})\}^2$$

where all the cross-product terms will be 0 by calculations like those discussed above, so that there remains

$$\sum_i \sum_j (y_{ij} - \bar{y})^2 + r \sum_i (\bar{y}_i - \bar{y})^2 + v \sum_j (\bar{y}_j - \bar{y})^2$$

The three terms are separate from (independent of) each other; the first and second are the total sum of squares and the sum of squares for treatments which we have already seen, and the third is a sum of squares for blocks exactly analogous to that for treatments. For computing the sum of squares for treatments

$$r \sum_i (\bar{y}_i - \bar{y})^2 = \sum_i \frac{T_i^2}{r} - \frac{G^2}{N}$$

and for blocks it is

$$v \sum_j (\bar{y}_j - \bar{y})^2 = \sum_j \frac{B_j^2}{v} - \frac{G^2}{N}$$

Defining one more *summation term*

$$S_B = \sum_j \frac{B_j^2}{v}$$

we can construct the Analysis of Variance table given below.

The arguments about expected values of mean squares are exactly similar to those detailed above for a completely randomized experiment, and as usual $\hat{\sigma}^2$ is given by the residual mean square.

The form of computing a simple randomized complete block design Analysis of Variance is as shown in the following table:

Source of variation	d.f.	Sum of squares	Mean square
Blocks	$(r-1)$	$\sum_{j=1}^{r} B_j^2/v - G^2/N = S_B - S_0$	$\dfrac{S_B - S_0}{(r-1)} = M_B$
Treatments	$(v-1)$	$\sum_{i=1}^{v} T_i^2/r - G^2/N = S_T - S_0$	$\dfrac{S_T - S_0}{(v-1)} = M_T$
Residual	$(r-1)(v-1)$	(by subtraction)	S.S. Resid/$\{(r-1)(v-1)\}$ $= M_R$
Total	$rv-1$	$\sum_i \sum_j y_{ij}^2 - G^2/N = S - S_0$	

Example 4.2. In an experiment to determine people's reaction times to a stimulus (a flashing light) under different environmental conditions A-E, five students from the same age group, all of whom had used the equipment before, recorded times (ms) as shown in the following table.

Student		I	II	III	IV	V	Total
Treatment	A	213	127	155	246	200	941
	B	178	143	147	210	192	870
	C	254	151	174	266	222	1067
	D	103	108	122	144	161	638
	E	177	199	212	168	182	938
Total		925	728	810	1034	957	4454

$$S_0 = \frac{G^2}{N} = \frac{4454^2}{25} = 793524.64$$

$$S = 839414.00$$

$S_T = \frac{1}{5}(941^2 + \cdots + 938^2) = 813551.60$ ('treatments' are conditions)

$S_B = \frac{1}{5}(925^2 + \cdots + 957^2) = 805342.80$ ('blocks' are students)

The details of this analysis are given in the table below.

Analysis of Variance for Example 4.2

Source of variation	d.f.	Sum of squares	Mean square	
Students	4	11818.16	2954.54	$F_{(4,16)} = 3.37^*$
Conditions	4	20026.96	5006.74	$F_{(4,16)} = 5.70^{**}$
Residual	16	14044.24	$877.765 = s^2$	
Total	24	45889.36		

The *F*-value for conditions (treatments) is significant at the 1% level (see Table II) and that for students (blocks) at the 5% level. A useful shorthand is to denote values that are significant at 5% by one asterisk *, those at 1% by two asterisks ** and those at 0·1% by three asterisks ***. Many computer programs can give the actual probability (*P*-value) of the *F* value found in the calculation, but for interpretation we still wish to know whether this *P*-value is greater than 5% (not significant), between 5% and 1% (*), between 1% and 0.1% (**) or less than 0.1% (***).

The blocking, whereby each student tested each set of environmental conditions (in random order, ideally on different days to avoid a speeding up, or learning effect, from one test to the next), has been useful: students do appear to vary from one to another. If this systematic variation had not been removed, the analysis taking the form for a completely randomized scheme, the student-to-student variation would have become part of the residual; besides increasing s^2 it would have made the assumption of random, normally distributed $\{\varepsilon_{ij}\}$ invalid because it contained a systematic component.

For the moment, we will assume that there is no particular structure in the set of treatments that would allow definite questions to be answered, and so we will follow the same lines as in Example 4.1. When all treatments have the same replication *r*, a simplification is possible. We may compute *least significant differences*, which are the values of differences between two means that would just achieve significance.

If the observed difference between two means divided by the standard error (standard deviation) of that difference is greater than or equal to *t* with the residual degrees of freedom at the 5% level (Table I), that pair of means may be declared significantly different at 5%; and we can repeat the process for 1% and 0.1% levels in the same way. In a randomized complete block, the standard error of the difference between any pair of means is $\sqrt{2s^2/r}$; hence the observed difference divided by $\sqrt{2s^2/r}$ is distributed as $t_{(\text{resid})}$. It follows that any pair that differ by $t\sqrt{2s^2/r}$ or more are 'significantly different', where *t* stands for the two-tail critical value (Table I) of *t* with the residual degrees of freedom.

In Example 4.2 we need $t_{(16)}$, whose critical values at 5%, 1% and 0.1% are 2.120, 2.921, 4.015, respectively. The standard error of the difference between any pair of means is

$$\sqrt{\frac{2 \times 877.765}{5}} = 18.738$$

Hence *least significant differences* are

$$18.738 \times \begin{cases} 2.120 = 39.72 \ (5\%) \\ 2.921 = 54.73 \ (1\%) \\ 4.015 = 75.23 \ (0.1\%) \end{cases}$$

In the table below the means are set out in increasing order of size, and as in the earlier example we bracket together those that do not differ significantly at a given level.

		Significance level		
Treatment	Mean	5%	1%	0.1%
D	127.6			
B	174.0			
E	187.6			
A	188.2			
C	213.4			

We can see that *D* is the only one that shows evidence of being different from the others. If least significant differences are used intelligently, in combination with any other information available on the experimental treatments and material, they can be quite helpful in interpreting the results of a preliminary experiment. We shall shortly (page 47) study methods which are much more suitable as a programme of experimental work develops.

4.6   Duncan's multiple range test

The full details and the necessary tables for this test can be found in Duncan (1955). Many statistical package programs include this test as an option, some as routine. The result of using it gives a display somewhat like that for least significant differences; but the purpose for which Duncan's test was designed is to correct for the error made if *all possible pairs* of means are compared using the least significant differences calculated above, namely that we violate the stated level (5%, etc.) at which we claim to be working. Strictly the least significant difference method only gives exactly the right answer when there are just two treatments, because there is only one *t*-test that can be made in that situation.

A typical layout for the output from a Duncan multiple range test is to label each group of treatments that do not differ by attaching the same letter to all in that group: in Example 4.2 the results for 5% and 1% are

$$5\%: \quad D \quad B \quad E \quad A \quad C \qquad 1\%: \quad D \quad B \quad E \quad A \quad C$$
$$ \quad a \quad b \quad b \quad b \quad b \qquad \quad a \quad ab \quad b \quad b \quad b$$

This shows that at 5% *D* differs from all the others (so it is the only one that has letter *a* with it) while *B*, *E*, *A* and *C* do not differ among themselves (so all carry letter *b*). At 1% there is a slight change, because now *B* has to be bracketed with *D* on one side, as well as with all the others above it, and we show this by writing *ab* against *B*. As we shall note later (page 90), the occasions on which this test is valid or useful are severely limited. O'Neill and Wetherill (1971) discuss multiple comparison methods.

4.7 Extra replication of important treatments

Provided every block contains the same selection of treatments, the normal equations lead to estimates of the treatment parameters $\{\tau_i\}$ and of μ that are independent of the block parameters $\{\beta_j\}$. A design in which this happens, and in which the block and treatment and residual sums of squares add to give the total sum of squares, is *orthogonal* in a mathematical sense and is particularly simple to analyse in practical use. Chapter 11 shows the additional complexity introduced when block designs are not simple, and become non-orthogonal.

The design with A twice, and BCD once each, in *every* block is still orthogonal and the analysis is almost the same as for a simple randomized complete block. The only difference is in the treatments sum of squares:

$$S_T = \frac{T_A^2}{2r} + \frac{T_B^2 + T_C^2 + T_D^2}{r}$$

When comparing means, we need to realize that any pair of means not including A will still have $\sqrt{2s^2/r}$ as the estimated standard error of their difference; but the comparison of A with any other mean will have estimated standard error $\sqrt{s^2/2r + s^2/r}$. If comparisons with A are particularly important, perhaps because it is a standard against which new treatments are being compared, this additional replication is a useful and simple way of increasing the precision of these comparisons without affecting orthogonality or reducing the precision of others.

Occasionally two treatments may need extra replication, one of which may be a standard A as the basis of comparison for others, B, C, D, and the other a *control*, O, often untreated in agricultural trials, which is included to examine the background conditions in which the experiment is being done (are there any of the pests, which we think we are trying to eradicate, actually present?); each block could then consist of $OOAABCD$, suitably randomized for each block. The treatments sum of squares now has two terms divided by $2r$ and three divided by r, while there are three standard errors to consider according to whether O and A are compared, *or* O or A with B, C, or D, *or* B, C and D among themselves.

The randomized complete block is quite a flexible design, the only restriction being that we must retain an adequate number of degrees of freedom for residual, so that t- and F-tests have good precision. In field experiments with moderate variability, 10 to 12 d.f. is considered a working minimum, though in very well-controlled studies in industry or in the laboratory where units do not vary much among themselves, this number may be reduced slightly.

4.8 Contrasts among treatments

When an experiment is designed to answer specific questions, as it should always be when possible (certainly in the later stages of any research programme or study) we can nominate particular comparisons or *contrasts* among the means or totals for the treatments; these comparisons enable the questions to be answered and are the only ones that will be tested or examined in the analysis. We will assume in our discussion of contrasts that all treatments have equal replication r, although the method can of course be used without this restriction.

If an experiment involves three treatments A, B, C, and A is different in character from the other two, it will be useful to examine the following contrasts:

- the difference between A and the average (mean) of B and C;
- the difference between B and C.

The first of these requires $\bar{y}_A - \frac{1}{2}(\bar{y}_B + \bar{y}_C)$ in the simple case of equal replication. This is equivalent to $2\bar{y}_A - \bar{y}_B - \bar{y}_C$, or in terms of treatment totals it is $2T_A - T_B - T_C$, provided we make proper allowance for the units in which the numbers are expressed. It is usual to express these contrasts with integer coefficients $(2 \ -1 \ -1)$ [rather than $(1 \ -\frac{1}{2} \ -\frac{1}{2})$] because when computing sums of squares the units are taken care of in the method of analysis.

An expression like $2T_A - T_B - T_C$ is a *linear contrast* among the treatment totals.

We could study contrasts simply by carrying out the same type of Analysis of Variance as before, finding s^2 and then using this to estimate the standard error of the linear contrast we are interested in. For example,

$$\text{var}\,(\bar{y}_A - \tfrac{1}{2}\bar{y}_B - \tfrac{1}{2}\bar{y}_C) = \frac{\sigma^2}{r} + \frac{\sigma^2}{4r} + \frac{\sigma^2}{4r} = \frac{3\sigma^2}{2r}$$

provided the experiment has been properly randomized so that the treatment estimates are independent of one another. The standard error of this contrast is therefore estimated by $\sqrt{3s^2/2r}$ and so

$$\frac{\bar{y}_A - \tfrac{1}{2}(\bar{y}_B + \bar{y}_C)}{\sqrt{3s^2/2r}}$$

is distributed as t with the residual degrees of freedom. This could be used to carry out a significance test of the hypothesis '$\mu_A = \frac{1}{2}(\mu_B + \mu_C)$', or to set confidence limits for the true value of this contrast.

A better approach for testing the hypotheses represented by the contrasts is to divide the sum of squares for treatments into single degrees of freedom, one for each contrast, giving $F_{(1,\text{resid})}$ tests for each. These are much more useful than the overall F-test we have used so far. There are restrictions on how this subdivision can be done, because the contrasts for the $(v-1)$ single degrees of freedom must all be mutually orthogonal if the sum of squares for treatments is to be completely divided up in this way. Sometimes there are not as many as $(v-1)$ separate contrasts that need to be examined. In such a case, as many (orthogonal) contrasts as needed may be extracted, and the remaining d.f. to complete $(v-1)$ are 'other treatment differences' (which we do not wish to study).

Formal definition of a linear contrast

When treatments are equally replicated, with totals $\{T_i\}$, $i=1, 2, \ldots, v$, and $\{c_i\}$ are a set of constants (coefficients), then $\sum_{i=1}^{v} c_i T_i$ is a linear contrast among the treatment totals, provided that $\sum_{i=1}^{v} c_i = 0$. A second contrast $\sum_{i=1}^{v} d_i T_i$, with $\sum_{i=1}^{v} d_i = 0$, is *orthogonal* to the first if and only if $\sum_{i=1}^{v} c_i d_i = 0$. The sum of squares, with 1 degree of freedom, associated with a contrast is $(\sum_{i=1}^{v} c_i T_i)^2 / (r \sum_{i=1}^{v} c_i^2)$, $\sum_{i=1}^{v} c_i T_i$ being the *value* of the contrast and $r \sum c_i^2$ its *divisor*.

Treatment structure and choice of contrasts are closely related.

Example 4.3. In a completely randomized design, four treatments are used. *A* is a 'standard'; *B* uses an additive *X* from one supplier and *C* uses *X* from another supplier; *D* uses additive *Y*. The results of a test are:

$$A \quad 34, 37, 40, 29, 29 \qquad C \quad 31, 35, 36, 36, 32$$
$$B \quad 38, 44, 36, 40, 47 \qquad D \quad 48, 51, 48, 56, 52$$

Examine the contrasts

(i) standard against other treatments,
(ii) substance *X* against *Y*,
(iii) one supplier of *X* against the other.

Treatment totals are: *A*, 169; *B*, 205; *C*, 170; *D*, 255. $r = 5$.
   In terms of contrasts, we require

(i) $\bar{y}_A - \frac{1}{3}(\bar{y}_B + \bar{y}_C + \bar{y}_D)$
(ii) $\frac{1}{2}(\bar{y}_B + y_C) - \bar{y}_D$
(iii) $\bar{y}_B - \bar{y}_C$.

As usual, we convert these to a form in which all coefficients are integers before applying them to the totals $\{T_i\}$. The calculation may be set out as follows.

Treatment	A	B	C	D			Sum of
Total	169	205	170	255	Value	Divisor	squares
(i) *A* versus (*B, C, D*)	3	−1	−1	−1	−123	12 × 5	252.15
(ii) (*B, C*) versus *D*	0	1	1	−2	−135	6 × 5	607.50
(iii) *B* versus *C*	0	1	−1	0	35	2 × 5	122.50
							982.15

Check that

$$S_T = \frac{1}{5}(169^2 + 205^2 + 170^2 + 255^2) = 164511/5 = 32902.20$$

and

$$S_0 = 799^2/20 = 31920.05$$

giving

$$S_T - S_0 = 982.15$$

the 3-degree-of-freedom sum of squares for treatments. All pairs of contrasts satisfy $\sum_{i=1}^{v} c_i d_i = 0$, so are mutually orthogonal, and therefore the three single-degree-of-freedom 'sums of squares' will add up to the 3 d.f. sum of squares for treatments. Unless contrasts are orthogonal, this will not happen.

To complete the analysis, $S = 34^2 + 37^2 + \cdots + 56^2 + 52^2 = 33143$ and $S - S_0 = 1222.95$. The Analysis of Variance results are given in the following table.

Source of variation	d.f.	Sum of squares	Mean square	
A versus (B, C, D)	1	252.15	252.15	$F_{(1,16)} = 16.75$***
(B, C) versus D	1	607.50	607.50	$F_{(1,16)} = 40.37$***
B versus C	1	122.50	122.50	$F_{(1,16)} = 8.14$*
Treatments	3	982.15		
Residual	16	240.80	$15.05 = s^2$	
Total	19	1222.95		

From this and from the treatment means we see that

(i) A is significantly lower than the mean of B, C, D,
(ii) D is significantly higher than the mean of A, C,
(iii) B is significantly higher than C.

If confidence limits to these contrasts were required, expressed in terms of means per unit plot, we should require the variances:

(i) $\mathrm{Var}\,[\bar{y}_A - \frac{1}{3}(\bar{y}_B + \bar{y}_C + \bar{y}_D)] = \dfrac{\sigma^2}{5} + \dfrac{1}{9}\left(\dfrac{\sigma^2}{5} + \dfrac{\sigma^2}{5} + \dfrac{\sigma^2}{5}\right) = \dfrac{4\sigma^2}{15}$

   estimated by $\frac{4}{15} \times 15.05 = 4.013$;

(ii) $\mathrm{Var}\,[\bar{y}_D - \frac{1}{2}(\bar{y}_B + \bar{y}_C)] = \dfrac{\sigma^2}{5} + \dfrac{1}{4}\left(\dfrac{\sigma^2}{5} + \dfrac{\sigma^2}{5}\right) = \dfrac{3\sigma^2}{10}$

   estimated by $\frac{3}{10} \times 15.05 = 4.515$;

(iii) $\mathrm{Var}\,[\bar{y}_B - \bar{y}_C] = \dfrac{2\sigma^2}{5}$ estimated by $\frac{2}{5} \times 15.05 = 6.020$.

Since s^2 has 16 d.f., $t_{(16)}$ is used in calculating confidence intervals. For example, a 95% confidence interval for the true value of $(\mu_D - (\mu_B + \mu_C)/2)$ is

$$\{\bar{y}_D - \tfrac{1}{2}(\bar{y}_B + \bar{y}_C)\} \pm t_{(16)} \sqrt{3s^2/10} \text{ or } 13.5 \pm 2.12 \times 2.125 \text{ or } 13.5 \pm 4.50$$

We can therefore say with 95% confidence that the interval (9.0; 18.0) contains the true value of this contrast. This interval is rather wide, showing that the variability in these data is rather high.

Note that in the analysis of variance, there was no need to test the significance of $F_{(3,16)}$ for treatments; this would examine the Null Hypothesis 'all $\{\tau_i\}$ are 0', in which we are not interested because specific treatment contrasts are given to be examined. Note also that in this example there does not appear to be any need to quote individual treatment means or their standard errors.

Example 4.4 Choice of contrasts. Suppose that in an agricultural trial O is an untreated 'control', to measure the infestation y of a certain crop-pest if it is not treated, and S is a standard method of treating it. Experimental treatments A, B, C, D are new (and possibly improved) methods of controlling the pest. *A* and *B* use one compound in different physical forms, while *C* and *D* are another compound applied at different strengths. Appropriate questions are:

(i) Is *O* different from the others (is any infestation present)?
(ii) Is *S* as good as the new treatments?
(iii) Are *A* and *B* different (is there any effect of physical form)?
(iv) Are *C* and *D* different (is there any strength effect)?
(v) Are *A* and *B* different from *C* and *D* (is there a compound effect)?

There are six treatments *O, S, A, B, C, D* and therefore five independent possible comparisons. Provided that every treatment *O, S, A, B, C, D* is replicated the same number of times, *r*, the sets (i)–(v) are orthogonal, as we can check from the table of coefficients below, and they will therefore account exactly for the 5 d.f. sum of squares between the treatment totals.

	O	*S*	*A*	*B*	*C*	*D*	Divisor
(i) *O* versus (*S, A, B, C, D*)	5	−1	−1	−1	−1	−1	30*r*
(ii) *S* versus (*A, B, C, D*)	0	4	−1	−1	−1	−1	20*r*
(iii) *A* versus *B*	0	0	1	−1	0	0	2*r*
(iv) *C* versus *D*	0	0	0	0	1	−1	2*r*
(v) (*A, B*) versus (*C, D*)	0	0	1	1	−1	−1	4*r*

These contrasts satisfy one of the practical rules for achieving orthogonality: first make comparisons between groups and then remain within groups that have already been compared. After (i), further comparisons must be within the group (*S, A, B, C, D*); after (ii) we must remain within (*A, B, C, D*); (v) may be considered next and after that we must remain within (*A, B*) and within (*C, D*), which lead automatically to (iii) and (iv).

Including a standard for comparison

For certain experiments, it is necessary to compare each new treatment *A, B, C* . . . against a standard *S*. Here *S* will be an established method of carrying out a particular operation and one of the new treatments will only be adopted if it is clearly better than *S*. *S* versus *A*, *S* versus *B*, *S* versus *C*, etc., are *not* orthogonal because they do not satisfy $\sum_i c_i d_i = 0$; but there are only $(v - 1)$ comparisons in this set so it is acceptable to carry out each of these tests as a simple difference between two means, using s^2 estimated by ANOVA.

Example 4.5. In a test of engine performance with three different fuels *F, G, H* and four additives 1, 2, 3, 4, it is not possible to include all additives with all fuels. The treatments possible are *F*1, *F*2, *G*2, *G*3, *G*4, *H*1, *H*2, *H*3, *H*4. Possible contrasts are:

(i) *F* against *G* (i.e. *F*1, *F*2, against *G*2, *G*3, *G*4),
(ii) (*F, G*) against *H* (all the above against *H*1, *H*2, *H*3, *H*4),
(iii) *F*1 against *F*2,

(iv) within the group (*G*2, *G*3, *G*4),
 (v) within the group (*H*1, *H*2, *H*3, *H*4).

We could *not* do '*F* against *G*' and '*F* against *H*' as these are not orthogonal; but '(*F*, *G*) against *H*' *can* follow '*F* against *G*'. The table below illustrates this.

	F1	F2	G2	G3	G4	H1	H2	H3	H4
F against *G*:	3	3	−2	−2	−2	0	0	0	0
F against *H*:	2	2	0	0	0	−1	−1	−1	−1
[(*F*, *G*) versus *H*:	4	4	4	4	4	−5	−5	−5	−5]

The d.f. for (iv) and (v) are 2 and 3, respectively, and there may not be any particular structure to these treatments leading to single d.f. contrasts. We could compute (iv) as a sum of squares between the totals for *G*2, *G*3, *G*4, ignoring all other treatments.

If a set of treatments has not been thought out sufficiently carefully, one or more useful treatments may have been omitted from what would be needed to construct a full orthogonal set of contrasts; in this case as many as possible are extracted and the rest of the treatment d.f. are left as 'remainder'. Consider these fertilizer treatments:

A, no added fertilizer;
B, nitrogen alone;
C, nitrogen + phosphorus;
D, nitrogen + lime;
E, nitrogen + phosphorus + potassium (standard mixture);
F, nitrogen + phosphorus + potassium (special mixture).

Possible contrasts are shown below.

	A	*B*	*C*	*D*	*E*	*F*
Addition of nitrogen	−5	1	1	1	1	1
N only against N plus others	0	−4	1	1	1	1
N + P against N + P + K	0	0	2	0	−1	−1
Different mixtures *E*/*F*	0	0	0	0	1	−1

There is now *no* orthogonal contrast that can be added to this set to examine the effect of lime: we cannot use

$$0 \quad 1 \quad 0 \quad -1 \quad 0 \quad 0$$

4.9   Latin squares and other orthogonal designs

A randomized block design removes one systematic source of variation in addition to treatments. There may be two such sources of variation to be removed; this is not unusual in industry. If different groups or 'shifts' of workers take part in a production process there may be different levels of quality in the resulting product. If more than one set of machinery is used, the same may occur. The actual treatments being studied could be the supply of raw materials or the amounts of particular compounds included in a mixture in a chemical reaction. These treatments are likely to be independent of machinery or workers in the effects they produce. (If they are *not* independent, see Chapter 5.)

A mathematical pattern, the Latin square, allows two systematic sources of variation to be removed before treatments A, B, C, . . . are compared. This pattern has each letter occurring once in each row and once in each column. Squares are usually listed in 'standard order' with the first row and the first column in alphabetical order (Fig. 4.1).

$$
\begin{array}{cccc}
\text{A B} & \text{A B C} & \text{A B C D} & \text{A B C D E} \\
\text{B A} & \text{B C A} & \text{B D A C} & \text{B C E A D} \\
2 \times 2 & \text{C A B} & \text{C A D B} & \text{C A D E B} \\
& 3 \times 3 & \text{D C B A} & \text{D E B C A} \\
& & 4 \times 4 & \text{E D A B C} \\
& & & 5 \times 5
\end{array}
$$

Fig. 4.1 Latin squares

There is only one standard 2×2 or 3×3 square; there are more for larger sizes. In order to prepare a square for use, after choosing one at random from a standard list such as Table XV of Fisher and Yates (1963), the rows are permuted at random and so are the columns. For the 5×5 square shown in Fig. 4.1, this may give

$$
\begin{array}{ccccc}
B & E & C & A & D \\
A & C & B & D & E \\
E & A & D & B & C \\
C & D & A & E & B \\
D & B & E & C & A
\end{array}
$$

The treatments should then be allocated to letters at random. Clearly the number of treatments has to be the same as the number of rows and the number of columns: this can be an inconvenient restriction. If five different machines are to be used, on five days of the week, to compare five different treatments A, B, C, D, E, we may make machines correspond to rows and days to columns. On day 1, machine 1 uses treatment B; machine 2, A; machine 3, E, etc. On day 2, machine 5 uses treatment B; and so on to complete 25 experimental units, each observation being the production from one machine on one day using one treatment.

The linear model for analysis is

$$y_{ijk} = \mu + \tau_i + \rho_j + \kappa_k + \varepsilon_{ijk}$$

in which μ and τ_i are as before, and ρ_j, κ_k are both terms like blocks. (Although there are three suffixes i, j, k, there are only v^2 units, because once i and j are given, in the ranges $i = 1$ to v and $j = 1$ to v, then k must follow uniquely from the design used. For example, $i = 5$ and $j = 5$ above requires $k = 3$ for that unit: treatment $E(5)$ on row 5 lies in column 3.) The residual term ε_{ijk} is assumed to be distributed as $\mathcal{N}(0, \sigma^2)$ as usual. All terms in the model are orthogonal: each treatment occurs once in each row and once in each column, and every row/column combination has an experimental unit in it.

The analysis (shown in the following table) is an exact extension of a randomized block.

Source of variation	d.f.	Sum of squares
Rows	$v-1$	$S_R - S_0$
Columns	$v-1$	$S_C - S_0$
Treatments	$v-1$	$S_T - S_0$
Residual	$(v-1)(v-2)$	(by subtraction)
Total	$v^2 - 1$	$S - S_0$

Residual degrees of freedom are $(v-1)(v-2)$:

$$v = 2 \quad 3 \quad 4 \quad 5 \quad 6 \quad 7$$
$$\text{d.f.} = 0 \quad 2 \quad 6 \quad 12 \quad 20 \quad 30$$

In a well-planned experiment, we have already remarked that the residual d.f. should usually be at least 10, although in a carefully controlled industrial or laboratory experiment 6 may be adequate. *F*- and *t*-tests will not be very sensitive with few d.f. The 5×5 square is commonly used, as 25 units can usually be handled satisfactorily in a single experiment. But if there are either only four days, *or* only four machines, to be used, an orthogonal row-and-column arrangement can still be found by using two 4×4 squares together, provided that *either* eight machines can be used on four days *or* four machines on eight days. An example of such a scheme is:

$$
\begin{array}{cccccccc}
B & D & A & C & A & D & C & B \\
D & A & C & B & C & B & A & D \\
C & B & D & A & D & C & B & A \\
A & C & B & D & B & A & D & C
\end{array}
$$

We should now be restricted to comparing only 4 treatments A, B, C, D, and the Analysis of Variance would contain the terms shown in the table below.

Source of variation	d.f.
Rows	3
Columns	7
Treatments	3
Residual	18
Total	31

When calculating sums of squares, we must remember that each row total is based on 8 observations, so $S_R = \sum_j R_j^2/8$, while $S_C = \sum_k C_k^2/4$. Treatments also have 8 replicates, so $S_T = \sum_i T_i^2/8$ and the estimated standard error of a difference between two means is $\sqrt{2s^2/8}$, s^2 as usual coming from the residual mean square (with 18 d.f.).

Example 4.6. Analysis of a single 4×4 square. Four companies A, B, C, D supply a component for a motor vehicle. The component can be used in four different models of engine for four different versions of car; all models of engine can be fitted to all versions of car. A standard test is applied, and the lengths of time for which the component works satisfactorily in the test are recorded (the units in the table have been coded for simple analysis).

Model of engine	(1)	(2)	(3)	(4)	Total
Version of car (1)	B: 25	A: 31	C: 40	D: 26	122
(2)	A: 32	C: 33	D: 32	B: 34	131
(3)	D: 24	B: 30	A: 36	C: 41	131
(4)	C: 37	D: 28	B: 28	A: 37	130
Total	118	122	136	138	G = 514

$N = 16$. Supplier totals are A, 136; B, 117; C, 151; D, 110.

$$S_0 = \frac{514^2}{16} = 16512.25$$

For engines (columns of design),

$$S_E = \tfrac{1}{4}(118^2 + 122^2 + 136^2 + 138^2) = 16587.00$$

For cars (rows of design)

$$S_C = \tfrac{1}{4}(122^2 + 131^2 + 131^2 + 130^2) = 16526.50$$

For suppliers (A, B, C, D)

$$S_S = \tfrac{1}{4}(136^2 + 117^2 + 151^2 + 110^2) = 16771.50$$

The design assumes that the effects of engines, cars and suppliers are all independent. For example, engine (1) is tested only once on car (1), using B's component; this engine–car combination never uses material from A, C or D. In this way the number of experimental units is kept down to 16. The Analysis of Variance is shown in the table below.

Source of variation	d.f.	Sum of squares	Mean square	
Engines	3	74.75	24.917	$F_{(3,6)} = 2.79$ n.s.
Cars	3	14.25	4.750	$F < 1$
Suppliers	3	259.25	86.417	$F_{(3,6)} = 9.69^*$
Residual	6	53.50	8.917	
Total	15	401.75		

n.s. = not significant

The only differences appear to be those among suppliers. In the absence of any specific hypotheses for testing, we may calculate the least significant difference between a pair of means; it is

$$t_{(6)}\sqrt{\frac{2 \times 8.917}{4}} = 2.112 t_{(6)}$$

The critical values of $t_{(6)}$ are 2.45 (5%), 3.71 (1%) and 5.96 (0.1%) and the least significant differences are therefore 5.17 (5%), 7.84 (1%), 12.59 (0.1%). Treatment means (in order of size) are D, 27.50; B, 29.25; A, 34.00; C, 37.75. As a guide to possible hypotheses for testing in future experiments, we may note that C is greater than both D and B at the 1% level, while A is greater than D, this time at the 5% level.

Example 4.7. Contrasts in a Latin square. Five different groups of apples are stored under different conditions, and the percentage weight losses in each group during storage are recorded. Groups A and B are the same variety, and groups C, D and E are another variety; A and C are stored for a short period, B and E for a long period, while D is stored for the long period in a protective covering. A Latin square design (shown below) is used to take out possible variations in environmental conditions in the store, rows being vertically above one another and columns going from a door to a wall of the store.

Column (dist. from door)		1	2	3	4	5
Row (vertical height)	1	C: 18.30	B: 35.25	D: 30.32	A: 16.08	E: 42.85
	2	D: 28.05	E: 36.16	A: 17.25	C: 25.90	B: 31.98
	3	A: 25.12	D: 28.55	B: 37.10	E: 38.27	C: 23.68
	4	B: 40.25	C: 22.60	E: 41.15	D: 31.68	A: 22.15
	5	E: 34.24	A: 26.42	C: 15.05	B: 36.52	D: 33.20

Row totals are 142.80, 139.34, 152.72, 157.83, 145.43.
Column totals are 145.96, 148.98, 140.87, 148.45, 153.86.
$N = 25$, $G = 738.12$. $S_0 = 738.12^2/25 = 21792.8454$; $S = 23375.9838$;
$S_R = \frac{1}{5}(142.80^2 + \cdots + 145.43^2) = 21838.2136$; $S_R - S_0 = 45.3682$;
$S_C = \frac{1}{5}(145.96^2 + \cdots + 153.86^2) = 21810.8042$; $S_C - S_0 = 17.9588$; $S - S_0 = 1583.1384$.

Treatment totals are (A) 107.02, (B) 181.10, (C) 105.53, (D) 151.80, (E) 192.67. $S_T = \frac{1}{5}(107.02^2 + \cdots + 192.67^2) = 23110.4080$, $S_T - S_0 = 1317.5626$.

Suitable contrasts to extract from the 4 d.f. sum of squares for treatments would appear to be

(i)   varieties, (A, B) versus (C, D, E);
(ii) storage time for first variety, A versus B;
(iii) storage time for second variety, C versus (D, E);
(iv) effect of protection (second variety, long time), D versus E.

If we try to take out long versus short storage time, it becomes difficult to complete a useful orthogonal set of four contrasts:

Treatment Total	A 107.02	B 181.10	C 105.53	D 151.80	E 192.67	Value	Divisor	Sum of squares
(i) Varieties	-3	-3	2	2	2	35.64	30×5	8.4681
(ii) Time (V.1)	-1	1	0	0	0	74.08	2×5	548.7846
(iii) Time (V.2)	0	0	-2	1	1	133.41	6×5	593.2743
(iv) Protection	0	0	0	-1	1	40.87	2×5	167.0357
								1317.5627

Analysis of Variance

Source of variation	d.f.		Sum of squares	Mean square	
Rows	4		45.3682	11.342	$F < 1$
Columns	4		17.9588	4.490	$F < 1$
(i)	1	8.4681		8.468	$F < 1$
(ii)	1	548.7846		548.785	$F_{(1,12)} = 32.56^{***}$
(iii)	1	593.2743		593.274	$F_{(1,12)} = 35.20^{***}$
(iv)	1	167.0357		167.036	$F_{(1,12)} = 9.91^{**}$
Treatments	4		1317.5627		
Residual	12		202.2487	$16.854 = s^2$	
Total	24		1583.1384		

Although the Analysis of Variance above shows that there is no evidence of environmental differences, and no mean difference between varieties according to these data, storage time makes a large difference in percentage weight loss for both the varieties (contrasts (ii) and (iii)) and protection also has a considerable effect (contrast (iv)).

There is no need for *t*-tests, since an *F*-test with 1 d.f. is equivalent to t: $F_{(1,f)} = t_{(f)}$ for any chosen probability level, and any value f of the degrees of freedom for t.

If confidence intervals are required for any contrasts, they are easy to find by the methods already discussed.

Example 4.8. Two 3 × 3 squares. An experiment is carried out over six days (columns in the design) using three machines *A*, *B*, *C* running at three speeds (rows in the design) while making a mass-produced item. The number of sound items is counted in samples of a fixed size:

Day	1	2	3	4	5	6	Speed total
Speed: HIGH	*B*: 39	*C*: 31	*A*: 32	*C*: 58	*A*: 40	*B*: 53	253
MEDIUM	*A*: 52	*A*: 29	*B*: 37	*B*: 41	*C*: 80	*C*: 88	327
LOW	*C*: 89	*B*: 61	*C*: 62	*A*: 63	*B*: 99	*A*: 43	417
Day total	180	121	131	162	219	184	997 = G

$$N = 18, \ S = 63123, \ S_0 = 997^2/18 = 55222.722$$

Totals for machines are: *A*, 259; *B*, 330; *C*, 408. Hence

$$S_M = \tfrac{1}{6}(259^2 + 330^2 + 408^2) = 57074.167$$

$$S_S = \tfrac{1}{6}(253^2 + 327^2 + 417^2) = 57471.167 \text{ for speeds}$$

$$S_D = \tfrac{1}{3}(180^2 + 121^2 + \cdots + 184^2) = 57421.000 \text{ for days}$$

The ANOVA is shown below.

Source of variation	d.f.	Sum of squares	Mean square	
Days	5	2198.28	439.66	$F_{(5,8)} = 2.20$ n.s.
Treatments	2	1851.45	925.72	$F_{(2,8)} = 4.62$*
Speeds	2	2248.44	1124.22	$F_{(2,8)} = 5.61$*
Residual	8	1602.11	200.26	
Total	17	7900.28		

n.s. = not significant.

There is no real evidence that days vary significantly, but speeds and treatments both do so. Clearly low speed on machine *C* gave the best results in this experiment after allowing for day variation. Although the means could be studied further by the methods already discussed, we are obviously looking for the best combination in this experiment and '*C*, low' is that combination.

4.10   Graeco-Latin squares

There may be as many as three sources of systematic variation to remove before studying treatments. An extension of the Latin square, called the Graeco-Latin square, can do this. A typical pattern is

$$
\begin{array}{ccccc}
A\alpha & B\beta & C\gamma & D\delta & E\varepsilon \\
B\gamma & C\delta & D\varepsilon & E\alpha & A\beta \\
C\varepsilon & D\alpha & E\beta & A\gamma & B\delta \\
D\beta & E\gamma & A\delta & B\varepsilon & C\alpha \\
E\delta & A\varepsilon & B\alpha & C\beta & D\gamma
\end{array}
$$

in which rows, columns and Greek letters can represent the systematic variation removed before examining the treatments, which are denoted by the Latin letters *A–E*.

For example, five different people ($\alpha, \beta, \gamma, \delta, \varepsilon$) on five days of the week (rows) using five different word processors (columns) can carry out five different types of work (*A* = letters, *B* = technical report, *C* = manuscript of research paper, ...) and record the number of errors made. The main interest is in the differences between the types of work, which we therefore regard as the 'treatments', but the other sources of variation may also be interesting, especially the word processors which form the columns of the design.

Probably 5×5 is the most useful size of Graeco-Latin square; it is not possible to construct a 6×6 square, and residual degrees of freedom which are $(v-1)(v-3)$ are too small for single squares of less than 5×5.

v	=	2	3	4	5	6	7
Degrees of freedom		*	0	3	8	*	24

* = no square exists.

As with a Latin square, after selecting a standard square at random from those available using, e.g. Fisher and Yates' (1963) table, the order of rows and columns is permuted, and the treatments (Latin) and people (in this example Greek) are allocated at random to letters. Also as with a Latin square, it is necessary that the sources of variation represented by rows, columns, Latin and Greek letters shall not interact (see Chapter 5 if they do).

Example 4.9. An example of a design based on two 4×4 squares is as follows. Rows represented four positions on a multiple-head machine for making glass bottles; columns were eight different days of running the experiment; Latin letters were different mixes of material; Greek letters were different colouring additives. The numbers of flaws in bottles during an hour's production were recorded. The design was:

$$
\begin{array}{cccccccc}
C\gamma & A\alpha & D\delta & B\beta & D\gamma & A\beta & C\delta & B\alpha \\
D\alpha & B\gamma & C\beta & A\delta & C\beta & B\gamma & D\alpha & A\delta \\
A\beta & C\delta & B\alpha & D\gamma & B\delta & C\alpha & A\gamma & D\beta \\
B\delta & D\beta & A\gamma & C\alpha & A\alpha & D\delta & B\beta & C\gamma
\end{array}
$$

Each column contains every Latin letter once and every Greek letter once; each row contains every Latin letter twice and every Greek letter twice; and every Greek letter appears twice with every Latin letter. The design is completely orthogonal, and no records were lost during the experiment, so the full set of data (see below) can be analysed by a method that is an extension of that in Section 4.9.

Days		(1)	(2)	(3)	(4)	(5)	(6)	(7)	(8)	Total
Position	(1)	23	18	14	23	8	11	10	15	122
	(2)	14	35	24	19	15	6	12	9	134
	(3)	31	11	24	9	25	20	14	15	149
	(4)	26	20	10	15	8	7	22	19	127
Total		94	84	72	66	56	44	58	58	$G = 532$

Mixes A–D gave totals: A, 120; B, 176; C, 137; D, 99; therefore

$$S_M = \tfrac{1}{8}(120^2 + 176^2 + 137^2 + 99^2) = 9243.25 \text{ for mixes}$$

The totals for colours α–δ were: α, 126; β, 161; γ, 124; δ, 121, giving

$$S_C = \tfrac{1}{8}(126^2 + 161^2 + 124^2 + 121^2) = 8976.75 \text{ for colours}$$

Row totals give $S_P = \tfrac{1}{8}(122^2 + 134^2 + 149^2 + 127^2) = 8896.25$ for positions
Column totals give $S_D = \tfrac{1}{4}(94^2 + 84^2 + \cdots + 58^2) = 9308.00$ for days

$$S = 10476 \text{ and } S_0 = 532^2/32 = 8844.50$$

The Analysis of Variance is given in the following table:

Source of variation	d.f.	Sum of squares	Mean square	
Rows (positions)	3	51.75	17.25	$F < 1$
Columns (days)	7	463.50	66.21	$F_{(7,15)} = 1.70$ n.s.
Latin (mixes)	3	398.75	132.92	$F_{(3,15)} = 3.41^*$
Greek (colours)	3	132.25	44.08	$F_{(3,15)} = 1.13$ n.s.
Residual	15	585.25	$39.02 = s^2$	
Totals	31	1631.50		

n.s. = not significant.

We may examine means for mixes (each based on 8 replicates) using least significant differences (since no more information is given about mixes which may lead to more specific hypotheses). The means are

$$\begin{array}{cccc} D & A & C & B \\ 12.375 & 15.000 & 17.125 & 22.000 \end{array}$$

The variance of a difference is

$$\sqrt{\frac{2s^2}{8}} = \sqrt{\frac{39.02}{4}} = 3.123$$

and the least significant difference is therefore $3.123t_{(15)}$. The critical points of $t_{(15)}$ are 2.13 (5%), 2.95 (1%), 4.07 (0.1%), and the least significant differences are therefore 6.65 (5%), 9.21 (1%), 12.71 (0.1%). Mix B seems to give more flaws, because $B > D$ at the 1% level and $B > A$ at 5%; on the evidence of this experiment there are no real differences between A, C and D. Unless mix B can be improved it will be dropped. Any future experiment should concentrate on studying A, C and D more precisely, together with any useful variations on these.

4.11 Two fallacies

'Pooling' degrees of freedom to obtain more for the residual

In a randomized block design where there are not very many degrees of freedom for the residual, some people would advocate making an F-test for blocks first, and if this gave a non-significant result would then propose combining blocks with the residual. This makes the analysis look like a completely randomized analysis and of course increases the number of degrees of freedom for the residual. (We shall find in Chapter 8 that quite a variety of tricks are used when an experiment has not originally been designed to give enough residual degrees of freedom.)

An experiment is conducted as a randomized block if it seems likely that some control of systematic variation between groups of units will be needed, often because past experience of similar work with similar material has proved it so. Experimental conditions are hardly ever repeated in every detail and therefore it is very likely that there will be occasions when blocking proves to be unnecessary, even in situations where usually it has been useful. Also it is quite possible that in analysing two measurements from the same experiment there will apparently be a significant block effect for one set of data but not for the other; and on inspection the reason for this may be due to no more than one or two divergent measurements from one or two experimental units.

Just because on one particular occasion, when analysing one particular measurement, the mean square for blocks proves non-significant, that does not give firm ground for changing our Null Hypothesis, and altering our statistical model not only after seeing the data – which is bad enough – but even after completing the analysis. This is very like the habit of deciding whether to do a one-tail or a two-tail significance test *after* seeing the data, assuming that any mean that is higher than average in this experiment must always be so! If we did not know enough about the system beforehand when setting up the model we do not know now, and cannot leave out arbitrarily those terms that were included to remove systematic variation. What is designed as a randomized block must be analysed as a randomized block. Obviously the result of this experiment will be added to our store of knowledge when designing future experiments, but that is another matter.

Two or more measurements of the same thing

Some research workers realize the need for a reasonable amount of replication but try to achieve this by taking more than one measurement from each unit, instead of having enough units in the design of the experiment. When this happens, the variance of the repeat observations on the same unit will usually be numerically less than the variance between the units, often considerably so. In Chapter 13, we will look in more detail at the relevant theory, but Example 4.10 shows the correct way of dealing with data like this, and the result of the analysis in that example is typical. 'Within-unit' variation, whether the units are people, animals, plants or samples of industrial material, is less than 'between-unit' variation because the latter contains the former together with extra components.

Example 4.10. Two samples of plant material were taken from each field plot in a completely randomized layout, and laboratory analyses carried out on each sample to determine the amount of a chemical element in the plant tissue; results are given in the table below. Construct an Analysis of Variance dividing the total sum of squares into treatments, between-plots residual and within-plots residual. How should we test the Null Hypothesis that there are no treatment differences, or find an estimate s^2 to be used in constructing confidence intervals?

Treatment	A	B	C	D	E
Plot (1)	14; 15	16; 19	15; 18	18; 19	13; 12
(2)	12; 10	15; 18	14; 14	17; 16	12; 12
(3)	16; 12	15; 14	16; 15	17; 18	14; 12
(4)	15; 18	19; 17	18; 17	15; 20	15; 13

$G = 615$, $N = 40$. $S_0 = 615^2/40 = 9455.625$; $S = 9689$.
   The plot totals for each treatment, the sums of the pairs of laboratory determinations, are given in the following table.

	A	B	C	D	E
(1)	29	35	33	37	25
(2)	22	33	28	33	24
(3)	28	29	31	35	26
(4)	33	36	35	35	28
Treatment totals	112	133	127	140	103

$$S_T = \tfrac{1}{8}(112^2 + 133^2 + 127^2 + 140^2 + 103^2) = 9571.375$$

Using the 20 plot totals, $S_{plots} = \tfrac{1}{2}(29^2 + 35^2 + \cdots + 35^2 + 28^2) = 9638.500$; $S - S_0 = 233.375$; $S_T - S_0 = 115.750$; $S_{plots} - S_0 = 182.875$.

The Analysis of Variance is:

Source of variation	d.f.	Sum of squares	Mean square	
Treatments	4	115.750	28.938	$F_{(4,15)} = 6.47$**
Between-plots residual	15	67.125	$4.475 = s^2$	
	19	182.875		
Within-plots residual	20	50.500	2.525	
Total	39	233.375		

Treatments were applied to plots as units, so the between-plots residual 4.475 is the appropriate basis for testing treatment differences. The correct test is $F_{(4,15)}$ [*not* $F_{(4,20)}$]. The within-plots residual is usually smaller, though in this experiment the difference is not significant, as $F_{(15,20)} = 1.77$. Calculations of confidence intervals should use s^2 and $t_{(15)}$.

4.12 Assumptions in analysis: using residuals to examine them

We must consider the assumptions made when setting up a linear model and using it for the basis of an analysis. For a randomized block (page 42), the model is

$$y_{ij} = \mu + \tau_i + \beta_j + \varepsilon_{ij}$$

  (i) It is *additive*: we add terms, rather than (for example) multiply any of them together.
 (ii) *All* systematic variation is included in $\{\tau_i\}$, $\{\beta_j\}$, leaving $\{\varepsilon_{ij}\}$ as the random term.
(iii) The $\{\tau_i\}$ and $\{\beta_j\}$ do not interact: the response to a treatment is the same in each block.
(iv) All $\{\varepsilon_{ij}\}$ are independent, normally distributed from the same distribution: $\mathcal{N}(0, \sigma^2)$, σ^2 constant for all plots.

Assumption (i) is often adequately satisfied, at least as a good approximation, and (ii) implies that in this experiment only one blocking system was needed, otherwise we would use the appropriate model for a more complicated scheme such as a Latin square. Assumption (iii) is quite important: if, for example, several hospitals take part in a trial of different drug treatments for a particular illness, they may not all administer the drugs in quite the same way, and the comparison of any pair of drugs may differ from one hospital to another. This is *interaction* between treatments and blocks (drugs and hospitals) and some of the methods of Chapter 5 will be needed. Another example is in an agricultural trial where blocks differ considerably in general fertility, so that although the better treatments perform reasonably well in every block, the poorer ones may do very badly indeed in the poorest of the blocks. We can detect this by checking whether some treatments give much more variable observations than others. All these assumptions can be studied to some extent by looking at *residuals*, and (iv) is all about residuals.

Residuals

The $\{\varepsilon_{ij}\}$ are the *residuals*. Once a model has been fitted and the parameters estimated we can find estimates of the yield on each unit.

$\mu + \tau_i$ is estimated by T_i/r; $\mu + \beta_j$ by B_j/v; μ by G/N in the simple randomized block with each treatment appearing once in each block. The *fitted value* $\hat{y}_{ij}$ is then

$$\hat{y}_{ij} = \hat{\mu} + \hat{\tau}_i + \hat{\beta}_j$$

and so $y_{ij} - \hat{y}_{ij}$ is the *residual*, or estimate of ε_{ij}.

This process will be familiar to those readers who have studied regression analysis (see, for example, Draper and Smith (1981)), and computer programs will often give the set of values (observed, fitted, residual) for each unit.

These residuals should be independent, which proper randomization helps to ensure; but there may be patterns in them that point to various problems.

1. In an agricultural field plan, a systematic pattern shows that an extra term is needed in a linear model. A similar situation in an industrial trial might be if the first, or last, operation each day differed from the others, or if there was a time trend.

Example 4.11. An experiment was conducted as randomized blocks, with 5 replicates of the 6 treatments *A–F*. Data are provided below.

Block		I		II		III		IV		V
Observations										
(y_{ij})	A	3.5	C	5.0	F	11.5	E	8.5	B	11.0
	C	2.5	D	8.5	B	9.0	A	8.0	D	12.5
	E	3.0	A	5.0	C	4.5	C	6.0	F	16.5
	B	5.0	B	8.5	D	11.0	F	13.5	E	9.0
	F	8.0	E	5.0	E	6.0	B	12.5	C	7.5
	D	8.0	F	11.5	A	7.0	D	13.0	A	10.5
Block totals		30.0		43.5		49.0		61.5		67.0

Treatment totals are:

$$A, 34.0; B, 46.0; C, 25.5; D, 53.0; E, 31.5; F, 61.0$$

Hence the grand total = 251.0. Therefore

$$\hat{\mu} = \frac{251.0}{30} = 8.37, \quad \hat{\mu} + \hat{\tau}_A = \frac{34.0}{5} = 6.80, \quad \text{so } \hat{\tau}_A = -1.57$$

similarly

$$\hat{\tau}_B = +0.83, \quad \hat{\tau}_C = -3.27, \quad \hat{\tau}_D = 2.23, \quad \hat{\tau}_E = -2.07, \quad \hat{\tau}_F = 3.83$$

Block parameters are estimated in the same way:

$$\hat{\mu} + \hat{\beta}_I = \frac{30.0}{6} = 5.0, \quad \text{so } \hat{\beta}_I = -3.37$$

similarly

$$\hat{\beta}_{II} = -1.12, \quad \hat{\beta}_{III} = -0.20, \quad \hat{\beta}_{IV} = +1.88, \quad \hat{\beta}_V = +2.80$$

Because of equal replication, $\sum_{i=A}^{F} \hat{\tau}_i = 0$ (= −0.02 because of rounding to 2 decimal places), and $\sum_{j=I}^{V} \hat{\beta}_j = 0$ (= −0.01). The estimates of residuals are $y_{ij} - \hat{y}_{ij} = y_{ij} - \hat{\mu} - \hat{\tau}_i - \hat{\beta}_j$

and are given below.

	I	II	III	IV	V
$\hat{\varepsilon}_{ij}$	$A+0.07$	$C+1.02$	$F-0.50$	$E+0.32$	$B-1.00$
	$C+0.77$	$D-0.98$	$B\ \ \ 0.00$	$A-0.68$	$D-0.90$
	$E+0.07$	$A-0.68$	$C-0.40$	$C-0.98$	$F+1.50$
	$B-0.83$	$B+0.42$	$D+0.60$	$F-0.58$	$E-0.10$
	$F-0.83$	$E-0.18$	$E-0.10$	$B+1.42$	$C-0.40$
	$D+0.77$	$F+0.42$	$A+0.40$	$D+0.52$	$A+0.90$
	(0.02)	(0.02)	0	(0.02)	0

Residuals sum to zero (to within a small rounding error) in each block, because there is a block term included in the model, and we may check that the same is true for treatments, where the sums of residuals are: A, (0.01); B, (0.01); C, (0.01); D, (0.01); E, (0.01); F, (0.01).

The sums of residuals in order down the list are: row 1, -0.09; row 2, -1.79; row 3, -0.49; row 4, -0.49; row 5, -0.09; row 6, $+3.01$. This suggests that row 6 is different from the others; there is no regular pattern among the first 5 rows, but perhaps (in agricultural work) there was a source of shelter such as a hedge or a fence at that side of the field or (in industry) a distinct effect of being the last one in each session. Either we must not use as many as 6 units per block or we must find a row-and-column design which will be suitable for the required experiment.

2. Is σ^2 independent of treatment, and of block? If so, no pattern must exist among the $\{\varepsilon_{ij}\}$ that is related in any way to blocks or treatments. With small experiments this is difficult to show. The residuals for E, for example, seem basically smaller than those for C, but statistical tests for comparing variances are not very sensitive. Greater variability of poorer treatments, as discussed above, may suggest that poorer unit plots are very seriously at risk if these treatments are used.

Possibly the residuals for the larger observations will be greater, implying that the ratio of variance to mean is more stable than the variance itself, and this suggests that a *transformation* (page 67) should be considered. Fig. 4.2 illustrates such a set of residuals.

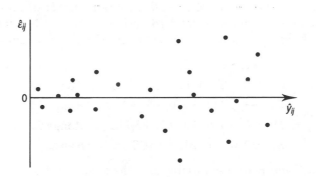

Fig. 4.2 Residuals proportional to the sizes of observations

3. Is the distribution of $\{\varepsilon_{ij}\}$ approximately normal? If so, the residuals should cluster around 0, symmetrically, with relatively few large positive or negative values. A skew distribution is suggested if all the large residuals have the same sign, as in Fig. 4.3.

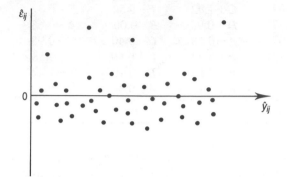

Fig. 4.3 Distribution of residuals is skew

A standard way of showing residuals, available in computer programs, is to plot estimated residuals $\hat{\varepsilon}_{ij}$ against fitted values $\hat{y}_{ij}$. A set of residuals satisfying the assumptions would appear as in Fig. 4.4.

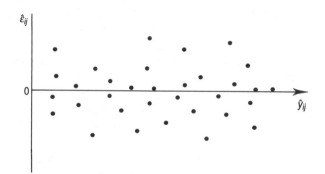

Fig. 4.4 Set of normally distributed residuals

A quick way of looking at the variability in each treatment is to find the ranges of the residuals, because in a sample of a relatively small number of observations from a normal distribution, the range and standard deviation are closely related.

Suppose that the set of residuals for 7 replicates of each treatment A–E were:

$$
\begin{array}{lll}
A & +2, -3, 0, -2, +4, +1, -2 & \text{Range 7} \\
B & -5, -2, +1, 0, -2, +3, +5 & \text{Range 10} \\
C & -3, +2, 0, +1, 0, -2, +2 & \text{Range 5} \\
D & -6, -4, 0, +3, +2, +3, +2 & \text{Range 9} \\
E & +2, 0, -1, +1, -2, 0, 0 & \text{Range 4}
\end{array}
$$

It is possible that C and E are less variable than the others, and σ^2 may not be constant for all observations. If $\hat{\varepsilon}_{ij}$ is proportional to $\hat{y}_{ij}$ we shall obtain a graph like Fig. 4.2.

4. We may look for an 'outlier', a large residual (positive or negative) which may suggest that that unit should be treated as a missing observation. For example:

$$A \quad +1, -2, +4, 0, +2, -5$$
$$B \quad -3, -5, 0, +1, -2, \mathbf{+9}$$
$$C \quad -2, +3, -1, 0, +2, -2$$

The value +9 in B is suspiciously high. If it is possible to check the record again that should be done, but otherwise we may prefer to discard that observation.

If only one residual is suspicious, as here, an 'outlier' may be a good explanation, but a situation as in Fig. 4.3 suggests general skewness rather than merely several outliers.

We will return to the problem of missing observations in block designs on page 69, but first we consider *transformations*.

4.13 Transformations

A graph like Fig. 4.2 suggests a relationship between the size of y_{ij} and the residual ε_{ij}, which we can often remedy by changing the scale of the measurement used in the analysis, i.e. by making a *transformation* of the data.

The *logarithmic transformation*, analysing $\log y_{ij}$ instead of y_{ij}, will usually stabilize the variance in the situation shown in Fig. 4.2. It also changes a multiplicative model into an additive one; in a multiplicative model it is the *proportional* response, not the actual response, which best measures the effect of treatments. Again the log of the response is the best figure to analyse because $\log(u/v) = \log u - \log v$. Many biological situations involve exponential growth laws, and so are naturally multiplicative rather than additive; there is nothing 'unnatural' about transforming such data for analysis. Although the model has become additive, we now have to measure residuals in logarithmic units, and assume a normal distribution in those units. We usually aim to stabilize variance, and assume that this will at the same time improve normality and additivity.

The analysis is more seriously affected by the variance not being constant than it is by mild departures from normality, so it is wise to deal with the variance first.

The theory for finding variance–stabilizing transformations when the relationship between mean and variance is known is explained below; this leads to $\log y_{ij}$ when the standard deviation is proportional to y_{ij}, to $\sqrt{y_{ij}}$ when the variance is proportional to y_{ij} (for example in Poisson-distributed data) and to arc sin $\sqrt{p_{ij}}$ for data expressed as percentages. (In the case of percentages or proportions, no transformation is needed when all p_{ij} lie between about 15% and 85%: $0.15 < p_{ij} < 0.85$.) Fisher and Yates (1963) tabulate this *inverse sine* (arc sin) transformation, but computer packages will calculate it directly from the data.

All analysis must be carried out in the transformed scale of measurement, but it is best to give results in the written report transformed back into the original units because these will be most easily understood. Careful explanation is needed so that readers will not try to make significance tests or find confidence intervals in the wrong units.

4.14 Theory of variance stabilization

Suppose that $\sigma(x)$ is a known function of $\mu(x)$, i.e. $\sigma(x) = g\{\mu(x)\}$; for example, we know that $\sigma^2(x) = \mu(x)$ in a Poisson distribution. The data x are transformed to

$y = f(x)$. The (approximate) expressions for the mean and variance of $f(x)$ are found from the Taylor series expansion:

$$f(x) = f(\mu) + (x - \mu)f'(\mu) + \frac{1}{2!}(x - \mu)^2 f''(\mu) + \cdots$$

so that

$$E[f(x)] = f(\mu) + f'(\mu)E[(x - \mu)] + \frac{1}{2!}f''(\mu)E[(x - \mu)^2] + \cdots$$

$$= f(\mu) + 0 + \tfrac{1}{2}\sigma^2 f''(\mu)$$

$$= f(\mu) \qquad \text{to second order}$$

$$\text{Var}\,[f(x)] = E[(f(x) - E\{f(x)\})^2] \approx E[\{f(x) - f(\mu)\}^2]$$

$$\approx E[\{(x - \mu)f'(\mu)\}^2] \approx \{f'(\mu)\}^2 E[(x - \mu)^2]$$

$$\approx \sigma^2 f'(\mu)^2$$

Hence Var $(y) = $ Var $[f(x)]$ will be constant if

$$f'(x) = \frac{k}{\sigma(x)}$$

so that

$$f(x) = \int \frac{k\,dx}{g(x)}$$

For the Poisson distribution, $g(x) = \sqrt{x}$, so

$$f(x) = \int \frac{k\,dx}{\sqrt{x}} = c\sqrt{x}$$

the square-root transformation.

When the variance increases roughly as the square of the observation, $\sigma(x) = ax$ and

$$f(x) = \int \frac{k\,dx}{x} = c \log_e x$$

For percentage data, we apply the binomial distribution which gives

$$\text{Var}\,(\hat{p}) = \frac{p(1 - p)}{n}$$

and leads to $f(p) = \arcsin \sqrt{p}$.

When a logarithmic transformation is used but there are zeros among the data, we apply $\log_e (x + 1)$ to avoid the indeterminacy of $\log_e 0$.

Sometimes $\sqrt{x + \frac{3}{8}}$ is used rather than $\sqrt{x}$, because it gives the minimum variance possible among the class of square-root transformations; $\sqrt{x + \frac{1}{2}}$ is nearly as good and

$\sqrt{x+1}$ may also be used. Similarly, if $\hat{p} = r/n$, then

$$\text{arc sin} \sqrt{\frac{r+\frac{3}{8}}{n+\frac{3}{4}}}$$

has optimum properties among its class.

The family of power transformations $y = x^\lambda$ was considered by Box and Cox (1964), λ being estimated from the data by a maximum likelihood method. Initially very popular, it now seems to be used less, because it is heavily dependent on the data and may suggest widely different values of λ in similar experimental situations, which is difficult to justify.

4.15 Missing data in block designs

When data are not collected from a unit, this will affect the whole analysis. If an agricultural experiment has some plots where plants have died, we have to decide whether this was due to the experimental treatment or to *natural* causes not connected with the experiment, such as damage by animals or by farm machinery. If the treatment is responsible a response 0 will usually be appropriate; but if several plots from the same treatment are missing it may be best to leave that treatment out of the analysis and consider it separately.

Provided that losses of data do not appear to be related to treatment, a *missing value* analysis is done. In a completely randomized layout, replication need not be the same for each treatment, so we simply analyse what is left.

In a randomized block layout, or a Latin square, the orthogonality is lost because all blocks are no longer the same: the estimated effect of a treatment is biased according to whether it was lost from a good block or from a poor one. Computing packages allow us to carry out the non-orthogonal analysis of the data that remain, and Chapter 11 describes how this is done. But the older method of *missing values* is still useful. The missing item is replaced by an estimate, based on the remaining data, which makes the residual on the affected plot 0.

An alternative way of reaching the same estimate is shown in the following table. m is the lost datum, and we minimize the residual sum of squares considered as a function of m.

Block	I	II	III	IV		
Treatment A	m	21	21	18	:	$60+m$
B	13	15	19	13	:	60
C	12	14	17	13	:	56
D	11	18	15	20	:	64
Total	$36+m$	68	72	64		$240+m$

Summation terms for analysis are:

$$S_B = \tfrac{1}{4}(13904 + (36+m)^2); \qquad S = 3998 + m^2$$
$$S_T = \tfrac{1}{4}(10832 + (60+m)^2); \qquad S_0 = (240+m)^2/16$$

The residual sum of squares (S.S) is $S - S_B - S_T + S_0$

$$= m^2 - 2186 - \tfrac{1}{4}(36 + m)^2 - \tfrac{1}{4}(60 + m)^2 + \tfrac{1}{16}(240 + m)^2$$

The derivative of this with respect to m

$$= 2m - \tfrac{1}{2}(36 + m) - \tfrac{1}{2}(60 + m) + \tfrac{1}{8}(240 + m)$$

Minimize the residual sum of squares (S.S.) by setting this derivative $= 0$ to give

$$\frac{9m}{8} = 18 \quad \text{or} \quad m = 16$$

(Check that the second derivative is >0, for a minimum.) With $m = 16$, the minimum value of the residual S.S. is 46.

Block and treatment terms are not orthogonal, so it is not exactly correct to complete the analysis with $m = 16$; however, it is a good approximation. For the exact analysis, consider the within-blocks S.S. $= S - S_B$, which is $522 + m^2 - \tfrac{1}{4}(36 + m)^2$.

This is minimized when $2m^* = \tfrac{1}{2}(36 + m^*)$, or $m^* = 12$. The minimum sum of squares is 90. The within-blocks analysis is given as follows:

Source	d.f.	Sum of squares	Mean square	
Treatments	3	44	14.67	$F_{(3,8)} = 2.55$
Residual	8	46	5.75	
Within blocks	11	90		

One observation having been lost, the within-blocks S.S. and the residual S.S. each lose 1 d.f. from what they would have in a full analysis.

The variance of $\bar{y}_A - \bar{y}_B$ is approximately (because of non-orthogonality)

$$\sigma^2 (\tfrac{1}{3} + \tfrac{1}{4})$$

Comparisons not involving A are not affected.

The mean of treatment A is found by setting $m = 16$;

$$\bar{y}_A = \tfrac{1}{4}(76) = 19$$

Two missing values

If two items are lost in a randomized block, a good method for estimating them is computer iteration. The above method for a single missing value is equivalent to the formula

$$m = \frac{rB' + vT' - G'}{(r-1)(v-1)}$$

in which B', T', G' are the incomplete totals of the block and treatment affected, and the incomplete grand total, respectively. In the example above,

$$m = \frac{4 \times 36 + 4 \times 60 - 240}{3 \times 3} = \frac{144}{9} = 16$$

If two estimates m and p have to be made, we must first guess a value for p, a *first approximation*. This may be the mean of the remaining replicates of the same treatment. An estimate m_1 of m can now be found by the formula.

With m set equal to m_1, remove the first approximation to p and use the formula to estimate p_1. Now put p_1 in the table, remove m_1 and use the formula to estimate m_2. In a similar way obtain p_2, m_3, p_3, ... until m and p each converge. The analysis can then continue as for the case of a single missing value.

Another method is based on regression (Analysis of Covariance), and is mentioned in Chapter 10.

The corresponding formula for a single missing value in a Latin square is

$$m = \frac{vT' + vR' + vC' - 2G'}{(v-1)(v-2)}$$

where R' and C', the incomplete row and column totals, are used.

4.16 Exercises

1. In a comparison of the effectiveness of four therapy programmes, each of 23 subjects was assigned to a programme. The score recorded was the number of words spoken by the subject over a given period of time. The results were as follows:

Therapy	Scores							Total
A	30	74	46	58	62	38		308
B	50	38	66	62	44	58	80	398
C	18	56	34	24	66	52		250
D	88	78	60	76				302

Note: The sum of the squares of the scores is 76444.

(i) It is first suggested that the therapy programmes may be of equal effectiveness. Using an appropriate test, and stating any assumptions you make, examine this suggestion and report your conclusions.

(ii) Test whether the effects of therapies A and B can be regarded as equal, and report your conclusions. A colleague notices that D gave the best average result and C the worst, and tries to undertake a similar test comparing those two therapies. State the objections you would have to this procedure.

(iii) The small experiment reported above is viewed as a pilot. Therapy A is standard, and B, C and D are experimental. Therapy C, which gave the worst result, is to be discarded. A further experiment is planned, with 140 subjects available to be allocated (on a suitably randomized basis) between therapies A, B and D, with B and D having the same numbers of subjects.

Comparisons between A and B and between A and D are of particular interest. If the aim is to minimize the sum of the variances of the contrasts between A and B and between A and D, how would you allocate the subjects to the three therapies?

2. Four different suppliers A, B, C, D provide raw material for a manufacturing process; before use, the manufacturer carries out a test to assess the percentage impurity of each. The results are:

A (10 samples)   12.8, 13.4, 11.2, 11.6, 9.4, 10.3, 14.1, 11.9, 10.5, 10.4
B (8 samples)   8.1, 10.3, 4.2, 7.8, 8.1, 12.7, 6.8, 6.4
C (8 samples)   9.8, 10.6, 9.1, 4.3, 11.2, 8.3, 9.2, 6.4
D (10 samples)   16.4, 8.2, 15.1, 10.4, 7.8, 9.2, 12.6, 11.0, 8.0, 9.8

Examine these data for possible differences in mean impurity, and also consider residuals and their variances.

3. An experiment was carried out to determine whether there was any significant difference between the strength of four materials. The data collected are shown below (measured in standard units). The design used was randomized blocks, with different days acting as blocks.

Material	1	2	3	4
Block I	20.7	23.8	21.5	25.2
II	19.1	21.5	22.9	25.5
III	21.4	23.2	22.4	24.8
IV	22.5	22.5	22.2	24.6
V	19.2	23.1	23.2	23.1

Analyse the data and determine whether the strengths of the materials are significantly different from each other. Put 95% confidence limits on the average strength of material 4.

Comment upon the results of your analysis.

4. During the manufacture of an organic chemical, five different blends of the original mixture (A–E) were compared and four batches of the chemical were prepared from each blend. The measurement of interest is the percentage loss from the theoretical yield. The results are:

Batch	I	II	III	IV
Bend A	14.9	14.5	15.5	15.0
B	16.2	17.2	15.1	16.4
C	15.0	16.1	15.3	15.8
D	13.1	14.0	15.8	14.3
E	16.3	16.3	17.8	16.9

Test whether there are differences among the blends. Is the analysis improved by a suitable transformation of the observations?

5. An agricultural engineering company has made two new types of lighting and housing for chickens. These make it possible to increase the amount of 'daylight' which the birds receive. The two new treatments are

      E: extended day (14 h)
      F: flash lighting (natural day plus two 20-s flashes per night);

and the control, O: natural daylight, is also included. Six farms are chosen, forming the blocks of a randomized block design.

    Carry out an analysis, extracting suitable contrasts.

Block	I	II	III	IV	V	VI
O	330	288	295	313	304	276
E	372	340	343	341	336	321
F	359	337	373	302	342	320

The data shown in the table are the numbers of eggs laid by groups of six birds in a period of 12 weeks.

6. An organization uses four schemes available for teaching word-processing keyboard skills. Scheme A is an established method, B is a new method that is similar to A; C and D are new methods but different in character. Forty people taking courses at the same time are allocated, randomly, ten to each scheme, and given an initial test. At the end of the course they are tested again and the difference in scores is used to assess the success of the scheme. The differences are:

    A: 33, 44, 39, 38, 29, 41, 39, 30, 42, 44
    B: 31, 33, 40, 34, 31, 41, 34, 28, 25, 33
    C: 32, 29, 34, 41, 27, 26, 43, 25, 35, 26
    D: 39, 42, 46, 42, 42, 46, 39, 43, 41, 38 $(\Sigma y = 1445, \Sigma y^2 = 53757)$

Examine whether the new methods differ on average from the old; and whether the methods C, D differ from the other new one (B). Obtain the residuals for this analysis, and comment on any pattern in them.

7. When the experiment described in Example 4.4 (page 50) is carried out, the research worker observes symptoms that convince her that she does not need to use O, and she considers she will have a better experiment if each block contains (S, S, A, B, C, D). Also, she is not interested in all the orthogonal contrasts but only in comparing each new treatment with S. She collects these data on an assessment of percentage damage to the crop; results are given in the table.

Block					
I	S 22, 30;	A 14;	B 16;	C 8;	D 10
II	S 24, 26;	A 11;	B 16;	C 5;	D 6
III	S 18, 14;	A 9;	B 12;	C 8;	D 6
IV	S 16, 12;	A 8;	B 6;	C 4;	D 5

Carry out significance tests, and construct confidence intervals, which will help her attain her objective.

8. Five varieties of corn were grown at each of four out-stations during a programme of development in a country. Two sample crop-cuts, each of 1 m², were taken from each variety at each station and the grain yields were recorded. The results are summarized (in coded units) as follows:

Station		P	Q	R	S
Variety	*a*	575, 545	655, 640	665, 645	690, 686
	b	625, 632	700, 685	650, 640	738, 744
	c	570, 510	600, 580	655, 640	678, 685
	d	515, 520	630, 650	585, 590	648, 666
	e	780, 790	695, 660	605, 620	720, 748

Analyse and comment on these results.

9. A randomized block experiment using *b* blocks is to be carried out using a standard treatment *S* and *v* other treatments *A*, *B*, ... Each block is to contain $(v+c)$ plots, consisting of *c* replicates of *S* and one of each of *A*, *B*, ...

 Write down a suitable mathematical model, state clearly the assumptions to be made, and find the least-squares estimate of the difference between treatment *S* and any other treatment. Find also the variance of this difference, and obtain a rule for choosing *c* so that this variance is minimized for the given total number of plots.

10. An agricultural research worker in a developing country wishes to test two locally available products with insecticidal properties. One product, *N*, may come from either the leaves or the seeds of a plant (we may label these *NL*, *NS*) and the other, *R*, from leaves only. Each of *R*, *NL*, *NS* is to be tested at three concentrations, and it is required to test linear spacing on a natural scale (0, 1, 2) *or*, as an alternative, on a logarithmic scale (1, 2, 4). The experiment must also contain a standard insecticide, *S*, and a control, *O*, making 11 treatments in all. Suggest a set of mutually orthogonal contrasts for subdividing the sum of squares for treatments, and explain what each one is comparing.

11. A supermarket buys a particular product from four suppliers, *A*, *B*, *C*, *D*, and regular tasting tests by expert panels are carried out as the product is sold in their food halls. Various characteristics are scored and an analysis of the totals of these scores is made. Four tasters *a*, *b*, *c*, *d* obtained these results at four sessions 1–4. Analyse and comment on them.

Taster		*a*	*b*	*c*	*d*
Session	1	*A*: 21	*B*: 17	*C*: 18	*D*: 20
	2	*B*: 20	*D*: 22	*A*: 23	*C*: 19
	3	*C*: 20	*A*: 24	*D*: 22	*B*: 19
	4	*D*: 22	*C*: 21	*B*: 22	*A*: 26

12. The effects of three different packaging materials for a particular product, the type of store or supermarket in which it is sold and the position on a row of the

shelves are being studied by a marketing company. The numbers sold in a week are recorded. Packaging materials are *A, B, C*.

Type of store	1	2	3	4	5	6
Row on shelf 1	*A*, 31	*B*, 24	*C*, 42	*C*, 36	*A*, 35	*B*, 30
2	*B*, 25	*C*, 39	*A*, 20	*B*, 26	*C*, 37	*A*, 27
3	*C*, 37	*A*, 26	*B*, 27	*A*, 43	*B*, 28	*C*, 43

Analyse and report on these data.

13. An experiment was conducted to compare four methods (*A, B, C, D*) of propagating lettuce in the early stages of growth. One of the treatments, *A*, was new and comparisons of *A* with the other three older methods were of greatest importance. For this reason twice as many plots of lettuces received *A* as received each of the other methods. In order to reduce the influence of positional effects, a 5×5 Latin square design was used, with the fifth treatment replaced by *A*, giving 10 observations on *A* in all. The results of the experiment are shown in Fig. 4.5, in kg of lettuce per plot.

C 20.1	*A* 14.3	*B* 25.1	*D* 21.6	*A* 17.8
B 22.0	*C* 21.0	*A* 12.9	*A* 16.2	*D* 19.1
A 14.0	*D* 21.5	*C* 22.4	*B* 20.6	*A* 19.1
A 12.8	*A* 16.4	*D* 19.9	*C* 21.5	*B* 22.8
D 21.3	*B* 20.0	*A* 15.5	*A* 13.4	*C* 21.8

Fig. 4.5 Results for Exercise 13

Construct the appropriate analysis of variance table and test the null hypothesis of no effects due to different propagating methods. Estimate the treatment contrast

$$A - (B + C + D)/3$$

together with its standard error. Does this contrast account for all the treatment differences?

14. (i) Discuss the importance of randomization in experimental design. Explain how you would apply randomization to a Latin square design.
    (ii) In a certain pyschological experiment, four different tests are to be given to each of 12 subjects. The subjects may improve their scores on the tests as the experiment progresses, so the order in which the four tests are applied should be balanced in some way.

Describe in detail how you would design and set about analysing this experiment; you should incorporate a discussion of the graphs and tables, etc., you would include.

15. Four laboratories A, B, C, D are carrying out chemical analyses of each of four compounds α, β, γ, δ. The study is spread over eight days (1–8) and the compounds are supplied from four sources (I–IV). The allocation of which material is studied on which day is made according to the scheme:

Day		1	2	3	4	5	6	7	8
Source	I	$D\beta$	$A\alpha$	$C\alpha$	$B\delta$	$D\delta$	$C\gamma$	$A\gamma$	$B\beta$
	II	$B\gamma$	$B\delta$	$A\delta$	$D\alpha$	$C\alpha$	$D\beta$	$C\beta$	$A\gamma$
	III	$A\alpha$	$C\beta$	$B\beta$	$C\gamma$	$B\gamma$	$A\delta$	$D\delta$	$D\alpha$
	IV	$C\delta$	$D\gamma$	$D\gamma$	$A\beta$	$A\beta$	$B\alpha$	$B\alpha$	$C\delta$

Analyses are made of an element that should have a known theoretical value in each compound, and the (absolute) departure from that is calculated. Results, coded to whole numbers for ease of analysis, are:

Day		1	2	3	4	5	6	7	8
Source	I	52	40	4	51	95	30	44	2
	II	45	81	10	44	56	34	55	5
	III	10	73	7	56	63	26	88	11
	IV	53	92	33	28	44	6	33	30

Examine the effects of the 4 classifications Laboratories, Compounds, Days and Sources, assuming they are all mutually independent.

16. The differences between observers in a laboratory examination using microscopes were compared in an experiment. Five slides were prepared and four observers each counted the number of a certain type of cell seen on the slides through the microscope. Due to a mistake, two of the slides were destroyed before all the observers had seen them. The observed counts and the missing values, m_1 and m_2, are given below.

Slide number		1	2	3	4	5
Observer	A	m_1	51	44	75	63
	B	62	58	57	96	75
	C	57	54	55	83	78
	D	39	m_2	44	70	66

Obtain estimates of the two missing values and analyse these data to determine whether the observers produce different mean counts. Using a suitable approximation find the standard errors of the observer differences. Hence construct approximate 95% confidence intervals for the differences in mean counts between observers A and B, A and D and B and C. Comment on the results.

17. Four drivers (Adams, Brown, Carter, Davis) drove four cars on four different roads and the level of brake wear was assessed. The data were as shown below.

	Road 1	Road 2	Road 3	Road 4
Ford Escort	B: 44	A: 46	D: 39	C: 52
Vauxhall Cavalier	C: 51	B: 37	A: 43	D: 40
Rover Metro	A: 42	D: 39	C: 46	B: 34
Peugeot 205	D: 45	C: 52	B: 36	A: 42

(The names of the drivers have been abbreviated to the initials A, B, C, D.)

Obtain an analysis of the data to discover whether there is evidence of differences between the drivers and indicate what design has been used. Comment on the way that drivers should have been allocated to cars and roads for your analysis to be valid.

Adams and Carter drove in the morning, the other two in the afternoon. Adams and Brown drove at the weekend, the other two on weekdays. Investigate the data for differences between times of day and times in the week and any other relevant comparisons. Discuss whether it is reasonable to attribute these differences to time.

18. Suppose that in a randomized block experiment the observation on the kth treatment in the lth block is missing. Explain how you could 'estimate' this missing value in order for the analysis to proceed conveniently, and indicate briefly how such a procedure can be justified. Suggest how this method might be extended to the case of several missing values.

In an investigation of the effects of diet on mice, three diets D_1, D_2 and D_3 were tested on five litters of three mice. (In the event, two litters contained only two mice that could be used, and for those litters one diet, selected at random, was omitted.) The observations in the table below are the growth weights in grams of the mice over the experimental period.

Litter		1	2	3	4	5
Diet	D_1	152	93	110	x	143
	D_2	106	y	72	92	127
	D_3	109	55	58	59	112

By calculating treatment and block totals in terms of x, y and the numbers in the table, obtain formulae for the residuals for the observations in the positions of x and y. Hence obtain estimates of x and y, using as estimates those values that make the corresponding residuals zero.

Write down the 'source of variation' and 'degrees of freedom' columns of an Analysis of Variance table for this experiment, and describe how you would test the hypothesis that the diets have identical effects. (No calculations are required.)

19. A small experiment involving 3 treatments is arranged in two randomized blocks, and a model of the form

$$E(y_{ij}) = \mu + \tau_i + \beta_j \qquad (i = 1, 2, 3; j = 1, 2)$$

(where y_{ij} represents the observation on treatment i in block j) is proposed for the analysis. Suppose now that two observations are lost. Investigate the effect of this loss on the ability to estimate treatment differences. For those differences that remain estimable, find the efficiency of the appropriate estimators relative to those that would be used if all 6 observations were available. You should consider three cases:

   (i) when the missing observations lie in the same block;
  (ii) when the missing observations are on the same treatment;
 (iii) when the missing observations are on different treatments in different blocks.

20. Four patients were asked to perform a certain task which required concentration and skill under a variety of conditions. These conditions involved the hour of day and whether the patient had been given a particular drug shortly before the test. The times in seconds are recorded as follows:

Treatment	No drug			Drug		
Hour	Morning	Afternoon	Evening	Morning	Afternoon	Evening
Patient 1	34	37	43	37	38	46
Patient 2	38	36	40	39	45	42
Patient 3	28	32	36	30	35	36
Patient 4	42	45	43	40	42	46

Analyse the data and report the results of your analysis; in particular comment on whether there is evidence that

   (i) the drug causes patients to take longer over the task;
  (ii) the effects of treatment and hour are additive;
 (iii) there is a difference in performance times between the morning and other times of day on average;
 (iv) there is a difference between afternoon and evening performance;
  (v) there is a difference in performance between patients.

(You may disregard any interaction involving patients.)

APPENDIX 4A   Cochran's Theorem on Quadratic Forms

There is a general theorem, originally due to Cochran (1934) and studied further by Lancaster (1954), which justifies the many different subdivisions we make of the total sum of squares. It uses standard results from matrix theory.

    Kendall *et al.* (1994) in Section 15.16 of Vol. I, give further details. Scheffé (1957) also contains proofs.

THEOREM. Let $\{x_j\}$ be n independent standard normal ($\mathcal{N}(0, 1)$) random variates. Let there be k quadratic functions $\{Q_i\}$ such that $\sum_{j=1}^{n} x_j^2 = \sum_{i=1}^{k} Q_i$, with all $\{Q_i\} \geq 0$ and rank $(Q_i) = n_i$.

Then a necessary and sufficient condition that the $\{Q_i\}$ are *independent* χ^2 variates, with degrees of freedom $\{n_i\}$, is $\sum_{i=1}^{k} n_i = n$.

COROLLARY 1. If $\{x_j\}$ are $\mathcal{N}(0, \sigma^2)$, then it is Q_i/σ^2 that will follow $\chi^2_{(n_i)}$.

COROLLARY 2. If $\{x_j\}$ are $\mathcal{N}(\mu, \sigma^2)$, then $\sum_{j=1}^{n} (x_j - \bar{x})^2/\sigma^2$ is $\chi^2_{(n-1)}$. So if $\sum_{j=1}^{n} (x_j - \bar{x})^2 = \sum_{i=1}^{k} Q_i$, with rank $(Q_i) = n_i$, a necessary and sufficient condition for the $\{Q_i/\sigma^2\}$ to be independent $\chi^2_{(n_i)}$ variates is $(n-1) = \sum_{i=1}^{k} n_i$.

PROOF. Every quadratic form in n variables $y_1, y_2, \ldots, y_n$ can be written as $\sum_{i=1}^{n} c_i Y_i^2$, the rank of which may be $\leq n$. The rank of a sum of quadratic forms is less than or equal to the sum of their ranks.

If a quadratic form Q can be written in the form $\sum_{i=1}^{k} L_i^2$ where each L_i is linear in $x_1, x_2, \ldots, x_n$, and there are just h independent functional relationships among $\{L_i\}$, then rank $(Q) = k - h$, which is $\leq n$.

If $Q = X'AX$ and $Y = BX$, then $Q = Y'A_1 Y$ in which $|A| = |A_1||B|^2$ since $Y'A_1 Y = X'B'A_1 BX$.

Finally, if $Q_1, Q_2, \ldots, Q_k$ are quadratic forms in $x_1, x_2, \ldots, x_n$ of ranks $n_1, n_2, \ldots, n_k$, and if $\sum_{i=1}^{k} Q_i = \sum_{j=1}^{n} x_j^2$, then the condition $n = \sum_{i=1}^{k} n_i$ is a necessary and sufficient condition that there is a non-singular linear transformation $Y = CX$ such that

$$Q_1 = y_1^2 + \cdots + y_{n_1}^2$$

$$Q_2 = y_{n_1+1}^2 + \cdots + y_{n_1+n_2}^2, \ldots$$

$$Q_k = y_{n_1+n_2+\cdots+n_{k-1}+1}^2 + \cdots + y_n^2$$

Note also that $\chi^2_{(f)} + \chi^2_{(g)}$ is distributed as $\chi^2_{(f+g)}$ when the two variates are independent.

Applying these results of Cochran's Theorem,

(1) if Q_i is $\chi^2_{(n_i)}$ then $\sum_i Q_i$ is $\chi^2_{(\Sigma n_i)}$; but by definition $\sum_i Q_i$ is $\chi^2_{(n)}$; therefore $n = \sum_i n_i$ is a necessary condition;

(2) if $n = \sum_i n_i$, there exists $Y = CX$ such that $Q_1 = y_1^2 + \cdots + y_{n_1}^2$, etc; each y is linear in the x's and so is normally distributed (as are the x's); var $(y_i) = \sigma^2 \sum_{j=1}^{n} c_{ij}^2 = \sigma^2 = 1$ (an idempotent orthogonal transformation); and cov $(y_i, y_k) = \sigma^2 \sum_{j=1}^{n} c_{ij} c_{kj} = 0$. Thus $\{y_i\}$ are $\mathcal{N}(0, 1)$ and independent; so the $\{Q_i\}$ are $\chi^2_{(n_i)}$ and are independent.

Note. Lancaster proved the generalized form of the theorem, that any of the following (a), (b), (c) implies the other two:

(a) the ranks of $\{Q_i\}$ add to n;
(b) each $\{Q_i\}$ is χ^2;
(c) all $\{Q_i\}$ are independent.

5
Factorial experiments

5.1 Introduction

There is one important type of experiment in which well-defined contrasts have long been used. This is when the treatments consist of combinations of levels (i.e. amounts) of different *factors*, and one major interest is to see whether the factors operate independently of one another or whether there is *interaction* between them. Having decided this, we shall then be in a position to recommend which treatments – i.e. which combinations of levels of factors – are producing good responses and which are not so good.

In one of the original examples from agricultural research, different fertilizers are used on growing crops such as wheat, in the hope of finding a fertilizer that will give high crop yield. Experiments of this sort are still very important in developing regions or where there is a rapid population increase. It is well known that some chemical elements are essential to the healthy growth of plants, most particularly the elements nitrogen (N), phosphorus (P) and potassium (K); but the exact amounts of these elements to use for any growing crop, in a new area not previously studied, will have to be determined by experiment.

The elements are the *factors*, and the experimental *treatments* will be various combinations of levels of the factors. But it is unlikely that all the factors will operate independently of one another: the response to increasing the level of N may well be greater when P is supplied at an adequate level than when P is in short supply. Figure 5.1 shows the possible results when two factors N and P are used at just two levels each, low (labelled 0) and high (labelled 1). For present purposes we will call the actual treatment combinations $N_0 P_0$, $N_0 P_1$, $N_1 P_0$, $N_1 P_1$, which are all the possible combinations of the two levels of the two factors. (There is a simplified notation used when factors only have two levels, and we introduce this on page 82.)

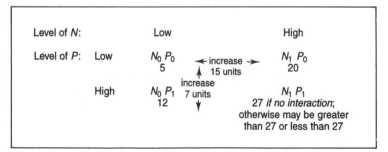

Fig. 5.1 Possible results of a 2×2 experiment

If the yield with $N_0 P_0$ is, say, 5 units, we may find it increases to 20 units when the fertilizer is changed to $N_1 P_0$; thus there is an increase of 15 units due to adding extra N without changing P. Suppose also that the yield with $N_0 P_1$ is 12, so that the increase due to adding extra P without changing N is 7 units. What will be the yield under $N_1 P_1$? When N and P are independent, it will be

base yield $(N_0 P_0)$ + increase due to N alone + increase due to P alone

i.e.

$$5 + 15 + 7 = 27 \text{ units}$$

But unless we already know that the two factors really are independent, the only way of actually finding out what is the yield under treatment $N_1 P_1$ is to have that combination of levels in the experiment as well. Often, especially with fertilizer elements like N and P, there is a substantial lack of independence; the yield under $N_1 P_1$ may, for example, be 40 units, so that we have an *interaction* of + 13 units in addition to the effects of the two factors. We may call this a 'positive' interaction; but a 'negative' interaction is also possible when a higher level of one factor depresses the response to the other.

Situations like this happen in all areas of study: the response to one factor depends on the precise level of the other factors in the experiment. Examples in this chapter and later provide plenty of illustrations. This well-known existence of interactions between factors in experiments seems, strangely, to have been overlooked in some types of experiment recently proposed for use in engineering and industry (see Chapter 8), which is all the more surprising in view of the previous large contributions of industrial statisticians to the design and analysis of experiments. It is akin to keeping a balanced diet for humans or animals: change the amount of one constituent and some other (or others) will most likely need changing too to keep a healthy balance.

Graphical study of interactions

Suppose that the yield of a chemical reaction depends on the amount of a factor A which is included in a mixture, and also on the presence or absence of a catalyst C. The means found from a properly designed factorial experiment are:

	A	Low	High
C	Absent	18.4	25.5
	Present	21.6	32.0

In addition to the calculations that will be made, a graph showing these four means helps to make the results much clearer. Mean yield is plotted vertically against the levels of one of the factors on the horizontal scale; Fig. 5.2 has A on the horizontal axis, and the graph is completed by joining up the two points for C absent, and also the two points for C present. Clearly, on this scale, the two lines for C are not parallel: the change in yield when A is altered from its low to its high level is smaller when C is absent than it is when C is present. We shall find that this is the pattern when interaction is present; if the two lines are parallel (or very nearly so) this is a sign that there is no

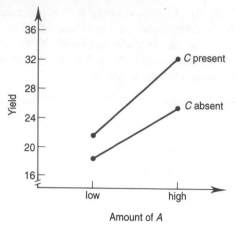

Fig. 5.2 Graphical study of interaction

interaction, since the response to changing one factor is about the same whatever the level of the other.

5.2   Notation for factors at two levels

For 2 factors, each at 2 levels, there are $2^2 = 4$ possible combinations. The usual notation is to label the *factors* by capital letters A, B. The actual treatments (which in earlier chapters have been denoted by capital letters) consist of the combinations of levels of A and B, shown in the following way:

> *ab*   is the combination with both factors at high level;
> *a*   has A at high level and B at low level;
> *b*   has A at low level and B at high level;
> (1)   has both factors at low level

We may replace the descriptions 'high level' and 'low level' by 'present' and 'absent' when appropriate, and then the notation shows which factors are present in each treatment. The symbol (1) is used in preference to zero (0) for the lowest combination of factor levels.

5.3   Definition of main effect and interaction

The *main effect* of factor A is defined as the *average* response to a change in the level of A, this average being taken over both levels of B. It is therefore the average of $(a - (1))$ and $(ab - b)$, because the first of these is the response when B is at its low level and the second is the response when B is at its high level. The corresponding definition of the main effect of B is the average of $(b - (1))$ and $(ab - a)$.

The *interaction* of the 2 factors, always denoted by the symbol AB, is the *difference* between the response to change in A at the high level of B and the response to change in A at the low level of B: this is $(ab - b) - (a - (1))$.

These contrasts can be set out in a table that is valid for any experiment in which each treatment combination is replicated the same number of times:

	(1)	*a*	*b*	*ab*
Main effect *A*	−	+	−	+
Main effect *B*	−	−	+	+
Interaction *AB*	+	−	−	+

Strictly we should write −1 and +1; but since every coefficient is 1 this is not necessary. The main effect for *A* has a + sign wherever *a* appears in the treatment combination, and a − sign when it does not; there is a similar pattern for the main effect *B*. The interaction is found by 'multiplying' the two rows for *A* and *B* in the sense of an inner product: in the column for (1) there is a − sign in the *A* row and also in the *B* row, so the sign in the *AB* row is *minus* times *minus*, which is *plus*. The *a* column has one minus sign and one plus, so the product is minus; and similarly in the *b* column. Finally there are plus signs in both the main effect rows under *ab*, and therefore the interaction row will also contain a plus. This rule is extremely helpful when there are more than two factors, and more complicated interactions have to be defined.

These three contrasts are all orthogonal to one another, and so together they will account for the three degrees of freedom for treatments in an analysis; they lead to a complete subdivision of the sum of squares for treatments in the way described in Chapter 4.

By this process we have of course only been concerned with one part of the design of an experiment, namely the choice of the treatments that go into it. We still have to consider what sources of variation there are in the material (the units) available for the experiment, and so to use a completely randomized layout, or a randomized complete block, or a Latin square or other design according to need. Every treatment combination must be included once in each block, if a randomized block is used; and equal replication is required in a completely randomized layout in order to keep the analysis to its usual, relatively simple, form.

Example 5.1. A 2^2 experiment in 5 randomized blocks. Factor *S* is the source of supply of a particular material. There are two sources, so we might call these S_0 and S_1; but in order to use the simple notation we shall write *s* when the first source is used and the absence of *s* will indicate that the second source was used. Factor *M* is the speed of running a machine which makes the material into a component for use in electronics; there are two speeds, and we shall write *m* whenever the higher of these is used.

The experiment was carried out on five separate days, and these form the blocks in a randomized block arrangement. The average numbers of faulty items per batch, based on examination of random samples from 10 batches, are given in the following table.

Treatment	I	II	III	IV	V	Treatment totals
		Block				
(1)	5.3	5.7	5.1	5.3	5.6	27.0
m	11.8	13.0	12.6	12.1	11.5	61.0
s	20.0	19.0	20.3	19.5	20.2	99.0
ms	26.7	24.1	25.7	26.0	25.5	128.0
	63.8	61.8	63.7	62.9	62.8	$315.0 = G$

Summation terms are:

$$S = 5.3^2 + 5.7^2 + \cdots + 25.5^2 = 6133.520$$

$$S_T = \tfrac{1}{5}(27.0^2 + 61.0^2 + 99.0^2 + 128.0^2) = 6127.000$$

$$S_B = \tfrac{1}{4}(63.8^2 + \cdots + 62.8^2) = 4961.905$$

$$S_0 = G^2/20 = 4961.250$$

The Analysis of Variance is:

Source	d.f.	Sum of squares	Mean square	
Blocks	4	0.655	0.16375	$F_{(4,12)} < 1$
Treatments	3	1165.750	388.58333	
Residual	12	5.865	0.48875	
Total	19	1172.270		

Often V.R. ('variance ratio') is made the heading of the final column, instead of using the symbol F, since it is only on the null hypothesis that the calculated value follows the F distribution. Also, because the blocks in the experimental layout are not selected at random, it is strictly not correct to do a test; the most we should do is look at the size of the ratio and compare it roughly with tabulated F-values.

Block differences appear unimportant; treatments are very large indeed. We have a treatment structure that permits proper contrasts to be defined, and this is the only satisfactory way to complete the analysis, which we do by using the method of Chapter 4 to split the treatments sum of squares into three contrasts.

Treatment Total	(1) 27	m 61	s 99	ms 128	Value	Divisor	Sum of squares
M	−	+	−	+	63	4×5	198.45
S	−	−	+	+	139	4×5	966.05
MS	+	−	−	+	−5	4×5	1.25
							1165.75

The Analysis of Variance table can be rewritten thus:

Source	d.f.	Sum of squares	Mean square	
Blocks	4	0.655	0.164	$F < 1$
M	1	198.45	198.45	$F_{(1,12)} = 406.0$
S	1	966.05	966.05	$F_{(1,12)} = 1977$
MS	1	1.25	1.25	$F_{(1,12)} = 2.50$
Treatments	3	1165.750		
Residual	12	5.865	0.4888	
Total	19	1172.270		

The interaction must be studied first. The 5% point of $F_{(1,12)}$ is 4.75, so we have no evidence of interaction. It is therefore valid to examine the main effects and to express the results in terms of these. Both main effects give very large F-values, significant at the 0.1% level. The means are calculated at each level of each factor; they are based on figures from 10 plots in this experiment. For high M we have $(61.0+128.0)/10=$ 18.9, and likewise for low M 12.6; while for high S the mean is $(99.0+128.0)/10=$ 22.7, and for low S it is 8.8.

No t-tests are needed, because the F-tests with a single degree of freedom in the numerator are equivalent to them. We are looking at the numbers of faulty products in this experiment, so clearly we want this to be as small as possible; we shall achieve this by using 'low S', which is the second source of supply, and 'low M', which is the lower machine speed.

A graph of the four individual treatment means for these data shows that the response to changing machine speed is similar for both the sources of material (Fig. 5.3).

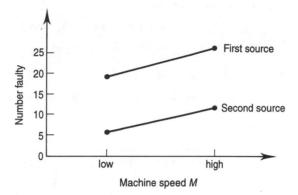

Fig. 5.3 Means for source/machine speed data

Example 5.2. Using a Latin square design. An experiment to compare two methods of applying fertilizer to a corn crop was laid out in a Latin square design. The two methods were (A) to 'broadcast' the fertilizer all over a plot, and (B) to 'place' it carefully in bands near the seeds. Two alternative fertilizers, X and Y, were used. The weights of a sample of plants from each plot were obtained after a fixed period of growth, as follows (g):

Column	1	2	3	4
Row 1	BY, 121	AX, 168	AY, 155	BX, 140
2	AY, 131	BX, 142	AX, 165	BY, 138
3	AX, 143	BY, 136	BX, 151	AY, 140
4	BX, 126	AY, 148	BY, 135	AX, 139

Because there may be an interaction between the method of application and the actual fertilizer used, we should treat this as a factorial experiment. One factor is 'Fertilizer X versus fertilizer Y' and the other is 'Broadcasting versus Placing'. In fact it is rather a small experiment (page 47), but provided the samples were of a fair number of plants, the total weights should not be too variable.

For the first factor, write x where fertilizer X was applied, and let the absence of x indicate fertilizer Y; and likewise for the second factor write a where the broadcast method was used and let the absence of a denote placing. Totals are:

Row	1	2	3	4	Column	1	2	3	4
	584	576	570	548		521	594	606	557
Treatment	AX	AY	BX	BY					
Code	ax	a	x	(1)	Grand total $G=2278$				
	615	574	559	530					

Summation terms:

$$S_0 = 2278^2/16 = 324330.25$$

$$S = 121^2 + 131^2 + \cdots + 140^2 + 139^2 = 326756.00$$

$$S_R = \tfrac{1}{4}(584^2 + 576^2 + 570^2 + 548^2) = 324509.00$$

$$S_C = \tfrac{1}{4}(521^2 + 594^2 + 606^2 + 557^2) = 325440.50$$

$$S_T = \tfrac{1}{4}(615^2 + 574^2 + 559^2 + 530^2) = 325270.50$$

The Analysis of Variance is:

Source	d.f.	Sum of squares	Mean square	$F_{(3,6)}$
Rows	3	178.75	59.58	1.82 n.s.
Columns	3	1110.25	370.08	11.30**
Treatments	3	940.25	313.42	9.57**
Residual	6	196.50	32.75	
Total	15	2425.75		

n.s. = not significant.

Subdivision of treatment sum of squares:

Treatment	(1)	a	x	ax			
Total	530	574	559	615	Value	Divisor	Sum of squares
A	−	+	−	+	100	4×4	625.00
X	−	−	+	+	70	4×4	306.25
AX	+	−	−	+	12	4×4	9.00
							940.25

Each of A, X, AX has 1 degree of freedom; testing AX against the residual mean square gives $F_{(1,6)} < 1$ so there is no evidence of interaction. For main effect A, $F_{(1,6)} = 19.08$, significant at the 0.1% level; and for B, $F_{(1,6)} = 9.35$, significant at 1%.

Rows did not remove a significant part of the variation among units (plots) but columns did. Since there is no interaction, the results are summarized by the main effect means, each based on 8 observations:

$$A: \text{`broadcast'} = (574 + 615)/8 = 148.6; \text{ Fertilizer } X = (559 + 615)/8 = 146.8;$$

$$\text{`placed'} = (530 + 559)/8 = 136.1. \qquad Y = (530 + 574)/8 = 138.0.$$

These show that greater plant weights are obtained by broadcasting the fertilizer, and by using fertilizer X.

5.4 Three factors each at two levels

This is often called a 2^3 experiment, because there will be $2^3 = 8$ combinations altogether for the three factors A, B and C. In the same notation as above, these are:

(1)   where all factors are at their low levels;
a which has A at high level, and B, C both at low level;
b which has B at high level, and A, C both at low level;
c which has C at high level, and A, B both at low level;
ab which has A and B both at high level, and C at low level;
ac which has A and C both at high level, and B at low level;
bc which has B and C both at high level, and A at low level;
abc which has all three factors at their high levels.

Seven orthogonal contrasts can be defined:

	(1)	a	b	c	ab	ac	bc	abc
A	$-$	$+$	$-$	$-$	$+$	$+$	$-$	$+$
B	$-$	$-$	$+$	$-$	$+$	$-$	$+$	$+$
C	$-$	$-$	$-$	$+$	$-$	$+$	$+$	$+$
AB	$+$	$-$	$-$	$+$	$+$	$-$	$-$	$+$
AC	$+$	$-$	$+$	$-$	$-$	$+$	$-$	$+$
BC	$+$	$+$	$-$	$-$	$-$	$-$	$+$	$+$
ABC	$-$	$+$	$+$	$+$	$-$	$-$	$-$	$+$

As with the 2-level factors, main effects are defined by placing a $+$ wherever the relevant letter appears in a treatment combination, and a $-$ otherwise. The interaction AB is found by 'multiplying' the row for A by that for B in the same way as described for 2-factor experiments; and similarly we may find AC and BC. The 3-factor interaction ABC may be found in any of three ways: by 'multiplying' AB by C, or AC by B, or BC by A.

The basic *definitions* of the interactions are these: AB is the difference between the A effect at high B and that at low B, averaged over both the levels of C which are used in the experiment, with corresponding statements for the other 2-factor interactions. The ABC interaction is the difference between the AB interaction computed at high C and that at low C. The reader should check through some of these, to see that they do indeed lead to the contrasts given in the table above. These basic definitions can often be useful in explaining interactions found in the analysis of experimental data and in deciding which tables of means to present in a report.

Example 5.3. The strength of a component (measured in a standard test, and expressed in suitable units) depends on three factors: T, the temperature at which it is made; P, the percentage purity of one of its raw materials; and R, the reaction time allowed for one part of the process. An experiment is carried out on 4 days, the 8 TPR combinations being run once each day. T is (250°C, 300°C); P is (85%, 90%); R is (2 h, $2\frac{1}{2}$ h). In the usual notation, results are

	(1)	t	p	r	tp	tr	pr	tpr	Day totals
Day I	7	8	11	10	12	12	15	16	91
II	8	10	11	11	15	16	16	15	102
III	6	8	13	12	14	12	13	15	93
IV	7	8	12	10	14	14	17	16	98
	28	34	47	43	55	54	61	62	384

Summation terms are:

$$S_0 = 384^2/32 = 4608.00$$

$$S_D = \tfrac{1}{8}(91^2 + 102^2 + 93^2 + 98^2) = 4617.25$$

$$S_T = \tfrac{1}{4}(28^2 + \cdots + 62^2) = 4876.00$$

$$S = 4912.00$$

The Analysis of Variance is

Source	d.f.	Sum of squares	Mean square	
Days	3	9.25	3.083	$F_{(3,21)} = 2.42$
Treatments	7	268.00		
Residual	21	26.75	1.274	
Total	31	304.00		

The days *F*-test is not significant (the 5% point is 3.07). The 7 d.f. for 'treatments' require to be split into 7 single d.f.

	(1)	t	p	r	tp	tr	pr	tpr	Value	Divisor	S.S./M.S.	$F_{(1,21)}$
Total	28	34	47	43	55	54	61	62				
T	$-$	$+$	$-$	$-$	$+$	$+$	$-$	$+$	26	32	21.125	***
P	$-$	$-$	$+$	$-$	$+$	$-$	$+$	$+$	66	32	136.125	***
R	$-$	$-$	$-$	$+$	$-$	$+$	$+$	$+$	56	32	98.000	***
TP	$+$	$-$	$-$	$+$	$+$	$-$	$-$	$+$	-8	32	2.000	n.s.
TR	$+$	$-$	$+$	$-$	$-$	$+$	$-$	$+$	-2	32	0.125	<1
PR	$+$	$+$	$-$	$-$	$-$	$-$	$+$	$+$	-14	32	6.125	*
TPR	$-$	$+$	$+$	$+$	$-$	$-$	$-$	$+$	-12	32	4.500	n.s.
											268.000	

Each sum of squares has 1 d.f., and so the mean squares are the same as the sums of squares: tests of each against the residual mean square with 21 d.f. give the significance levels shown. A summary of the results therefore needs P and R to be examined together because they interact, and T means quoted for the main effect.

$$T \text{ low } (250°C) \text{ mean} = 11.19$$
$$T \text{ high } (300°C) \text{ mean} = 12.81$$

and by the F-test we know these are significantly different at the 0.1% level. The higher temperature is therefore recommended.

	P low	high
R low	(1), t	p, tp
high	r, tr	pr, tpr

so means are

	P low	high
R low	7.75	12.75
high	12.13	15.38

The standard error of the difference between any two of these means is

$$\sqrt{\tfrac{2}{8} \times 1.274} = 0.564$$

The interaction can be studied in (at least) two ways. We may look for differences in the table; least significant differences are

$$t_{(21)} \times 0.564 = 0.564 \times \begin{cases} 2.080 \ (5\%) = 1.17 \\ 2.831 \ (1\%) = 1.60 \\ 3.819 \ (0.1\%) = 2.15 \end{cases}$$

Hence a summary is

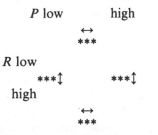

which in this particular example is not very helpful.

A graph (Fig. 5.4) of the means shows the presence of interaction because the responses to P are not parallel when R is compared for low and high levels.

Multiple Comparison tests, usually Duncan's, which are often found in computer packages perform very badly indeed for factorial experiments – this example could be used to show how badly – because they were never designed for this purpose and they fail to make use of the structure in the treatments. The information present in the results is not extracted.

If the TPR interaction had been significant, interpretation would have required a three-way table of the $T \times P \times R$ table of means, and this could be approached in either of three ways according to which is most helpful in explaining the reason for the interaction: *either* a PR table can be presented at each level of T, *or* a TP table at each level of R, *or* a TR table at each level of P. The main effects and two-factor interactions would have given little useful information in this case, because they average out the

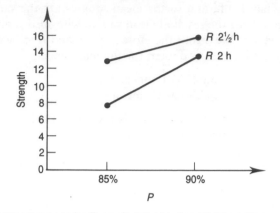

Fig. 5.4 Graph of treatment means for Example 5.3, showing *PR* interaction

interesting differences between individual treatment combinations which cause the three-factor interaction.

5.5   A single factor at more than two levels

In the early stages of a research programme, factors are often used at two levels to discover which of them interact, because those that do interact will need to be studied together in subsequent experiments. But in the later stages of the programme, it is not sufficient to study factors at only two levels. When treatments are different levels (amounts) of the same factor, that is they are *quantitative*, we will be interested in fitting a straight line or a curve to the graph of mean response y against level of factor applied, x. A good idea of the shape of the response curve is often obtained by using about four or five levels of a factor; when studying a single factor about five replicates of each factor level gives a reasonably sized experiment.

We first consider quantitative factors having their applied levels *equally spaced*, such as temperatures of 275, 300, 325, 350°C in operating an industrial process, or amount of nitrogenous fertilizer 10, 12, 14, 16 or 18 kg per unit plot. The responses to the set of levels should be studied together; much information about the shape of the response is lost if the means are compared individually.

In a 3-level experiment, the main effect of a factor has 2 degrees of freedom; its sum of squares is found in the usual way from the total yields of r replicates at each of the 3 levels. These 2 d.f. can be split into two components (contrasts), each with 1 d.f., the *linear* and *quadratic* components of the main effect (Fig. 5.5).

A graph of response against level of factor may be (approximately) a straight line; if so, and if the levels are equally spaced on a quantitative scale 0, 1, 2, the total responses should increase by the same step whenever the level increases by 1 unit.

$$T_2 - T_1 \text{ should equal } T_1 - T_0$$

If so $T_2 - T_1 = T_1 - T_0$, i.e. $T_2 - 2T_1 + T_0 = 0$.

The contrast $(1, -2, 1)$ is the *quadratic* component, not the linear one: because if it is 0 there is no curve. The linear component is estimated by the average of $(T_2 - T_1)$ and $(T_1 - T_0)$, which is

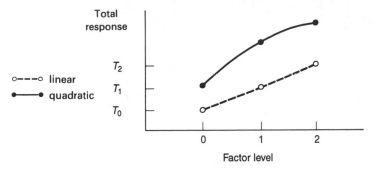

Fig. 5.5 Graph of response against level of factor

$$\tfrac{1}{2}\{(T_2 - T_1) + (T_1 - T_0)\} = \tfrac{1}{2}(T_2 - T_0)$$

and in the usual notation for contrasts is written $(-1, 0, 1)$. (The factor levels are always written in increasing order $(0), (1), (2)$.)

The two contrasts are summarized in the usual form of a table for the Analysis of Variance:

Treatment (factor level)	0	1	2			
Total	T_0	T_1	T_2	Value	Divisor	S.S.
Linear	-1	0	$+1$	$T_2 - T_0$	$2r$	
Quadratic	$+1$	-2	$+1$	$T_2 - 2T_1 + T_0$	$6r$	

where each level is replicated r times. These are two orthogonal contrasts, as defined in Chapter 4, and will therefore account for all the sum of squares for treatments. Each contrast may be tested for significance against the residual mean square. If the quadratic contrast is not significant, we may say that the response is effectively linear; when there is a quadratic component there is usually a linear one also, because a rather special pattern like Fig. 5.6(a) would be needed for pure quadratic, with the response at level

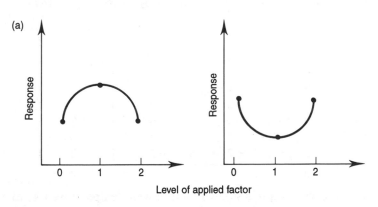

Fig. 5.6 (a) Patterns of response that are pure quadratic with no linear component

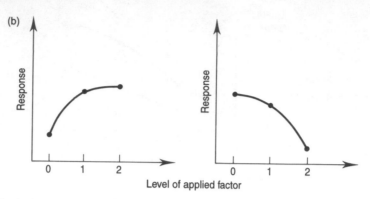

Fig. 5.6 (b) Typical responses with both linear and quadratic components

0 being the same as at level 2. Many responses take the form of Fig. 5.6(b), where there is a steady increase or decrease, in a curve rather than a straight line.

When there are two 3-level factors in an experiment, there are 9 treatment combinations: a_0b_0, a_0b_1, a_0b_2, a_1b_0, a_1b_1, a_1b_2, a_2b_0, a_2b_1, a_2b_2. There must be 8 d.f. among these. The main effects of A and B will take up 2 d.f. each: in computing the linear component of A (which we shall call A_L) every treatment having A at level 0 must be given a coefficient -1, every a_1 has a coefficient 0, and every a_2 has a coefficient $+1$. This is shown in the following table. For the quadratic component of A (A_Q) we proceed in a similar way: attach coefficients $+1$ to all the a_0's and a_2's, and -2 to all the a_1's. The main effect of B is split up in the same way, and the divisors for all these components follow the usual rule.

This leaves 4 d.f. for the interaction AB: interactions always have d.f. which are the product of the two main effects involved. These 4 d.f. can be split into four single d.f. by the 'multiplication' process using the linear and quadratic components of A and B to give 'linear $A \times$ linear B', which we call $A_L B_L$, and the other three contrasts $A_L B_Q$, $A_Q B_L$, $A_Q B_Q$ can be found in similar ways. The example that follows the table indicates how this subdivision is useful in an analysis.

Treatment Total	a_0b_0 .	a_0b_1 .	a_0b_2 .	a_1b_0 .	a_1b_1 .	a_1b_2 .	a_2b_0 .	a_2b_1 .	a_2b_2 .	Value	Divisor
A_L	-1	-1	-1	0	0	0	1	1	1	.	$6r$
A_Q	1	1	1	-2	-2	-2	1	1	1	.	$18r$
B_L	-1	0	1	-1	0	1	-1	0	1	.	$6r$
B_Q	1	-2	1	1	-2	1	1	-2	1	.	$18r$
$A_L B_L$	1	0	-1	0	0	0	-1	0	1	.	$4r$
$A_L B_Q$	-1	2	-1	0	0	0	1	-2	1	.	$12r$
$A_Q B_L$	-1	0	1	2	0	-2	-1	0	1	.	$12r$
$A_Q B_Q$	1	-2	1	-2	4	-2	1	-2	1	.	$36r$

Example 5.4. In a chemical process, two factors were temperature (A) at 80°C, 85°C and 90°C, and concentration of an acid used in a reaction (B) at 5%, 15% and 25%.

Five replicates of a completely randomized scheme gave the table of totals of process yield, in coded units:

A:	$a_0 = 80°C$	$a_1 = 85°C$	$a_2 = 90°C$		Total
B: b_0 (5%)	10	13	13	:	36
b_1 (15%)	12	18	21	:	51
b_2 (25%)	15	20	28	:	63
	37	51	62		150

$$N = 45, \qquad S_0 = 150^2/45 = 500.00, \qquad S = 594.00 \text{ (given)}$$

$$S_A = \frac{37^2 + 51^2 + 62^2}{15} = 520.93, \qquad S_B = \frac{36^2 + 51^2 + 63^2}{15} = 524.40$$

$$S_T \text{ (for all 9 treatments)} = \tfrac{1}{5}(10^2 + 13^2 + \cdots + 28^2) = 551.20$$

The Analysis of Variance is:

Source	d.f.	Sum of squares	Mean square	
A	2	20.93	10.465	$F_{(2,36)} = 8.80$***
B	2	24.40	12.200	$F_{(2,36)} = 10.26$***
AB	4	5.87	1.468	$F_{(4,36)} = 1.23$ n.s.
Treatments	8	51.20		
Residual	36	42.80	1.189	
Total	44	94.00		

n.s. = not significant

The interaction sum of squares is found by subtracting the sums of squares for *A* and for *B* from the 8-degree-of-freedom sum of squares for all 9 treatment combinations.

Although the overall *F*-tests indicate significant main effects and non-significant interaction, the full interpretation needs the sum of squares for treatments split into 8 single d.f. The totals in the contrasts table are, in the order above,

$$10 \quad 12 \quad 15 \quad 13 \quad 18 \quad 20 \quad 13 \quad 21 \quad 28$$

and $r = 5$. *Values* therefore are:

A_L	25	$A_L B_L$	10
A_Q	-3	$A_L B_Q$	-2
B_L	27	$A_Q B_L$	6
B_Q	-3	$A_Q B_Q$	6

and using the divisors given above, the sums of squares are

A_L	20.833	$A_L B_L$	5.000
A_Q	0.100	$A_L B_Q$	0.067
B_L	24.300	$A_Q B_L$	0.600
B_Q	0.100	$A_Q B_Q$	0.200

When each is tested against the residual mean square, A_L and B_L are *** i.e. have $P<0.001$, $A_L B_L$ is *, or $P<0.05$, and all others n.s. ($P>0.05$). The results show that only linear components are important and a graph of the 9 treatment-combination means illustrates this (Fig 5.7).

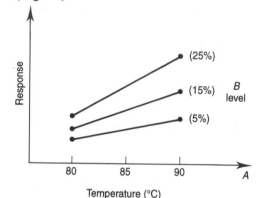

Fig. 5.7   Treatment means showing linear $A \times$ linear B interaction

In this example, both factors were *quantitative*, and it would have been possible to present the results with B plotted in the x-direction and lines drawn for each level of A. Tables of means can be used as previously.

Qualitative factors may be used (see below) and more than 3 levels will sometimes be desirable. If there are more than 3 levels we can still define linear, quadratic and also additional components of main effects, which will have different coefficients from 2-level factors. For equally spaced levels, single d.f. contrasts are:

3 levels: L $(-1, 0, 1)$; Q $(1, -2, 1)$

4 levels: L $(-3, -1, 1, 3)$; Q $(1, -1, -1, 1)$;   Cubic $(-1, 3, -3, 1)$

5 levels: L $(-2, -1, 0, 1, 2)$; Q $(2, -1, -2, -1, 2)$; $C(-1, 2, 0, -2, 1)$;   Quartic $(1, -4, 6, -4, 1)$

These contrasts are all *orthogonal* to one another, and so give complete subdivisions of the treatment sum of squares computed from the totals of the responses at each level of the factor. They are called *orthogonal polynomials*, and are tabulated for equally spaced factor levels by Fisher and Yates (1963) and in other reference books.

The general method for computing orthogonal polynomials is described on page 96; it can be applied (as illustrated there) when factor levels are not equally spaced, although the coefficients are less simple than for equally spaced levels.

Example 5.5. An experiment was carried out to measure the effects of operating temperature (10°C, 16°C, 22°C, 28°C) and the material used in making plates (A or B) on the maximum output voltage of a battery. Voltages, to the nearest 5 V, were recorded in 3 replicates of the experiment carried out in completely randomized order.

Temperature		10°C	16°C	22°C	28°C
Plate material:	A	150	120	135	60
		135	125	125	60
		145	140	110	50
	B	165	170	75	45
		150	185	100	45
		170	190	95	60

$r = 3$

Totals:		10°C	16°C	22°C	28°C		
	A	430	385	370	170	:	1355
	B	485	545	270	150	:	1450
		915	930	640	320		2805

Contrasts

		A10	A16	A22	A28	B10	B16	B22	B28		
Totals		430	385	370	170	485	545	270	150	Value	Divisor
Temperature	L	−3	−1	1	3	−3	−1	1	3	−2075	120
	Q	1	−1	−1	1	1	−1	−1	1	−335	24
	C	−1	3	−3	1	−1	3	−3	1	275	120
Materials A/B		1	1	1	1	−1	−1	−1	−1	−95	24
Temps × Mats	L	−3	−1	1	3	3	1	−1	−3	485	120
	Q	1	−1	−1	1	−1	1	1	−1	25	24
	C	−1	3	−3	1	1	−3	3	−1	−705	120

As a check, compute summation terms:

$$S = 377175; \qquad S_0 = 2805^2/24 = 327834.375$$

$$S_{\text{MATS}} = 328210.4167; \qquad S_{\text{TEMPS}} = 369020.8333; \qquad S_{\text{T}} = 375525$$

Source		d.f.	Sum of squares	Mean square	$F_{(1,16)}$
Materials		1	376.042		3.65 n.s.
Temperature	L	1	35880.208		347.9***
	Q	1	4676.042		45.34***
	C	1	630.208		6.11*
		3	41186.458		
Temps × Mats	L	1	1960.208		19.01***
	Q	1	26.042		<1
	C	1	4141.875		40.16***
		3	6128.125		
Treatments		7	47690.625		
Residual		16	1650.000	$103.125 = s^2$	
Total		23	49340.625		

Graphs based on the table of means help in interpretation; the standard error of difference between two means in this table is $\sqrt{2s^2/3} = 8.29$, and least significant differences are

$$8.29t_{(16)} = 8.29 \times \begin{cases} 2.120 \ (5\%) = 17.58 \\ 2.921 \ (1\%) = 24.22 \\ 4.015 \ (0.1\%) = 33.29 \end{cases}$$

Means:	10°C	16°C	22°C	28°C
A	143.3	128.3	123.3	56.7
B	161.7	181.7	90.0	50.0

The differences between means are rather irregular, and the graph (Fig. 5.8) is, in this case, probably the most useful item. Because only temperature is qualitative, the graph must be drawn like this; we cannot sensibly have *A* and *B* on the horizontal axis. It is reasonable to use the least-significant-difference approach, carefully and with adequate thought about what the results mean, because in this example no specific contrasts have been nominated.

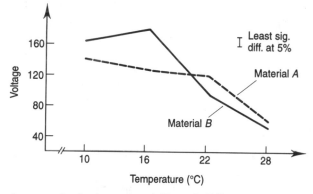

Fig. 5.8   Treatments means for 2×4 experiment (Example 5.5)

5.6   General Method for computing coefficients for orthogonal polynomials

The levels of a factor X need not be equally spaced; let them be $\{x_i\}$ and each level has the same replication r. Then the linear contrast has coefficients $\{l_i\}$, such that $l_i = a + bx_i$. Suppose that there are five levels of X, which are $x_i = 0, 2, 3, 4, 6$ units. Because a contrast must have $\sum_i l_i = 0$,

$$a + (a + 2b) + (a + 3b) + (a + 4b) + (a + 6b) = 0$$

Hence

$$5a + 15b = 0 \qquad \text{or} \qquad a + 3b = 0$$

We may choose any pair of values for a and b to satisfy this, and simple ones are $a = -3$, $b = 1$. The successive coefficients are then $l_0 = a = -3$; $l_2 = a + 2b = -1$; $l_3 = a + 3b = 0$; $l_4 = a + 4b = 1$; $l_6 = a + 6b = 3$.

The linear contrast is therefore $(-3, -1, 0, 1, 3)$.

The contrast is computed from the treatment totals T_i as $-3T_0 - T_2 + T_4 + 3T_6$, and its value is squared and divided by $20r$ for Analysis of Variance.

The quadratic contrast will have coefficients $q_i = f + gx_i + hx_i^2$, where the $\{q_i\}$ must add to 0; and $\sum_i l_i q_i = 0$ also if the linear and quadratic contrasts are to be orthogonal.

$$\sum_i q_i = f + (f + 2g + 4h) + (f + 3g + 9h) + (f + 4g + 16h) + (f + 6g + 36h) = 0$$

which reduces to

$$5f + 15g + 65h = 0$$

or

$$f + 3g + 13h = 0$$

$$\sum_i l_i q_i = -3f - (f + 2g + 4h) + (f + 4g + 16h) + 3(f + 6g + 36h) = 0$$

which reduces to

$$20g + 120h = 0$$

or

$$g + 6h = 0$$

Take $g = -6$, $h = 1$; then from the first equation $f = 5$. The successive coefficients are $q_0 = f = 5$; $q_2 = f + 2g + 4h = -3$; and similarly $q_3 = -4$, $q_4 = -3$, $q_5 = 5$.

The quadratic contrast is therefore $(5, -3, -4, -3, 5)$.

The contrast will be computed as $5T_0 - 3T_2 - 4T_3 - 3T_4 + 5T_6$, and the divisor for analysis is $84r$.

Because the divisor reduces the sum of squares to its correct dimension, it is only the ratio of $a : b$ and $f : g : h$ that is important, and this is why we may take simple integer values as appropriate during the calculation.

Since there are five levels in this example, cubic and quartic contrasts also exist. The coefficients $\{c_i\}$ must be of the form

$$c_i = p + rx_i + sx_i^2 + tx_i^3$$

and must satisfy $\sum_i c_i = 0$, and also $\sum_i l_i c_i = 0$ and $\sum_i q_i c_i = 0$ to give orthogonality. Likewise, the quartic must satisfy four relationships so as to make it a properly defined contrast which is orthogonal to the other three.

The most useful information is usually contained in the linear and quadratic components, however many levels a factor has. A 'cubic' component may sometimes be no more than a rather systematic, instead of purely random, deviation from linearity; and a 'quartic' can bear the same relationship to the quadratic. After removing linear and quadratic components, the remainder may be called 'deviations from quadratic', and be left as a single item in the analysis with more than 1 d.f.

5.7 Exercises

1. (i) Find all the orthogonal contrasts (linear, quadratic, ...) for a factor with six equally spaced levels.
   (ii) Find the linear and quadratic contrasts for a factor applied at levels 0, 1, 2 and 4 units.

2. Three different minerals A, B, C, each used at 2 levels (coded $-$, $+$) were added to molten steel during a manufacturing process. For each of these 8 treatment combinations, three steel ingots were cast and their breaking strengths (kN) determined in a standard test. The results were:

	B^-		B^+	
	C^-	C^+	C^-	C^+
A^-	12, 9, 9	14, 16, 17	9, 12, 11	16, 18, 15
A^+	16, 16, 11	17, 19, 19	13, 12, 13	16, 20, 17

$$\sum y^2 = 5269$$

Analyse these data to determine what differences these minerals may be producing, and write a brief report saying which of these minerals should be used in future and at what levels.

3. An experiment is carried out in two blocks, each of 10 units. Four factors A–D are studied, each at two levels (0, 1), and each block also contains *two* untreated units S. Construct an outline Analysis of Variance for this experiment, including the following components of the treatment sum of squares:

   (i) an estimate of the mean linear increase for all of A–D,
   (ii) an estimate of the difference in mean linear increases of (A, B) and (C, D).

4. In an experiment on penicillin, two levels of each of two factors C, E were used during manufacture. Yields were:

Block		I	II	III	IV
Treatment	(1)	112	99	118	116
	c	116	155	185	170
	e	58	76	110	71
	ce	83	83	85	100

Analyse these data, examine residuals and consider whether the data should have been transformed for analysis.

5. The data below show the gains in weight of male rats under six feeding treatments in a completely randomized experiment. The factors were:

   (3 levels)   Source of protein: beef, cereal, pork
   (2 levels)   Level of protein: high, low

The experimenter is particularly interested in the comparisons:

   (i) cereal versus the average of beef and pork
   (ii) beef versus pork

and how these comparisons differ between levels of protein. Analyse the data in a way that you think appropriate to answer the experimenter's questions.

Gains in weight (g) of rats under six diets

	High Protein			Low Protein		
	Beef	Cereal	Pork	Beef	Cereal	Pork
	73	98	94	90	107	49
	102	74	79	76	95	82
	118	56	96	90	97	73
	104	111	98	64	80	86
	81	95	102	86	98	81
	107	88	102	51	74	97
	100	82	108	72	74	106
	87	77	91	90	67	70
	117	86	120	95	89	61
	111	92	105	78	58	82
Totals	1000	859	995	792	839	787

6. (i) Fertilizer A gives a large increase in crop yield when it is applied, and also interacts (positively) with fertilizer B; but the main effect of B is not significant. Which combination of levels should be recommended for use?

 (ii) Can the main effects of A and B in a 2^2 experiment be non-significant while the AB interaction is highly significant? Illustrate by drawing a graph of treatment means.

 (iii) In a 2^3 experiment, using factors A, B, C, which of the following situations could give rise to an ABC interaction?

 (1) The combinations (1), a, b, c perform less well than the others.
 (2) One treatment combination behaves very differently from all the others.
 (3) The AB interaction is larger when C is present than when C is absent.
 (4) The combination (1) fails in one replicate.

7. In the production of a synthetic fibre, extrusion temperature of the material is important in determining strength. Five temperatures are used on each of four machines, and each combination is sampled twice after running. The results are as follows. Analyse and report on them.

Temperature (°C)		250	255	260	265	270
Machine	I	9, 7	13, 17	15, 7	13, 8	11, 11
	II	9, 7	7, 9	12, 14	9, 15	16, 13
	III	4, 0	6, 9	8, 8	11, 16	20, 19
	IV	0, 7	1, 3	6, 11	12, 10	19, 21

(For ease of arithmetic, 80 has been subtracted from every observation.)

8. An experiment was carried out to investigate the effects of three different types of heat treatment on the decarburization of steel. Two furnaces were used and four temperatures. The coded results are given below. Analyse the data and comment on your results.

		Temperature (°C)							
		800		850		900		950	
Furnace		1	2	1	2	1	2	1	2
Heat treatment	1	12.8	13.8	15.1	16.9	18.9	19.5	22.9	22.1
	2	13.0	14.1	15.2	16.4	21.2	19.6	23.7	24.2
	3	10.4	13.4	16.7	17.5	19.6	21.6	29.8	22.7

The third-order interaction is known to be negligible from previous work.

9. An experiment is conducted to test whether the concentration of a chemical in the leaves of a particular species of plant is linearly related to the quantity of applied nutrient in the soil. Five levels of nutrient, equally spaced in the range where a linear response is expected, are applied to two plants of each of three varieties P, Q, R, and the observed concentrations are given in the following table. Complete an analysis of these data, write a report on the results and illustrate it with suitable graphs. (There was no blocking in this experiment.)

Level of nutrient:		1	2	3	4	5
Variety	P	1.17, 1.22	1.18, 1.21	1.22, 1.20	1.23, 1.17	1.26, 1.22
	Q	1.09, 1.16	1.16, 1.22	1.24, 1.19	1.18, 1.21	1.24, 1.26
	R	1.12, 1.15	1.11, 1.15	1.18, 1.14	1.28, 1.22	1.22, 1.27

10. A method of testing materials for abrasion resistance consists of measuring the loss in weight or decrease in thickness of specimens after suitable intervals of time when placed in a machine in which they are rubbed against a standard abrasive under controlled conditions. The results below were obtained in an experiment with 24 specimens of coated fabrics, and refer to weight losses (mg) for each specimen after 3000 revolutions of the testing machine.

Two different fillers, F_1 and F_2, were tried in three different proportions, Q_1, Q_2 and Q_3, with and without a surface treatment T.

Wear of coated fabrics in milligrams

Surface Treatment	Filler	Proportion of filler		
		Q_1 (25%)	Q_2 (50%)	Q_3 (75%)
T_0	F_1	527, 561	621, 664	724, 743
(T absent)	F_2	456, 377	426, 476	460, 426
T_1	F_1	475, 466	561, 540	626, 682
(T present)	F_2	296, 325	301, 235	322, 304

By means of an Analysis of Variance, examine what conclusions may be drawn regarding the effects of the different experimental treatments.

11. A company manufacturing a foodstuff is interested in how the pH (acidity) of the final product depends on the temperature at which production takes place. Four strains of yeast, which is an important ingredient in the process, can be used. These are included in an experiment, together with three production temperatures. Each combination of strain and temperature is tested twice, and the 24 observations are obtained in random order. The following table gives the acidity level of the final product, in coded units. (The sum of the squares of all 24 observations is 237 799.)

Temperature (°C)	Yeast strain							
	I		II		III		IV	
12	113	130	115	100	52	43	61	74
16	118	105	106	116	80	70	101	92
20	105	88	131	121	93	107	96	108

Carry out an analysis of these data and write a report that will help the company to identify one or more of the yeasts as suitable for general production. You should bear in mind that it will not be as easy to control the temperature in general production as under experimental conditions. Also, the acidity level should not be too far from 100 units.

Illustrate the report with suitable graphs, and provide any information that will help the company manager to carry out some elementary significance tests.

12. The following data are the coded production rates for a chemical process. Four reagents and three catalysts were used and each combination was run twice, the 24 observations being made in random order. Analyse these data and make a report. (The sum of the squares of all the observations is 4240.)

In fact, the reagent labelled as 4 was only brought in at the last moment and the experimenter suspected that it might not act in the same way as the rest. Consider the merits of leaving reagent 4 out of the analysis.

		Reagent				Total
		1	2	3	4	
Catalyst	1	4	8	12	15	86
		6	10	16	15	
	2	11	12	17	15	106
		7	15	17	12	
	3	9	16	21	10	112
		15	14	17	10	
Total		52	75	100	77	304

13. A Latin square design gave the following results:

A, 23	E, 33	B, 28	C, 37	F, 45	D, 30
F, 44	A, 26	C, 41	B, 30	D, 27	E, 37
E, 34	F, 48	D, 26	A, 21	B, 28	C, 37
D, 28	B, 32	E, 35	F, 50	C, 43	A, 25
B, 33	C, 42	A, 22	D, 29	E, 34	F, 47
*	D, 32	F, 51	E, 36	A, 23	B, 34

The observation * was lost accidentally during testing. The data are the lengths of time for which an aircraft component lasted in a destructive test after having been made in a process that can run for three lengths of time $M = m_0$, m_1 and m_2 (equally spaced), with a catalyst L present (l_1) or absent (l_0).

$$A = l_0 m_0; \quad B = l_0 m_1; \quad C = l_0 m_2; \quad D = l_1 m_0; \quad E = l_1 m_1; \quad F = l_1 m_2$$

Complete the analysis of the data and write a report.

14. An experimenter theorizes that the concentration of a chemical in the root cells of a particular species of plant is linearly related to the quantity of an added soil nutrient, at least within a certain range of added nutrient quantities. He has performed an experiment, and asks you to use the resulting data to test his theory. He states that the experiment is a completely randomized design with four treatments and 24 plants as experimental units. The treatments consist of equally spaced, increasing quantities of added soil nutrient, which span the range for which a linear response is predicted. He presents you with the four treatment means and their estimated standard error. (You may assume that there are no arithmetic errors in the calculations.)

Nutrient level	1	2	3	4	Standard error
Treatment mean	14.500	16.667	19.333	20.167	1.409

(i) Use these results to test the experimenter's theory.

(ii) In discussing the results with the experimenter you learn that the 24 plants actually consisted of eight plants of each of three varieties. Two plants of each variety received each level of nutrient. The individual totals for each variety and each level of nutrient are as follows:

		Level of nutrient			
		1	2	3	4
Variety	1	34	39	44	38
	2	26	37	41	35
	3	27	24	31	48

Re-analyse the data in the light of this new information, stating clearly any changes that must be made to the conclusions from (i) above.

15. An experiment was carried out to compare the resistance to fatigue of three types of alloy subjected to six different vibration rates. The time (to the nearest half hour) taken to achieve a given fatigue state was recorded for three separate samples of

each alloy under each vibration rate. These times are given below. Complete an appropriate basic Analysis of Variance table. What conclusions would you draw from it?

Alloy	Vibration rate (cycles/s)						Total
	100	150	200	250	300	350	
A	32.5	31.5	34.0	33.0	32.0	29.5	
	33.0	33.5	32.5	31.0	30.5	30.0	
	35.0	34.0	31.0	30.5	30.5	28.0	
Total	100.5	99.0	97.5	94.5	93.0	87.5	572.0
B	33.0	33.0	32.5	30.5	28.5	28.5	
	31.5	29.0	30.0	29.0	30.5	28.5	
	31.0	32.5	28.5	30.0	26.5	27.0	
Total	95.5	94.5	91.0	89.5	85.5	84.0	540.0
C	29.0	27.0	28.0	29.0	26.5	28.0	
	28.5	27.0	26.5	28.0	27.5	26.0	
	25.5	27.5	27.0	26.5	26.5	28.5	
Total	83.0	81.5	81.5	83.5	80.5	82.5	492.5
Total	279.0	275.0	270.0	267.5	259.0	254.0	1604.5

Describe the further analyses you feel would be appropriate for the data. (You should indicate, in general terms, how the calculations would be done, but it is not necessary to undertake any arithmetic.)

6
Experiments with many factors: confounding and fractional replication

6.1 Introduction

When several factors are included in an experiment, the number of treatment combinations quickly becomes so large that they will not all go into blocks of homogeneous material. In agricultural experiments, except with small-scale crops, a block size of eight or nine plots is often the upper limit if proper control of heterogeneity is to be achieved. This implies a maximum of three 2-level factors (the block size is then $2^3 = 8$ units) or just two 3-level factors (the block containing $3^2 = 9$ units). In industry 8, or perhaps 16, units is a reasonable block size for many experiments; but the number of factors can easily be six or seven, and even more in the early stages of a research programme. Occasionally, even with a small number of factors, block sizes have to be very small – for example using the upper and lower surfaces of the same leaves as blocks of size 2.

Often factors are included at 2 levels only, to see whether they interact: if they do, they must be studied together in later experiments, but otherwise can be examined separately. We therefore hope to reduce the number of factors required in an experiment as the programme goes on, and eventually to use some of the methods described in Chapter 9 to find good combinations of the factor levels. But we must begin with 2 level experiments containing many factors.

Confounding for 2-level factors

This is a method for arranging a 2^p experiment (p factors each at two levels) in blocks of size 2^k, where k is less than p. For example, if p is 5, there are 32 treatment combinations and we may wish to limit block size to 16 or 8 (or even, sometimes, 4). The block size must be a power of 2 when this method is used.

Consider first an elementary example of three 2-level factors, A, B and C: as we saw on page 87 there are seven orthogonal contrasts that can be defined, namely three main effects, three two-factor interactions and the three-factor interaction. Ideally, we want to estimate all seven of these; but the last one, the ABC interaction, is likely to be the least important. We might therefore accept a scheme in which AB, AC, BC could all be studied, and if they were not significant then the main effects A, B and C could be estimated. Anything less than this scheme loses the essential factorial nature of the experiment.

If we can find satisfactory blocks of size 8 there is no difficulty. If, however, block size must be restricted to 4 units for proper control of heterogeneity, we can achieve this by *confounding* in such a way that the ABC interaction is not estimable but all the main effects and two-factor interactions are. The *ABC* interaction is the comparison of treatment combinations *abc*, *a*, *b*, *c* (those that carry a + sign in *ABC*, as on page 87) with the other treatment combinations *ab*, *ac*, *bc*, (1) (those that carry a − sign). Let us place *abc*, *a*, *b*, *c* in random order in half the blocks in the experiment, and *ab*, *ac*, *bc*, (1) in random order in the other half; Fig. 6.1 illustrates this. Now the comparison of *abc*, *a*, *b* and *c* against *ab*, *ac*, *bc* and (1) measures *block* differences, as well as the *ABC* interaction if there is one. Therefore when this contrast is computed, we cannot say whether it is, in reality, due to block differences or to the existence of an *ABC* interaction. We say that *ABC is confounded with blocks*. None of the other contrasts is affected by this confounding, because they all require two positive and two negative terms from each block: these contrasts are therefore orthogonal to blocks.

BLOCK									
I	b	abc	c	a	bc	(1)	ab	ac	IV
II	ab	bc	ac	(1)	abc	c	b	a	V
III	c	a	abc	b	ac	bc	(1)	ab	VI

Fig. 6.1 A 2^3 experimental layout confounding *ABC*

Example 6.1. In a forestry nursery, seedlings are planted out for their initial period of growth at two different spacings (factor *A*), with or without herbicide to control ground weeds (factor *B*) and with two different levels of fertilizer (factor *C*). Because it was difficult to find enough clear ground for the experiment, six blocks of four plots were used in different parts of the nursery, so as to confound the interaction according to the plan of Fig. 6.1. At the end of the period of growth, the nurseryman assessed general health by giving visually estimated scores for vigour, colour, height and other characteristics. The total scores for each plot were used as the measurement to be analysed, and the values in the following table have been 'coded' by subtracting 50 from each figure to simplify the analysis.

	(1)	a	b	c	ab	ac	bc	abc	Block total
Block I		9	12	13				23	57
II	5				15	16	21		57
III		11	14	13				25	63
IV	8				18	16	20		62
V		12	15	16				22	65
VI	7				18	20	19		64
Treatment total:	20	32	41	42	51	52	60	70	368 = G

Summation terms:

$$S = 6268$$

$$S_0 = 368^2/24 = 5642.67$$

$$S_B = 22632/4 = 5658$$

$$S_T = 18674/3 = 6224.67$$

Treatment Total	(1) 20	a 32	b 41	c 42	ab 51	ac 52	bc 60	abc 70	Value	Contrast
A	−	+	−	−	+	+	−	+	42	73.50
B	−	−	+	−	+	−	+	+	76	240.67
C	−	−	−	+	−	+	+	+	80	266.67
AB	+	−	−	+	+	−	−	+	−2	0.17
AC	+	−	+	−	−	+	−	+	−2	0.17
BC	+	+	−	−	−	−	+	+	−4	0.67

Each contrast is $(\text{Value})^2/24$.

As a check, the ABC contrast is 0.17, and the 7 d.f. sum of squares for treatments including ABC is 582.02; $S_T - S_0 = 582.00$, equal apart from rounding error. But we cannot include ABC in the analysis because it is confounded with blocks.

$$S - S_0 = 625.33 \quad \text{and} \quad S_B - S_0 = 15.33$$

The complete Analysis of Variance is:

Source	d.f.	Sum of squares	Mean square	
Blocks	5	15.333	3.07	$F_{(5,12)} = 1.31$
A	1	73.500		$F_{(1,12)} = 31.32$***
B	1	240.667		$F_{(1,12)} = 102.5$***
C	1	266.667		$F_{(1,12)} = 113.6$***
AB	1	0.167		$F < 1$
AC	1	0.167		$F < 1$
BC	1	0.667		$F < 1$
Residual	12	28.165	2.347	
Total	23	625.333		

The aim is to obtain a high score for general health. Since none of the interactions are significant, the results are summarized by looking for the levels of A, B and C that produce the higher scores as given by the main effects. These are high A (wider spacing), high B (with herbicide), high C (more fertilizer). Finally, there is no evidence of any block effect due to using different parts of the nursery; but if there had been such an effect, the confounding would have eliminated it without damaging any of the six contrasts in this analysis.

6.2 The principal block in confounding

Now consider a more likely practical problem, with 4 factors, and therefore $2^4 = 16$ treatment combinations, where block size must be reduced to 8. We might confound the 4-factor interaction *ABCD*; but suppose instead that we choose to confound *BCD*. We could set up the table of $(+, -)$ coefficients for 4 factors and pick out the row for *BCD*, then put all the + treatments into one block and all the − treatments into the other. However, there is a better way, which requires the setting up of a 'principal block'; this block always contains (1). The other members of it are those treatment combinations that have no letters or an even number (2, 4, . . .) of letters in common with the interaction that is confounded. We find the principal block is

$$(1), cd, abc, abd, a, bc, bd, acd \qquad [6.1]$$

The other block is

$$b, bcd, ac, ad, ab, c, d, abcd \qquad [6.2]$$

It is of course the remaining eight treatments that do not appear in the principal block; but again there is a better way of finding it. We can 'interact' with the principal block any treatment that has not appeared in it, say *b*; this means 'multiplying' the set of treatment combinations through by *b*, and striking out any letter that occurs twice. So $b \times (1) = b$; $b \times cd = bcd$; $b \times abc = ab^2c = ac$; $b \times abd = ab^2d = ad$; $b \times a = ab$; $b \times bc = b^2c = c$; $b \times bd = b^2d = d$; $b \times acd = abcd$. We should obtain the same set of treatments (in a different order) by choosing any treatment not in the principal block: for example using *abcd* would begin $(1) \times abcd = abcd$; $cd \times abcd = abc^2d^2 = ab$; $abc \times abcd = a^2b^2c^2d = d$; . . . . Two blocks like [6.1] and two like [6.2], i.e. two complete replicates of all the treatment combinations, would give an experiment of reasonable size. There would be 31 d.f. altogether, of which 3 represent blocks, 4 are main effects, 6 are 2-factor interactions, 3 are 3-factor interactions (all except *BCD* which is confounded with blocks), and there is also the 4-factor interaction *ABCD*. Residual thus has 14 d.f. which should be quite adequate. The *ABCD* interaction must be examined first, then (if it is not significant) the three-factor interactions and so on in the usual order.

Suppose now that in this 4-factor experiment blocks of 4 units each are all that is available. When we reduced block size to one half of the total number of treatment combinations in the examples discussed above, we did so by confounding one inter-action. The present problem, of reducing block size to one quarter (which we may think of as $1/2^2$) of the total number, requires two interactions to be confounded. Let us return to the arrangement above, where *BCD* was confounded in a 2^4 experiment and blocks like [6.1] and [6.2] were required. We could choose *ACD* for our second con-founded interaction. The principal block for that *alone* is

$$(1), cd, abc, abd, b, bcd, ac, ad \qquad [6.3]$$

and the other block is

$$a, bc, bd, acd, ab, c, d, abcd \qquad [6.4]$$

In order to confound both *BCD* and *ACD* together, we need to form four blocks which are

(I)   treatments common to [6.1] and [6.3]

(II)   treatments common to [6.1] and [6.4]

(III)   treatments common to [6.2] and [6.3]

(IV)   treatments common to [6.2] and [6.4]

The blocks are then

(1), *cd, abc, abd*   (the principal block)

a, bc, bd, acd   (II)

b, bcd, ac, ad   (III)

ab, c, d, abcd   (IV)

There are three comparisons possible between the four blocks, and we have chosen only two of them so far: (I and II) against (III and IV) is [6.1] against [6.2] and therefore confounds *BCD*; (I and III) against (II and IV) is [6.3] against [6.4] and therefore confounds *ACD*. The remaining comparison is (I and IV) against (II and III), which is

(1), *cd, abc, abd, ab, c, d, abcd*

against

a, bc, bd, acd, b, bcd, ac, ad

Treating the first of these as a 'principal block' it appears that *AB* is confounded by this comparison. Now *AB* is the 'generalized interaction' of *ACD* and *BCD*, found by the same 'interaction' method as we applied to treatment comparisons above: $ACD \times BCD = ABC^2D^2 = AB$. When we chose our two interactions *ACD* and *BCD* for confounding, we did not intentionally set out to confound a 2-factor interaction; but the choice we made forced us to confound *AB* even without intending to.

We may ask whether we can find a system for constructing a 2^4 experiment in blocks of 4 units without confounding any 2-factor interactions, because 2-factor interactions are usually important. If *ABCD* is confounded, we cannot also confound a 3-factor interaction, for example *ABC*, because the generalized interaction of these is *D*, a main effect, and that will also be confounded. So the only systems available are the one already described, two 3-factor interactions which must carry a 2-factor interaction with them, and an alternative (*AB, CD, ABCD*) which is less satisfactory because it uses two 2-factor interactions.

Confounding systems for reducing block size by more than one half can be constructed directly: there is no need to do it in steps as we did above.

Example 6.2. A 2^5 experiment is to be carried out in blocks of 8 units. *ABC* and *ABCDE* are chosen for confounding. This requires *DE* to be confounded also. The principal block contains (1) and all those treatment combinations that have 0, 2 or 4 letters in common with *all* of the confounded interactions *ABC*, *DE* and *ABCDE*:

(1), *de, ab, ac, abde, acde, bc, bcde*

The other three blocks are found by 'interacting' the elements of this principal block (Block I) with treatments not so far used: *a, e* and *ae* are one set of possibilities. We sometimes call these treatments the *generators* of the other blocks.

Block II: *a, ade, b, c, bde, cde, abc, abcde*

III: *e, d, abe, ace, abd, acd, bce, bcd*

IV: *ae, ad, be, ce, bd, cd, abce, abcd*

The analysis of a replicated confounded factorial design is straightforward. If two replicates of this system are carried out, in 8 blocks altogether, there will be 64 units and the degrees of freedom are assigned as follows:

Source of variation	d.f.	
Blocks	7	This includes *DE, ABC, ABCDE* which are confounded with blocks
Main effects *A, B, C, D* and *E*	5	
Two-factor interactions *except DE*, which is confounded	9	
Three-factor interactions *except* ABC	9	
Four-factor interactions	5	
Residual 'error'	28	
	63	

6.3 Single replicate

The experiment just discussed, in 64 units, is rather large, and 28 d.f. are more than we need for residual. Further, we do not usually expect to find important 4-factor interactions; 3-factor ones are sometimes important but not always. It is therefore useful to look at the form of analysis for a single complete replicate of the experiment (the four blocks once each). The 31 degrees of freedom are *all* accounted for by main effects and interactions (three of which are confounded with blocks).

Source of variation	d.f.	
Blocks	3	(including *DE, ABC, ABCDE*)
Main effects *A, B, C, D, E*	5	
Two-factor interactions *except DE*	9	
Three-factor interactions *except ABC*	9	
Four-factor interactions	5	
	31	

If this design is to be used, we must find an estimate of residual natural variation. It can only come from making some more assumptions. If we assume that no interaction above 3-factor is important (or statistically significant) there are 5 d.f. that we could use as 'residual'. In a well-controlled industrial experiment this may just be a sufficient number of d.f., though in a typical agricultural experiment it would not. Assuming that 3-factor interactions also are unimportant, a further 9 d.f. are available, making a total of 14 d.f. for residual which should be adequate in general. However, if unfortunately one or more of the 3-factor interactions were to be significant this would cause a bias (upward) in the estimate of σ^2: a systematic element of variability would then be present in addition to random residual variation. Anything we know already about the experimental material and the likely possible interactions will help in deciding whether such a system should be used. Also, such knowledge will help us to choose which factors shall be allocated the letters D and E in this arrangement so that their (2-factor) interaction is confounded: they should where possible be factors whose behaviour is already known so that an estimate of interaction is not needed, and the reason for using them is to study their relationships to A, B and C.

In the analysis, the blocks sum of squares should be computed in the usual way. Then those main effects and interactions that are not confounded should be calculated using the appropriate contrasts for a 2^5 experiment.

Example 6.3. Observations are made on the surgical equipment used at four different places to compare sterilization procedures: these are combinations of four factors, oxidants A and B (present or absent), time in autoclave C (low $= 5$ min; high $= 10$ min) and heat level D. Average bug counts per mm^2 are made at the four different sites (places) 1, 2, 3, 4 as follows:

Site 1		Site 2		Site 3		Site 4		
(1)	52.5	a	52.1	c	56.0	abc	42.1	
b	49.5	ab	44.7	bc	49.8	ac	51.1	
acd	50.3	cd	57.2	ad	52.1	bd	49.6	
abcd	36.6	bcd	51.1	abd	42.9	d	55.3	
	188.9		205.1		200.8		198.1	$G = 792.9$

Since we have a 2^4 design in blocks of four (i.e. 2^2), two interactions must have been confounded between sites, and their generalized interaction will also have been confounded. Putting 1 and 2 together and taking them as a principal block, CD is the interaction confounded; in the same way 1 and 3 give AD, and we can check this by seeing that 1 and 4 give AC. It is not immediately clear why 2-factor interactions have been chosen for confounding, unless we have some more information about the process.

In fact the experimenters knew from earlier work that the only factors likely to interact were A and B, so perhaps they deliberately retained ABC, ABD, $ABCD$ as well as AB.

In the usual notation

$$S = 39744.43, \qquad S_0 = \frac{729.9^2}{16} = 39293.15$$

$$S_B = \tfrac{1}{4}(188.9^2 + 205.1^2 + 200.8^2 + 198.1^2) = 39328.37$$

	(1)	a	b	c	d	ab	ac	ad	bc	bd	cd	abc	abd	acd	bcd	abcd	Value
	52.5	52.1	49.5	56.0	55.3	44.7	51.1	52.1	49.8	49.6	57.2	42.1	42.9	50.3	51.1	36.6	Value
A	−	+	−	−	−	+	+	+	−	−	−	+	+	+	−	+	−49.1
B	−	−	+	−	−	+	−	−	+	+	−	+	+	−	+	+	−60.3
C	−	−	−	+	−	−	+	−	+	−	+	+	−	+	+	+	−4.5
D	−	−	−	−	+	−	−	+	−	+	+	−	+	+	+	+	−2.7
AB	+	−	−	+	+	+	−	−	−	−	+	+	+	−	−	+	−18.3
BC	+	+	−	−	+	−	−	+	−	−	−	+	−	−	+	+	−9.7
BD	+	+	−	+	−	−	+	−	−	+	−	−	+	−	+	+	−9.1
ABC	−	+	+	+	−	−	−	+	−	+	+	+	−	−	−	+	−2.5
ABD	−	+	+	−	+	−	+	−	+	−	+	−	+	−	−	+	−3.9
ACD	−	+	−	+	+	+	−	−	+	+	−	−	−	+	−	+	−4.1
BCD	−	−	+	+	+	+	+	+	−	−	−	−	−	−	+	+	−0.1
ABCD	+	−	−	−	−	+	+	+	+	+	+	−	−	−	−	+	−5.7

The Analysis of Variance can now be constructed, with each main effect and unconfounded interaction computed as (Value)^2/16.

Source of variation	d.f.	Sum of squares	Mean square	
Blocks	3	35.22	11.74	$F_{(3,5)} = 13.34$**
A	1	150.68		$F_{(1,5)} = 171.22$***
B	1	227.26		$F_{(1,5)} = 258.25$***
C	1	1.27		$F_{(1,5)} = 1.44$ n.s.
D	1	0.46		$F_{(1,5)} < 1$
AB	1	20.93		$F_{(1,5)} = 23.78$**
BC	1	5.88		$F_{(1,5)} = 6.68$*
BD	1	5.18		$F_{(1,5)} = 5.89$ n.s.
ABC	1	0.39		
ABD	1	0.95		
ACD	1	1.05	0.88	(5 d.f.)
BCD	1	0.00$^+$		
ABCD	1	2.01		
Total	15	451.28		

The experimental design was not good: a single replicate of 2^4 does not give a sound basis for estimating residual variation. The 3- and 4-factor interactions will have to be used, giving 5 d.f. This is not enough, especially as BC and BD seem to be larger than the experimenters expected: we have no information on AC and AD, which is very unfortunate because A and B are both oxidants. The point of including C and D in the experiment has really been lost. There were block differences, which are differences between the places where the experiment was done; this is to be expected.

AB means are:

	A Absent	Present
B Absent	55.25	51.40
Present	50.00	41.58

Clearly the two oxidants *A* and *B* together do reduce bug counts, which is the aim of the treatment. There is some improvement when only one of them is used, either *A* or *B*, but in view of the unsatisfactory design it is not wise to work out 'least significant differences' using the 5 d.f. estimate of error. For a satisfactory experiment, 2 replicates of the set of treatments are needed, either by using 8 sites or by having each of the sites run 8 treatments instead of 4. One final question that might be asked about *C* and *D* is whether the levels used in the experiment were sufficiently far apart; perhaps high *C* should have been 15 min (or more) and high *D* hotter.

Example 6.4. How to find out what interactions have been confounded. In the following design for a 2^5 factorial experiment, the original plan has been lost but it is known which treatment combinations were in which block, as given below:

> Block   I   *c, cde, ae, bde, b, ad, abcd, abce*
>
> II   *acde, ab, e, d, bce, ac, bcd, abde*
>
> III   *bc, de, (1), bcde, abd, abe, ace, acd*
>
> IV   *a, be, ce, bd, abcde, ade, abc, cd*

Find out which interactions are confounded.

The principal block is III, since it contains (1). The other treatments in it must have 0, 2 or 4 letters in common with all confounded interactions. There will be 3 of these interactions (2 chosen for confounding and their interaction). The 4-letter combination *bcde* suggests trying *BCDE* as one; this is correct, as we see from looking at all the other treatments in III. Comparing the first four and the last four elements in III indicates *ABC* and *ADE* as the other two.

6.4   Small experiments: partial confounding

In an experiment with only very few factors, block size may still be a problem. A 2^3 experiment in blocks of 4 can be carried out by confounding *ABC*, as we saw in Example 6.1, so that the 2 basic blocks are (1), *ab, ac, bc* and *a, b, c, abc*. Three or four replicates of the whole scheme would be needed for a satisfactory experiment. There is a good alternative method for a small experiment such as this, which is to confound a different interaction between each pair of blocks that make up a complete replicate. So we could have

I: (1), ab, ac, bc and   II: a, b, c, abc,   confounding ABC

III: (1), a, bc, abc and   IV: b, c, ab, ac,   confounding BC

V: (1), b, ac, abc and   VI: a, c, ab, bc,   confounding AC

VII: (1), c, ab, abc and   VIII: a, b, ac, bc,   confounding AB

If 8 blocks of material were available, this plan would allow each main effect A, B, C, to be estimated from all the blocks, and each of the interactions AB, AC, BC, ABC from 6 of the 8 blocks. (This is sometimes described as having '$\frac{3}{4}$ information' on the interactions). Similarly, if we had only 6 blocks, we could choose which pairs of blocks to use so as to confound those interactions we were least interested in. The analysis is carried out in the usual way, except that some interactions are estimated from less than the full number of blocks.

Example 6.5. Crop weights of strawberries

Block	I	II	III	IV	V	VI	VII	VIII	
(I)	21		24		28		23		
a		40	37			37		33	
b		29		26	30			25	
ab	33			35		38	32		
c		37		40		42	38		
ac	47			45	41			42	
bc	39		38			35		36	
abc		48	39		42		44		
Block total	140	154	138	146	141	152	137	136	$G=1144, N=32$
Interaction confounded	*ABC*		*BC*		*AC*		*AB*		$S=42448$

$$S_0 = \frac{1144^2}{32} = 40898, \qquad S_B = \tfrac{1}{4}(163926) = 40981.5$$

Main effects and interactions

	(1)	a	b	ab	c	ac	bc	abc	Blocks	Value	Divisor	S.S.
A	−	+	−	+	−	+	−	+	All	122	32	465.125
B	−	−	+	+	−	−	+	+	All	−6	32	1.125
C	−	−	−	−	+	+	+	+	All	162	32	820.125
AB	+	−	−	+	+	−	−	+	I–VI	−17	24	12.042
AC	+	−	+	−	−	+	−	+	I–V, VII–VIII	−25	24	26.042
BC	+	+	−	−	−	−	+	+	I–II, V–VIII	−8	24	2.667
ABC	−	+	+	−	+	−	−	+	III–VIII	16	24	10.667

Analysis of Variance

Source of variation	d.f.		Sum of squares	Mean square	
Blocks	7		83.500	11.929	$F_{(7,17)} = 1.58$ n.s.
A	1	465.125		465.125	$F_{(1,17)} = 61.44$***
B	1	1.125		1.125	$F < 1$
C	1	820.125		820.125	$F_{(1,17)} = 108.32$***
AB	1	12.042		12.042	$F_{(1,17)} = 1.59$ n.s.
AC	1	26.042		26.042	$F_{(1,17)} = 3.44$ n.s.
BC	1	2.667		2.667	$F < 1$
ABC	1	10.677		10.667	$F_{(1,17)} = 1.41$ n.s.
	7		1337.793		
Residual	17		128.707	7.571 = estimate of σ^2	
Total	31		1550.000		

The main effects A and C are significant but there are no interactions. So it is correct to quote the means of A and C as a summary of the results.

$$A: \text{low} \quad 31.94 \qquad C: \text{low} \quad 30.69$$
$$\text{high} \quad 39.56 \qquad \text{high} \quad 40.81$$

Assuming that high yield is required, the high levels of A and C should be used, but it does not matter about B. [Note: If we require them, the standard errors of A means and C means are $\sqrt{\sigma^2/16}$, estimated by $\sqrt{7.571/16} = 0.69$; however, we already have all the information we can obtain, because the $F_{(1,17)}$ tests for A and C have shown significance. But the standard errors of the means in the two-way AC table

will be $\sqrt{\sigma^2/6}$, not $\sqrt{\sigma^2/8}$, because they are based on 3 replicates, not four. In this example we do not need them because there was no AC interaction. Similarly, the standard error of a mean of an individual treatment combination, such as *abc*, or (1), is $\sqrt{\sigma^2/3}$ and *not* $\sqrt{\sigma^2/4}$.]

6.5 Very large experiments: fractional replication

An experiment with six 2-level factors is quite likely to be needed early in a research programme, especially in industry. This needs $2^6 = 64$ units even for a single replicate.

A single replicate may in fact be quite adequate if we are prepared to use 5- and 6-factor interactions as the estimate of residual variation, and certainly if we use 4-factor interactions as well. But 64 units is a large number, and an experiment using 64 units may take a long time to complete. Suppose that no more than 32 units can be used; and suppose, for simplicity, that a completely randomized scheme is satisfactory. Then we require a method of choosing 32 treatment combinations which will allow us to estimate main effects and interactions in the most useful way possible; obviously we cannot estimate them all if we do not use all 64 treatment combinations.

The first step is to choose an interaction that does not need to be estimated: this is called the *defining contrast*. Next we decide whether to use all the treatment combinations with positive signs from this interaction, or to use all those with negative signs; the choice must be made at random (e.g. by tossing a coin) and now we have the set of 32 treatment combinations for the experiment. This is called a *half-replicate* of the 2^6 experiment, and the method in general is called *fractional replication*.

If *ABCDEF* is the defining contrast we might choose to use all the treatment combinations with the same sign as (1); these would be those with 2, 4 or 6 letters in common with *ABCDEF*. (The alternative choice of course takes in all the treatment combinations with 1, 3 or 5 letters in common with *ABCDEF*.) In Example 6.6 below, the set including (1) is used:

(1)	*ab*	*cd*	*ef*	*abcd*	*abce*	*abcf*	*acde*
ac	*ad*	*ae*	*af*	*acdf*	*acef*	*abde*	*abdf*
bc	*bd*	*be*	*bf*	*abef*	*adef*	*bcde*	*bcdf*
ce	*cf*	*de*	*df*	*bcef*	*bdef*	*cdef*	*abcdef*

In a half-replicate, each main effect and interaction has an *alias*, and when a significant result appears in an analysis we cannot tell which member of the alias pair is causing the significance. If we perform only half the experiment we must expect to obtain only half the information available from the full experiment. The loss of information takes the form that we cannot distinguish between pairs of effects.

Mathematically, the defining contrast is written I = *ABCDEF* (I stands for 'identity'). To find the alias of any main effect or interaction, we 'multiply' both sides of this by the appropriate letters and strike out any that occur in pairs: for example the alias of *A* is *AABCDEF* = *BCDEF*. Each main effect in this example has a 5-factor interaction as alias. The alias of *AB* is *ABABCDEF*, or *CDEF*, and each 2-factor interaction has a 4-factor alias. Clearly the 3-factor interactions will be aliased in pairs: *ABC* = *DEF* and so on. What this means is that for the interaction *ABC* we must use the same (+, −) combination of the 32 treatments actually included in the experiment as we use for estimating *DEF*: this (+, −) combination or contrast has two 'names', *ABC* and *DEF*. The same is true for all other alias pairs. (Sometimes a required effect is estimated by exactly the same terms as its alias, sometimes the terms are the same but with exactly opposite signs.)

Example 6.6. The list below shows in shorthand form the set of treatments in a $\frac{1}{2}$-replicate of a 2^6 experiment, using the 32 treatments including (1).

	def	de	df	ef	d	e	f	(−)
abc	6.1				8.6	7.2	8.8	
ab		12.4	6.3	5.6				10.1
ac		7.4	4.2	6.6				9.8
bc		9.2	4.4	8.2				8.4
a	5.3				7.5	8.9	6.2	
b	2.6				6.0	9.3	5.0	
c	0.4				5.6	6.6	4.9	
(−)		6.1	−0.1	5.5				5.3

For example, the combination *abcdef* gave 6.1; and (1) 5.3. The data are coded by subtracting 60 and are from a biological experiment. When listing the treatments in order to apply the usual (+, −) contrasts to calculate main effects and interactions, the best method is to write down the list of 32 combinations that would arise from a *complete* 2^5 experiment: (1), *a*, *b*, *c*, *d*, *e*, *ab*, *ac*, ..., *de*, *abc*, ..., *cde* and so on but then to replace those that have not been used by their interaction with *f*. For example, *abcde* was not in the experiment, so we put in *abcdef* instead; *a* was not used, so we put *af* in its place in the table of yields (Table 6.1). The 3-factor alias pairs $ABC = DEF$, $ABD = CEF$, ..., $AEF = BCD$ (10 pairs in total) can be found in the same way; all are fairly small except $ADE = BCF$ which is 14.8 in value, giving a sum of squares of 6.845. We have only a single $\frac{1}{2}$-replicate, so there are no residual degrees of freedom unless we use high-order interactions. We must assume (1) that the 3-factor interactions can be used as residual 'error', and (2) that no 4-factor or 5-factor interactions are present, so that any significance is due to their aliases: main effects are aliases of 5-factor interactions, and 2-factor interactions are aliases of 4-factor ones. By doing this we can complete the Analysis of Variance table, which will contain the following types of term:

Items	d.f.
Main effects, *A–F* (with aliased 5-factor interactions)	6
Two-factor interactions *AB–EF* (with 4-factor aliases)	15
'Residual' (consisting of 10 pairs of aliased 3-factors)	10
	31

We might argue that since all 3-factor interactions are to be used as 'residual', there is no need to compute them; but if one or two of them did prove to be large we might have misgivings about constructing a residual term in this way without, at least, a further look at the data.

The total S.S is found from $S - S_0$ in the usual way, and is 214.755 with 31 d.f. so that the residual S.S. is 18.815 with 10 d.f., giving a residual mean square of 1.8815 which is the estimate of σ^2. $F_{(1,10)}$ tests of each main effect and 2-factor interaction

Table 6.1.   Table of yields for Example 6.6

Treatment		A	B	C	D	E	F= ABCDE	AB	AC	AD	AE	AF= BCDE	BC	BD	BE	BF= ACDE	CD	CE	CF= ABDE	DE	DF= ABCE	EF= ABCD
(1)	5.3	−	−	−	−	−	−	+	+	+	+	+	+	+	+	+	+	+	+	+	+	+
a(f)	6.2	+	−	−	−	−	+	−	−	−	−	+	+	+	+	−	+	+	−	+	−	+
b(f)	5.0	−	+	−	−	−	+	−	+	+	+	−	−	−	−	+	+	+	−	+	−	−
c(f)	4.9	−	−	+	−	−	+	+	−	+	+	−	−	+	+	−	−	−	+	+	−	−
d(f)	−0.1	−	−	−	+	−	+	+	+	−	+	−	+	−	+	−	−	+	−	−	+	−
e(f)	5.5	−	−	−	−	+	+	+	+	+	−	−	+	+	−	−	+	−	−	−	−	+
ab	10.1	+	+	−	−	−	−	+	−	−	−	−	−	−	−	−	+	+	+	+	+	+
ac	9.8	+	−	+	−	−	−	−	+	−	−	−	−	+	+	+	−	−	−	+	+	+
ad	7.5	+	−	−	+	−	−	−	−	+	−	−	+	−	+	+	−	+	+	−	−	+
ae	8.9	+	−	−	−	+	−	−	−	−	+	−	+	+	−	+	+	−	+	−	+	−
bc	8.4	−	+	+	−	−	−	−	−	+	+	+	+	−	−	−	−	−	−	+	+	+
bd	6.0	−	+	−	+	−	−	+	−	+	−	+	+	−	+	−	−	+	+	−	−	+
be	9.3	−	+	−	−	+	−	−	+	+	−	+	−	−	+	−	+	−	+	−	+	−
cd	5.6	−	−	+	+	−	+	−	+	−	+	+	+	−	−	+	+	−	−	−	−	+
ce	6.6	−	−	+	−	+	+	−	+	−	+	+	−	+	−	+	−	−	+	−	+	−
de	6.1	−	−	−	+	+	−	+	+	−	−	+	+	−	−	+	−	−	+	+	−	−
abc(f)	8.8	+	+	+	−	−	+	+	+	−	−	+	+	−	−	+	−	−	+	+	−	−
abd(f)	6.3	+	+	−	+	−	+	+	−	+	−	+	−	+	−	+	−	−	+	−	+	−
abe(f)	5.6	+	+	−	−	+	+	+	−	−	+	+	−	−	+	+	−	−	+	−	−	+
acd(f)	4.2	+	−	+	+	−	+	−	+	+	−	+	+	−	−	+	−	+	−	+	−	+
ace(f)	6.6	+	−	+	−	+	+	−	+	−	+	+	−	+	−	+	−	−	+	−	−	+
ade(f)	5.3	+	−	−	+	+	+	−	−	+	+	+	+	−	−	+	−	−	+	+	+	−
bcd(f)	4.4	−	+	+	+	−	+	−	+	+	−	−	+	+	−	+	+	−	+	−	+	−
bce(f)	8.2	−	+	+	−	+	+	−	−	+	−	−	+	−	+	+	−	+	+	−	−	+
bde(f)	2.6	−	+	−	+	+	+	−	+	−	−	−	+	+	+	+	−	−	−	+	+	+
cde(f)	0.4	−	−	+	+	+	−	+	+	−	−	−	−	−	−	−	+	+	+	+	+	+
abcd	8.6	+	+	+	+	−	−	+	+	+	−	−	+	+	−	−	+	−	−	−	+	+
abce	7.2	+	+	+	−	+	−	+	+	−	+	−	+	−	+	−	−	+	−	−	+	−
abde	12.4	+	+	−	+	+	−	+	+	−	+	+	−	−	+	+	−	−	+	+	−	−
acde	7.4	+	−	+	+	+	−	−	+	+	+	−	+	+	+	−	+	+	+	−	+	−
bcde	9.2	−	+	+	+	+	−	−	−	−	−	+	+	+	+	+	−	+	+	−	+	−
abcde(f)	6.1	+	+	+	+	+	+	+	+	+	+	+	+	+	+	+	+	+	+	+	+	+
S.S. (Value²/32)		35.280	24.500	0.605	18.605	1.280	73.205	2.880	4.205	5.780	3.380	0.245	0.245	3.380	0.005	0.000	0.720	4.805	3.125	1.805	11.045	0.845
Value		33.6	28.0	4.4	−24.4	6.4	−48.4	−9.6	−11.6	13.6	−10.4	2.8	2.8	10.4	0.4	0.0	−4.8	−12.4	10.00	7.6	−18.8	−5.2

against this residual mean square give non-significant *F*-values *except* for:

$$A: \ F_{(1,10)} = 18.75^{**}$$

$$B: \ F_{(1,10)} = 13.02^{**}$$

$$D: \ F_{(1,10)} = 9.89^{*}$$

$$F: \ F_{(1,10)} = 38.91^{***}$$

$$DF: \ F_{(1,10)} = 5.87^{*}$$

The analysis is completed by quoting *A* means, *B* means and the means in a 2-way *DF* table.

6.6   Replicates smaller than half size

A $\frac{1}{4}$-replicate of a 2^6 experiment uses only 16 treatment combinations. Two interactions have to be included in the defining contrast to achieve this. The choice has to be made carefully, as in confounding, otherwise the alias pattern is unsatisfactory. Basing a scheme on $ABDE$ and $ACDF$, we also lose their interaction $BCEF$. The defining contrast now is

$$I = \quad ABDE = \quad ACDF = \quad\quad BCEF$$

so that alias sets are

$$
\begin{array}{cccc}
A= & BDE= & CDF= & ABCEF \\
B= & ADE= & ABCDF= & CEF \\
C= & ABCDE= & ADF= & BEF \\
D= & ABE= & ACF= & BCDEF \\
E= & ABD= & ACDEF= & BCF \\
F= & ABDEF= & ACD= & BCE \\
AB= & DE= & BCDF= & ACEF \\
AC= & BCDE= & DF= & ABEF \\
AD= & BE= & CF= & ABCDEF \\
AE= & BD= & CDEF= & ABCF \\
AF= & BDEF= & CD= & ABCE \\
BC= & ACDE= & ABDF= & EF \\
BF= & ADEF= & ABCD= & CE \\
ABC= & CDE= & BDF= & AEF \\
ABF= & DEF= & BCD= & ACE \\
\end{array}
$$

As regards main effects, the design is satisfactory because none are aliased with one another or with 2-factor interactions. But several 2-factor interactions are aliased in pairs, which is not satisfactory unless information is already available on some of these, so that careful choice of aliasing can be made. In practice, a $\frac{1}{4}$-replicate would not often be used with less than 7 or 8 factors, and then confounding may also be needed to control block size (see below).

Fractional replication is very useful in exploratory work where a very large number of factors need to be investigated. In the chemical industry, for example, as many as 20 factors are possible in a research programme. Many of the factors may produce no response, but interactions are possible and need to be explored. Fractional replication can be used to 'screen' factors, and when the important ones have been found they can be included in one or more of a set of experiments in greater depth (at 3 or more levels). The first 'screening' stage is conducted efficiently and economically in this way.

6.7   Confounding with fractional replication

We have assumed so far in discussing fractional replication that we can find enough homogeneous units to lay the experiment out as a completely randomized design. In experiments with many factors this is rather unlikely, and in order to reduce blocks to an acceptable size we need to employ confounding also. The following example illustrates how confounding and fractional replication are combined in the same experiment.

Example 6.7. A $\frac{1}{4}$-replicate of a 2^8 experiment (factors $ABCDEFGH$) which preserves all main effects and 2-factor interactions needs an identity relationship (defining contrast) whose terms all have at least 5 letters. For example

$$I = ABCDE = ABFGH = CDEFGH$$

This avoids confounding any pairs of 2-factor interactions, and also avoids main effects and 2-factor interactions being confounded. The 64 treatment combinations to be used may be the set containing (1), or any of the other 3 sets found by interacting this with a suitable letter (or letters) as generator. For simplicity we choose (1), *ab*, *cd*, *ce*, *de*, *fg*, *fh*, *gh*, . . . , and so on.

It is quite unlikely that we shall be able to run all 64 together in a completely randomized scheme; instead we shall need to split them into groups of 8, or perhaps 16, to provide convenient blocks for carrying out the treatments and recording the results.

(i) *Blocks of 16*. To split these 64 into 4 blocks of 16, we need to choose two more interactions for confounding, but we must remember that their aliases will also be confounded. $ACF = BDEF = BCGH = ADEGH$ and $BDG = ACEG = ADFH = BCEFH$ are possibilities. The principal block for this design will be found in the usual way, based on $ABCDE$, $ABFGH$, ACF and BDG, which were the *independent* interactions chosen to set up the complete scheme combining fractional replication with confounding (and which, of course, carry the others with them). The principal block is:

$$(1), \ ach, \ aef, \ cefh, \ bdh, \ abcd, \ abdefh, \ bcdef, \ beg, \ abcegh,$$

$$abfg, \ bcfgh, \ degh, \ acdeg, \ adfgh, \ cdfg$$

(We need only find four independent members of this set; the rest will be generated by interaction: *ach*, *aef*, *bdh* and *beg* are enough to set up the block.) The three remaining blocks must be from the *same* $\frac{1}{4}$-replicate; we can use *ab*, *ce* and *de* to generate them. We have also confounded the interaction of these two alias sets ($ACF \times BDG = ABCDFG$):

$$ABCDFG = EFG = CDH = ABEH$$

The analysis can be carried out by omitting *a* and *f* from the treatment combinations to obtain the main effects and interactions in terms of *B*, *C*, *D*, *E*, *G*, *H*, and then relabelling them with suitable aliases. The Analysis of Variance is:

	d.f.
Blocks	3
Main effects $A, B, \ldots, H$	8
Two-factor interactions $AB, \ldots, GH$	28
Residual 'error'	24
	63

(ii) *Blocks of 8*. To reduce block size to 8, some 2-factor interactions must be lost. A scheme that loses only two is based on the above design with

$$CDF = ABEF = ABCDGH = EGH \quad \text{also confounded}$$

This will carry with it the following interactions:

$$AD = \quad BCE = BDFGH = ACEFGH$$

$$BCFG = ADEFG = \quad ACH = BDEH$$

$$ABG = \quad CDEG = \quad FH = ABCDEFH$$

The Analysis of Variance is:

Source of variation	d.f
Blocks (including *AD*, *FH*)	7
Main effects	8
Unconfounded 2-factor interactions	26
Residual 'error'	22
Total	63

Again it can be carried out by omitting *a* and *f*, and relabelling. The principal block is (1), *abcd*, *cefh*, *abdefh*, *beg*, *acdeg*, *bcfgh*, *adfgh*, and the other 7 blocks (in the same $\frac{1}{4}$-replicate) can be found by interacting this block with *ab*, *ce*, *de*, *fg*, *gh*, *acf*, *bdh* in turn.

In this design, only two 2-factor interactions are lost, and these can be chosen carefully on the basis of any previous knowledge.

Example 6.8. There are 32 units of experimental material available for a 2-level factorial experiment. Block size must not exceed 8 units. Show that if 5 factors are used there is a confounding scheme, based on two 3-factor interactions, with no main effect or 2-factor interaction confounded. Write down the contents of each block in this scheme.

The experimenter wonders whether he might use the 32 units for a $\frac{1}{2}$-replicate of a 2^6 experiment instead. Find a scheme in which only one 2-factor interaction need be confounded, and write down the contents of the blocks for the $\frac{1}{2}$-replicate. Would you advise 2^5 or 2^6?

In a 2^5 experiment, to reduce block size to 8, we may confound *ABC*, *ADE*, *BCDE*. This will give as the principal block:

$$(1), bcde, abd, abe, acd, ace, bc, de$$

Then using *a*, *b*, *d* in turn as generators, we obtain the other three blocks:

Block II   *a*, *abcde*, *bd*, *be*, *cd*, *ce*, *abc*, *ade*

III   *b*, *cde*, *ad*, *ae*, *abcd*, *abce*, *c*, *bde*

IV   *d*, *bce*, *ab*, *abde*, *ac*, *acde*, *bcd*, *e*

In the Analysis of Variance, it would be necessary to use 3-factor interactions as well as the higher-order ones to find a suitable residual 'error' estimate.

Source of variation	d.f.	
Blocks	3	
A, *B*, *C*, *D*, *E*	5	
Two-factor interactions	10	
Three-factor (*except ABC*, *ADE*)	8	
Four-factor (*except BCDE*)	4 }	5 d.f. not sufficient for residual
ABCDE	1 }	'error' in most experiments
Total	31	

In 2^6, we take $ABCDEF$ as the *defining contrast* and use only one half of the treatments, *either* those with $+$ signs in this contrast or those with $-$ signs (it does not matter which). A scheme based on ABC, ABD and CD is then possible for confounding, but the 4-factor interaction will carry a 2-factor one as its alias. Aliases are: $A = BCDEF$, $B = ACDEF, \ldots, F = ABCDE$; $AB = CDEF, \ldots, EF = ABCD$ (but note that $CD = ABEF$ is confounded); $ABC = DEF$ and $ABD = CEF$ (both confounded), $ABE = CDF$, $\ldots, AEF = BCD$.

The principal block is found using ABC, ABD and $ABCDEF$:

$$(1), \; ab, \; abef, \; ef, \; bcde, \; acde, \; acdf, \; bcdf$$

Other blocks must be in the *same* half replicate, and so have 0, 2 or 4 letters in common with $ABCDEF$:

Block II   *cd, abcd, abcdef, cdef, be, ae, af, bf*

      III   *ac, bc, bcef, acef, abde, de, df, abdf* ·

      IV   *ad, bd, bdef, adef, abce, ce, cf, abcf*

Source of variation	d.f.	
Blocks	3	
A, B, C, D, E, F	6	(aliased with 5-factor interactions)
AB, ..., EF(*except CD*)	14	(aliased with 4-factor interactions)
ABE = CDF, ..., AEF = BCD	8	(8 alias pairs: other 2 pairs confounded)
Total	31	

In both 2^5 and 2^6 the 3-factor interactions will need to be used as residual 'error'; in 2^5 there are a few other d.f. available but these may not be enough except in a well-controlled industrial experiment. The decision depends on balancing this against the usefulness of including an extra factor in the particular experiment being planned.

6.8   Confounding three-level factors

We have already noted (page 90) that a factor A at 3 levels a_0, a_1, a_2 takes up 2 degrees of freedom for its main effect, which is the comparison of the 3 totals of the observations on a_0 and a_1 and a_2, respectively. If the levels are quantitative, we may want to extract linear and quadratic components, each with 1 d.f. Further, when 2 such factors A, B are included in the same experiment we can split the 4 d.f. for the AB interaction into four single degrees of freedom based on linear and quadratic components of A and B. Although the calculation of coefficients in the contrasts is more complicated, quantitative factors do not have to be equally spaced for this method to be applied (page 97), and so we will often want to subdivide main effects and interactions in this way when factors are quantitative.

Whether or not factors are quantitative, we can still identify some components of their main effects and interactions, and these form the basis of confounding when block sizes have to be less than the total possible number of treatment combinations for the experiment. Just as with 2-factor experiments we are able to reduce block sizes by factors of 2, so with 3-factor experiments we must work in blocks of $\frac{1}{3}$, $\frac{1}{9}$, etc., of the total number of treatment combinations.

For 2 factors A, B there are 9 treatment combinations, which may be written as pairs of numbers showing the A level first followed by the B level: 00, 01, 02, 10, 11, 12, 20, 21, 22. These may be written in a 3×3 layout and associated with the Latin and Greek letters of a 3×3 Graeco-Latin square as follows:

$$
\begin{array}{ccc} \quad\quad & \begin{array}{ccc} 00 & 10 & 20 \end{array} & \quad\quad & \begin{array}{ccc} a\alpha & b\beta & c\gamma \end{array} \\[4pt] & \begin{array}{ccc} 01 & 11 & 21 \end{array} & & \begin{array}{ccc} b\gamma & c\alpha & a\beta \end{array} \\[4pt] & \begin{array}{ccc} 02 & 12 & 22 \end{array} & & \begin{array}{ccc} c\beta & a\gamma & b\alpha \end{array} \end{array}
$$

The main effect of A is the comparison between the totals in the three columns, and that of B is the comparison of the three row totals; each accounts for 2 d.f. The remaining 4 d.f., for the interaction AB, can be split into two pairs, and these are usually called the I and J components of the AB interaction. I is the comparison α versus β versus γ, and J is the comparison a versus b versus c.

In order to lay a basis for examining interactions in large experiments where confounding will be necessary, we shall denote the full set of 9 treatment combinations by (x_1, x_2), where x_1 denotes the level of A applied to a unit and x_2 denotes its level of B. Thus each of x_1 and x_2 can take the values 0, 1 and 2.

The main effect of A is the comparison between all the 0s, all the 1s and all the 2s. We will use the idea from number theory of working 'modulo 3'; whenever a number greater than 3 is encountered, as many 3s as possible are taken from it and the residue, which can only be 0, 1 or 2, is actually used. The statement $x_1 = 0$ identifies the same units as $2x_1 = 0$; $x_1 = 1$ is the same as $2x_1 = 2$; and $x_1 = 2$ is the same as $2x_1 = 1$ because the actual result, $2x_1 = 4$, has to be reduced modulo 3 before it comes into the range of possible x-values. We shall call the main effect of A (A, A^2), because it is the comparison between the numbers that satisfy

$$x_1 = 0 \text{ or } 1 \text{ or } 2 \text{ (mod 3)} \qquad (A)$$

or, equivalently, between those that satisfy

$$2x_1 = 0 \text{ or } 2 \text{ or } 1 \text{ (mod 3)} \qquad (A^2)$$

The same argument can be used to express B as the two degrees of freedom (B, B^2). However, the main interest of this is in subdividing interactions; I is defined by

$$x_1 + 2x_2 = 0 \text{ or } 1 \text{ or } 2 \text{ (mod 3)}, \qquad \text{which we call } AB^2$$

or equally well by

$$2x_1 + x_2 = 0 \text{ or } 2 \text{ or } 1 \text{ (mod 3)}, \qquad \text{which we call } A^2B$$

This pair accounts for two of the 4 d.f. from the AB interaction. The other 2 d.f. are those for J, which is

$$x_1 + x_2 = 0 \text{ or } 1 \text{ or } 2 \text{ (mod 3)}, \qquad \text{which we call } AB$$

and

$$2x_1 + 2x_2 = 0 \text{ or } 2 \text{ or } 1 \pmod 3, \qquad \text{which we call } A^2 B^2$$

Example 6.9.   We may arrange a 3^2 experiment in blocks of 3 units each by confounding J. The relationship $x_1 + x_2 = 0$ is satisfied for $(0, 0)$, $(1, 2)$ and $(2, 1)$, so these will form one of the blocks. Another comes from $x_1 + x_2 = 1$, and is $(0, 1)$, $(1, 0)$ and $(2, 2)$; finally the third is defined by $x_1 + x_2 = 2$, giving $(0, 2)$, $(1, 1)$ and $(2, 0)$. The blocks therefore are:

$$\text{I}: \quad 00, \ 12, \ 21$$

$$\text{II}: \quad 01, \ 10, \ 22$$

$$\text{III}: \quad 02, \ 11, \ 20$$

We do not have to confound all 4 d.f. of the AB interaction, only part of it; this in fact is no help in interpreting results but can be useful in the design of larger experiments, where unconfounded parts of higher interactions may be used as residual variation in a single replicate.

If 2 replicates of the design above in blocks of 3 units have been run, the Analysis of Variance table will contain the following terms:

Source	d.f.
Blocks	5
A	2
B	2
$(A^2 B, AB^2)$	2
Residual	6
Total	17

The 2 d.f. for J can be found from the totals, from the whole experiment, of the treatment combinations that satisfy $2x_1 + x_2 = 0$; 1; 2 in turn: call these J_0, J_1, J_2. The sum of squares is $(J_0^2 + J_1^2 + J_2^2)/6 - G^2/18$.

Principal block

As we saw for 2-level factorial experiments, we can find a *principal block* and generate all the others from it in a suitable way: we do not have to work out each one separately as we have done above. With 3-level factors, the basic member of the principal block is the treatment combination which has every factor at the lowest level, i.e. 00, 000, ... according to how many factors are included in the experiment. We must therefore look at the equation with '$= 0$' on the right-hand side: in Example 6.9 we found that confounding AB, $A^2 B^2$ (J) led to a block 00, 12, 21 which satisfies $x_1 + x_2 = 0$; this is called the *defining congruence*.

There are two more blocks to be found, and these will arise by the same 'multiplication' method as in 2-factor designs, but this time applied modulo 3. First we have to find suitable generators. These generators will be treatment combinations not already used, so that with the principal block that we have we can take 10 as the generator for a second block; this leads to 10, 22, 01 using arithmetic modulo 3 so that $1+2=3$ becomes 0. The third block could be generated from the principal block using 11, which gives 11, 20, 02.

Example 6.10. In a 3^3 experiment, the 27 treatment combinations 000, 001, ..., 222 provide a treatments sum of squares with 26 degrees of freedom: 2 for each main effect, 4 for each 2-factor interaction and 8 for the 3-factor interaction. If we want to reduce this to blocks of 9 units, we need one pair (2 d.f.) out of the 3-factor interaction (or, indeed, out of any main effect or interaction, but the 3-factor one is the obvious choice). If we take ABC^2, A^2B^2C, these satisfy

$$x_1 + x_2 + 2x_3 = 0 \text{ or } 1 \text{ or } 2 \text{ (mod 3)}$$

$$2x_1 + 2x_2 + x_3 = 0 \text{ or } 1 \text{ or } 2 \text{ (mod 3)}$$

Clearly the principal block can be found from

$$x_1 + x_2 + 2x_3 = 0 \qquad (*)$$

since 000 satisfies this; so $(*)$ is the *defining congruence*.

The principal block will contain 9 treatment combinations:

$$000, 011, 022, 101, 112, 120, 202, 210, 221$$

The remaining blocks can be generated from it by using in turn 001 and 002:

Block II is 001, 012, 020, 102, 110, 121, 200, 211, 222

Block III is 002, 010, 021, 100, 111, 122, 201, 212, 220

We can check that these satisfy $x_1 + x_2 + 2x_3 = 2$ and 1, respectively.

In a single replicate of this experiment, the following analysis would be possible.

Source	d.f.
Blocks	2
A, B, C main effects	6
AB, AC, BC 2-factor interactions	12
Unconfounded 3-factor interactions	6
Total	26

Note: The six unconfounded 3-factor interactions are $(ABC, A^2B^2C^2)$, (AB^2C, A^2BC^2), (AB^2C^2, A^2BC), which could perhaps be used as residual.

Reduction of block size to one ninth of the full set of treatments

Consider a 3-factor experiment in which (ABC^2, A^2B^2C) is confounded, giving the defining congruence $x_1 + x_2 + 2x_3 = 0$. Another 2 d.f. from an interaction will be needed. When two independent pairs have been chosen, their generalized interaction is also confounded. Suppose we choose (AB^2C, A^2BC^2) for the second pair. We have to multiply *both* components of each pair together in order to obtain the full set of confounded interactions: they will be

$$ABC^2, A^2B^2C; \qquad AB^2C, A^2BC^2; \qquad ABC^2AB^2C=A^2;$$

$$ABC^2A^2BC^2=B^2C; \quad A^2B^2CAB^2C=BC^2; \quad A^2B^2CA^2BC^2=A.$$

Hence we have also confounded (A, A^2) and (BC^2, B^2C).

In fact we cannot take a second pair out of the 3-factor interaction without losing the 2 d.f. for one of the main effects, which would not usually be satisfactory.

Instead let us try (AC, A^2C^2) as the second pair, with (ABC^2, A^2B^2C). Generalized interactions now are

$$ABC^2AC=A^2B; \qquad ABC^2A^2C^2=BC;$$

$$A^2B^2CAC=B^2C^2; \qquad A^2B^2CA^2C^2=AB^2.$$

so the other two confounded pairs are (AB^2, A^2B) and (BC, B^2C^2), which will usually be acceptable. However, we have had to confound parts of each 2-factor interaction as well as 2 d.f. of the 3-factor interaction, so this is not an ideal situation – but not very surprising when we are seeking such a small block size.

The second part of the defining congruence for this system is $x_1 + x_3 = 0$. The principal block will be 000 together with all other treatment combinations that satisfy both of

$$x_1 + x_2 + 2x_3 = 0 \qquad (ABC^2)$$

$$x_1 + x_3 = 0 \qquad (AC)$$

since these automatically carry with them all the other confounded elements. Besides 000, the principal block will contain 112 and 221. The other 8 blocks may be found from it by using as generators 100, 010, 001, 011, 200, 020, 002, 022 in turn.

6.9   Fractional replication in 3-level experiments

Just as $\frac{1}{2}$-, $\frac{1}{4}$-, ..., etc., replicates are possible when using 2-level factors, so $\frac{1}{3}$-, $\frac{1}{9}$-, ..., etc., replicates can be found when factors are at 3 levels. To construct a $\frac{1}{3}$-replicate, choose an interaction for confounding and use *only* the set of combinations in the principal block, satisfying 'defining congruence = 0', *or else* use those in the block with '= 1' on the right of that equation, *or else* those from '= 2', *but not all three*. We will use '= 0' for illustration. With large experiments, confounding may also be needed to reduce block size.

Example 6.11. Construct a $\frac{1}{3}$-replicate of a 3^5 experiment in 9 blocks, of 9 plots each. If we take $(ABCDE, A^2B^2C^2D^2E^2)$ as the defining congruence, $x_1 + x_2 + x_3 + x_4 + x_5 = 0$ will give the members of the principal block. Then two more pairs will be needed for confounding to reduce the block size to 3^2. A possible scheme is to use (ABC^2, A^2B^2C) and (AB^2D, A^2BD^2). There are a considerable number of generalized interactions in this scheme, all of which will be confounded: they are (ACD^2, A^2C^2D), $(BCD, B^2C^2D^2)$, (CD^2E^2, C^2DE), (BC^2E^2, B^2CE), (BD^2E, B^2DE^2), (ABD^2E^2, A^2B^2DE), (AC^2DE^2, A^2CD^2E), $(AB^2C^2D^2E, A^2BCDE^2)$ and finally, unfortunately, (AE, A^2E^2), the J component of AE. It is not possible to avoid this in this system.

In the complete confounded system, the principal block has to satisfy

$$x_1 + x_2 + x_3 + x_4 + x_5 = 0$$
$$x_1 + x_2 + 2x_3 \qquad = 0$$
$$x_1 + 2x_2 + x_4 \qquad = 0$$

It contains 00000, 01110, 02220, 10122, 11202, 12012, 20211, 21021 and 22101. Possible generators for the other 8 blocks are 00012, 00021, 00111, 00102, 00120, 00201, 00210, 00222. Assuming there is just one replicate (which is already 81 units (plots)), the terms in the Analysis of Variance will be:

Source	d.f.	
Blocks	8	
Main effects (A, B, C, D, E)	10	
Two-factor interactions	38	(all *except* the J component of AE)
Three-factor and higher interactions	24	(can be used as 'residual')
Total	80	

Further discussion of this topic can be found in Box, Hunter and Hunter (1978) and Cochran and Cox (1992).

There are SAS computer programs available which will produce designs for confounding and fractional replication, in particular FACTEX and OPTEX (Tobias (1989)).

6.10 Exercises

1. An investigation was carried out to determine the effect on yield of two solvents, factor A, change in temperature, factor B, and change in machines, factor C. Two machines were used and two temperatures were used. The basic material for testing was produced in such a way that only 4 tests could be carried out on each batch of material. Since the 3-factor interaction was expected to be absent, the batch effect was confounded with the ABC interaction. The basic experiment was replicated in order to give an estimate for the residual error variance.

   The results are given below. Lower case letters denote the upper level of the factor, the lower level is denoted by 1.

				Batch				
	1		**2**		**3**		**4**	
abc:	2.6	1:	1.9	1:	2.6	a:	2.2
b:	2.3	bc:	2.1	ac:	2.7	c:	2.4
a:	2.1	ac:	2.2	bc:	2.9	abc:	2.8
c:	2.5	ab:	2.1	ab:	2.7	b:	2.4

Analyse the data and determine which effects are significant. Give your recommend-
ations for future work.

2. A 2^5 experiment is to be divided into 8 blocks of size 4 where the defining contrasts
are *ABC*, *BDE* and *BCD*.

   (i) Show how the experiment may be organized by arranging the treatments in
   the blocks.
   (ii) Write down the block comparison identified by the *BDE* contrast.
   (iii) Find the other confounded contrasts and criticize the suitability of the
   choice of contrasts in this experiment.
   (iv) Show that all main effects are estimable.
   (v) Describe how the Analysis of Variance is obtained.

3. The following design is for a 2^5 factorial experiment in 4 blocks of 8. Find the
confounded interactions, and indicate the form the analysis would take if higher-
order interactions could be ignored.

$$\text{I}: \; a, \, abc, \, ace, \, bde, \, bcd, \, cde, \, abe, \, d$$

$$\text{II}: \; ce, \, be, \, ad, \, (1), \, abcd, \, abde, \, bc, \, acde$$

$$\text{III}: \; c, \, bce, \, acd, \, e, \, abd, \, b, \, abcde, \, ade$$

$$\text{IV}: \; cd, \, de, \, ae, \, bd, \, ac, \, ab, \, bcde, \, abce$$

4. Sixteen lupin plants were grown in a similar soil which was treated with three
additives *A*, *B*, *C* in a 2^3 arrangement. After a few days the plant heights (y cm)
corresponding to the applied treatments were as shown:

Replicate 1		Replicate 2	
(1)	16.2	c	13.8
ac	16.6	bc	20.7
bc	20.0	ab	16.3
ab	19.2	a	17.2
b	15.1	ac	12.8
abc	21.4	b	14.0
a	18.3	abc	20.9
c	15.2	(1)	16.8

Analyse the data to investigate the differences between treatments if:

   (i) the experiment was divided into 2 blocks as indicated by the replicates
   (where the horizontal broken line is ignored);
   (ii) the experiment was divided into 4 blocks as indicated by all the lines of the
   table.

5. A 2^3 experiment with two replicates measured the yield of brussels sprouts (kg/plot) as follows, where the factors were A, B, C:

Replicate 1				Replicate 2			
ab	28	c	26	a	17	(1)	21
ac	27	a	22	bc	26	abc	32
(1)	20	abc	35	ab	26	b	24
bc	26	b	21	c	24	ac	29
(Block 1)		(Block 2)		(Block 3)		(Block 4)	

Obtain the analysis of variance for the above data, ignoring the block structure but accounting for replicates.

Given that the 4 columns in the table denote the 4 blocks as indicated in brackets, describe the confounding scheme and adjust your analysis accordingly. In this case discuss the importance of the treatment contrasts identified in your analysis.

6. Explain the meaning and uses of *partial confounding*. A 2^3 experiment with factors A, B, C, is to be arranged with three replicates in 6 blocks of size 4. All effects are to be estimable, but main effects and the AC interaction are of major importance. Find a suitable arrangement of the design and comment on the relative information on the effects.

7. A 2^4 experiment is divided into blocks as follows:

Block 1	Block 2	Block 3	Block 4
c	ab	bd	bc
bd	abd	(1)	d
abcd	(1)	ad	abd
b	d	abc	ac
cd	ac	bcd	a
ad	bcd	ab	b
abc	acd	c	abcd
a	bc	acd	cd

Show how confounding has been used in this design and indicate why this design may be better than an alternative where one contrast has been confounded into blocks of size 8 with a complete replication.

8. In a 2^6 experiment on factors $A, \ldots, F$ the ABC, $ACDE$ and BDF contrasts are confounded in order that a single replicate of the experiment may be divided into blocks of size 8. Find the principal block and find also what other contrasts are confounded. The experiment is now modified so that a half replicate is selected where ABC is the defining contrast and $ACDE$ and BDF are still confounded. Give the layout of the half replicate that contains the principal block and obtain the aliases of A. Show that the A effect is estimable in the resulting design and describe how the estimator for A is connected to an alias.

9. Discuss briefly the situations in which fractionally replicated factorial experiments may be useful, and explain the terms *defining contrast* and *alias* as used in designing

such experiments. (You may limit any illustrations in your discussion to experiments with 2-level factors only.)

In an industrial experiment there are 6 factors of interest and some of these are thought unlikely to interact. Each factor is to be used at 2 levels and there are 32 units of experimental material available, arranged in 4 blocks of 8 units each. Design a scheme for carrying out this experiment, list the contents of one of the blocks and say how to find the others, and write down the terms that will appear in the Analysis of Variance. How should the experiment be randomized when carried out in practice?

Point out any advantages and disadvantages of this scheme. How might any disadvantages be minimized?

10. An experiment was carried out to investigate 6 factors A, B, C, D, E and F. It was decided to use a $\frac{1}{4}$-replicate of a 2^6 design with E equated to ABC and F equated to BCD. The data are:

Treatment combination	(1)	ae	bef	abf	cef	acf	bc	abce
Response	8.1	11.0	9.9	15.1	11.2	11.5	10.0	9.9
Treatment combination	df	adef	bde	abd	cde	acd	bcdf	abcdef
Response	17.3	15.6	13.5	9.2	9.7	14.8	12.4	16.4

It was known that E and F did not interact with each other or with any of the other factors. It was also known that D and C did not interact with each other.

Analyse the data stating clearly what assumptions, if any, you make. Determine which effects, if any, are significant and comment upon your results.

11. (i) Explain carefully the function of fractional replication in the design of experiments, and say how it differs from confounding. Use examples to illustrate your answer.

(ii) It is desired to arrange a $\frac{1}{4}$-replicate of a 2^7 factorial experiment in a single block of 32 plots. The main effects are of most interest and 2-factor interactions involving A, B and C are of secondary interest. Two-factor interactions involving only D, E, F and G and interactions involving 3 or more factors are thought to be unimportant. Derive a suitable design and indicate the breakdown in the numbers of degrees of freedom. List the treatment combinations used in the block.

12. (i) Explain why fractional replication is useful in the design of experiments and indicate briefly how such an experiment would be analysed. A quarter replicate of a 2^7 experiment is to be arranged such that the defining contrasts are $ABCDE$ and $BDEFG$. Show that this scheme enables main effects and 2-factor interactions to be estimated. Describe problems that arise in the analysis, and discuss how they may be overcome.

(ii) A 3^4 experiment is to be arranged in 9 blocks of size 9 where the confounded contrasts are AB^2C^2 and ABC^2D. Obtain the principal block of the design and, without a complete enumeration of the whole design, show how the other blocks are obtained. Discuss the merits of the design in terms of the estimability of important contrasts.

13. (i) A 3^4 experiment is to be conducted. There is only sufficient material in each batch to provide 9 observations and so it is decided to confound the batch effect with the interactions AB^2C and ABD. State the group of defining contrasts and write out the treatment combinations for the principal block and one other block. State clearly how you arrived at these treatment combinations.

If it was decided to carry out an initial exploratory experiment using just one block of the above experiment, what would be the aliases for this reduced experiment? Choose a block from the above confounded experiment and use it to show clearly what effects can be investigated. State what assumptions are made. What additional information would be required in order to reach any conclusions about the factors? What are the drawbacks of this design?

(ii) A 3^2 experiment was carried out in which the observations were to be grouped in blocks of 3. It was decided to confound the block effect with the $AB(J)$ component of the interaction. The experiment was replicated once. The data are given below. Factor A is a quantitative factor with the levels spaced at equal increments. Factor B is a qualitative factor. Analyse the data and comment on your results.

	Replicate 1			Replicate 2		
	b_0	b_1	b_2	b_0	b_1	b_2
a_0	7.9	10.0	10.9	7.6	6.4	8.3
a_1	10.6	12.1	12.9	11.4	9.9	10.7
a_2	11.7	8.7	10.8	9.2	8.2	9.8

14. (i) In a 3^3 experiment, the AB^2C contrast is to be confounded with blocks. Show how the treatment levels may be arranged in 3 blocks of size 9 and describe how the analysis of variance is obtained.

(ii) In a 3^3 experiment find the arrangement of the treatments in 9 blocks of size 3, where the defining contrasts are AB^2C and AC. Which other contrasts are also confounded?

(iii) Discuss whether it is possible to arrange a 3^3 experiment in 9 blocks of size 3 in such a way that only three-factor interactions are confounded.

15. Describe what is meant by fractional replication and indicate why it is a useful technique. A 3^4 experiment is to be divided into 9 blocks of size 9 through the confounding of the ABC^2D and AB^2C contrasts. Obtain the principal block of the design and describe how the other blocks are formed. Given that interactions of order 3 and above are not considered to be important, supply the complete set of confounded contrasts and indicate any possible deficiences in the design.

If the above experiment is now to be arranged in blocks of size 3 where the ABD^2 contrast is also confounded, obtain the principal block and discuss whether any further important contrasts are affected.

16. A 3^4 experiment with factors A, B, C, D is to be divided into 9 blocks through the confounding of the AB^2CD and ABC^2D contrasts.

    Obtain the principal block for the design and show how all the other blocks may be obtained. (There is no need to enumerate all the blocks.)

    Find the complete list of confounded contrasts and show that some of the 2-factor interactions are not estimable.

17. An experiment with factors each at 2 levels is to be carried out, and a maximum of 64 experimental units can be found. The scientist in charge of the experiment has 5 factors A–E which it is necessary to include, and a sixth factor F which it would be interesting, though not essential, to study. You are asked for your advice on planning the experiment, and given the additional information that F is likely to interact – if at all – only with A, B and C, and not with D or E.

    (i) Assuming that the 64 units are reasonably homogeneous, compare the advantages and disadvantages of including the factor F in the experiment. Sketch out the Analysis of Variance for each possible form of experiment, and indicate how a satisfactory residual (error) term may be obtained.

    Explain also how the experiment would be laid out.

    (ii) On further enquiry you find that the experiment will be carried out over a period of 4 days, with 16 units being processed each day. How would each of your proposed forms of experiment in (i) be modified if a possible effect due to days is to be removed? Explain how these changes would affect the layout, and how blocks would be made up.

    Explain also whether, and if so why, you would want to alter your advice on the relative advantages and disadvantages of including F.

APPENDIX 6A   Methods of confounding in 2^p factorial experiments*

1. Block size restricted to 4 units (plots)

(a) With $p = 3$ factors, one interaction must be confounded. This may usually be ABC, but could be *any* interaction (or main effect) that was already known to be unimportant, or that had been studied in previous experiments.

(b) When $p = 4$ two interactions must be confounded, together with their generalized interaction. Two 3-factor interactions may be confounded, and their generalised interaction has to be a 2-factor interaction, e.g. ACD and BCD, together with AB. Label the factors so that AB is unimportant, or has already been studied. This is the most satisfactory scheme available.

(c) When $p = 5$, three independent interactions must be chosen; these will generate four more generalized interactions. The best scheme available is based on two 2-factor interactions with no common letter (e.g. AB and CD) together with one 3-factor interaction (e.g. ACE) chosen to have one letter from each confounded 2-factor interaction and completed by the fifth letter. The system confounds two 2-factor interactions, four 3-factor interactions and one 4-factor interaction; the factors should be labelled so that the two which are confounded are the least important of the 2-factor interactions.

* Factors are denoted as A, B, C, . . .

(d) When $p=6$, it is possible to confound the highest order interaction *ABCDEF*. Three more independent interactions must be chosen: two may be *AB* and *CD*, so that *EF* will be among the generalized interactions in this scheme. The remaining independent interaction is a 3-factor interaction that takes one letter from each of the first two 2-factors, e.g. *ACE*. The complete scheme confounds three 2-factor interactions, eight 3-factors, three 4-factors and the 6-factor. [It is not possible to construct a good scheme involving mainly 5-factor interactions.]

2. Block size restricted to 8 units (plots)

(a) With $p=4$, one interaction, usually *ABCD*, must be confounded: but any interaction (or main effect) may be chosen if there is good reason for not wanting to examine it.
(b) With $p=5$, two independent 3-factor interactions having one letter in common give the best scheme, because their generalized interaction will be a 4-factor one: e.g. *ABC*, *CDE* and therefore also *ABDE*.
(c) With $p=6$, choose two 3-factor interactions having one letter in common (e.g. *ABC*, *CDE*); the third chosen interaction must include the sixth letter (*F*) together with one letter – *not* the common one – from each of these first two (e.g. *ADF*). The scheme confounds four 3-factor and three 4-factor interactions.
(d) When $p=7$, use *ABCDEFG* together with two 3-factor interactions to which just one letter is common (e.g. *ABC*, *ADE* which imply also *AFG* as the generalized interaction). For the remaining independent interaction, choose a 3-factor interaction that has one letter – *not* the common one – from each of the three 3-factor interactions already confounded (e.g. *BDF*). The scheme confounds seven 3-factor and seven 4-factor interactions, together with the 7-factor interaction.

Other schemes may sometimes be satisfactory when we are prepared to sacrifice lower order interactions that no longer require to be studied.

3. Block size restricted to 16 units (plots)

(a) When $p=5$, confound one interaction, usually *ABCDE* but the remarks 1(a) and 2(a) above apply.
(b) With $p=6$, a good scheme uses three 4-factor interactions, the two independent ones having just two letters in common, e.g. *ABCD*, *ABEF* which carry with them *CDEF*.
(c) With $p=7$, a scheme is based on three independent 4-factor interactions which carry with them a further four 4-factor interactions. The first two chosen interactions have just two letters in common, e.g. *ABCD*, *ABEF*. This gives three 'pairs' of letters (*AB*, *CD*, *EF*). The remaining independent interaction uses the seventh letter together with one letter from each of these pairs, e.g. *ACEG*. The full scheme confounds seven 4-factor interactions.

Notes:
(i) A scheme that is possible for a 2^p-experiment in blocks of 2^k will also be possible for a 2^{p+q} experiment in blocks of 2^{k+q}.
(ii) Where alternative schemes exist for the same block size and the same number of factors, a partial confounding scheme may be useful.

7
Confounding main effects – split-plot designs

7.1 Introduction

When two or more factors are included in the same experiment, it may be difficult to apply one of them to very small-sized units but easy to do so for the other(s). One obvious example is in agricultural work where some treatments, especially if they involve machinery, can only be put on to quite large units of land: such things as cultivation methods which need ploughing machinery or the control of ground weeds which requires spraying equipment. Other treatments in the same experiment, such as the application of fertilizers, can be applied to much smaller units. In industry, treatments which require a change of operating temperature in a process will take time to set up because the system will have to heat up or cool down at every change of treatment, whereas other factors such as the amounts of different elements included in a mixture for making a product, or different lengths of running time for the process, can be changed very easily. A whole set of treatments making up the sub-plot factor can be run at one temperature, then the temperature changed and the sub-plot set run again – in a different random order, of course.

For simplicity we begin by thinking of just two factors in the experiment. The factor that needs applying to large units is the *main-plot* factor (sometimes called the *whole-plot* factor), and that which will require only small units is the *sub-plot* factor. The larger the units, the more variable we must expect them to be, and so the main-plot factor is often not very precisely examined. It is therefore all the more important to use the small units that are possible for the other factor, so that precision is not thrown away unnecessarily. The sub-plot factor and the interaction between the sub-plot factor and the main-plot factor are both studied with greater precision than the main-plot factor.

There are other occasions where a split-plot design is the obvious one to use: we may already know about the main effect of one of the factors, which is included in an experiment only to examine its interaction with the set of treatments making up the sub-plot factor. If it helps to simplify and speed up experimental work, the factor which is known about may be put on to main-plots where its main effect is not examined with any great precision; the interaction with the sub-plot factor, as well as the main effect of the sub-plot factor, will still have good precision.

Another occasion where splitting of plots is useful is when an experiment continues for some time, for example on a long-term crop like a fruit plant or tree; and after one phase of work (say the first season) we think of a new factor to include. If plots have been made large enough in the original experiment they may be split into sub-plots to accommodate a new factor on small units, the original treatments now being on main-plots.

7.2   Linear model and analysis

Suppose that the main-plot factor (or set of treatments – they may not be fully 'factorial' in the sense of Chapter 5) is laid out in a randomized complete block design; this factor is at m levels (i.e. the set contains m treatments), and there are r blocks, giving rm main plots in total. The variability at this level is represented by a main-plot variance σ_M^2.

Each main plot is split into s sub-plots, and the variance of sub-plots is σ_S^2. Main-plots are randomized as in a randomized complete block design, and within each main plot the s sub-plot treatments are randomized, with a fresh randomization for each main plot.

The linear model (Alternative Hypothesis) underlying the analysis is

$$y_{ijk} = \mu + \alpha_i + \beta_j + \varepsilon_{ij} + \tau_k + (\alpha\tau)_{ik} + \delta_{ijk} \qquad [1]$$

in which y is an observation, α_i a main-plot treatment effect, β_j a block effect, τ_k a sub-plot treatment effect, $(\alpha\tau)_{ik}$ an interaction of main-plot with sub-plot treatments, μ a general mean, ε_{ij} a normally distributed residual term representing variation between main plots and δ_{ijk} a normally distributed residual term representing variation between sub-plots. We have $i = 1$ to m; $j = 1$ to r; $k = 1$ to s.

The sub-plot residual will be just like a residual in any single-level layout we have looked at until now, and $\{\delta_{ijk}\}$ are independent $\mathcal{N}(0, \sigma_S^2)$. When considering main plots, we need to apply an argument which we shall develop further in Chapter 13. The observation on a main plot is composed of the total of the s observations that are taken from the sub-plots which form it, and the variance is the sum of two parts, σ_M^2 from the main-plot variation itself and σ_S^2/s from the mean of the s sub-plot observations. The main-plot residual ε_{ij} is the first of these two parts, and so is $\mathcal{N}(0, \sigma_M^2)$. It will also be useful to define $\sigma^2 = \sigma_S^2 + s\sigma_M^2$.

Constraints on the terms in this model will be $\sum_{i=1}^{m} \alpha_i = 0$, $\sum_{j=1}^{r} \beta_j = 0$, $\sum_{k=1}^{s} \tau_k = 0$; also the interaction terms $\sum_{i=1}^{m} (\alpha\tau)_{ik} = 0$ for each value of k and $\sum_{k=1}^{s} (\alpha\tau)_{ik} = 0$ for each value of i.

Totals of the observations are defined in similar notation to that used previously.

$G = \sum_i \sum_j \sum_k y_{ijk}$,   the total of all observations

$T_i = \sum_j \sum_k y_{ijk}$,   for each i, the total of observations on main-plot treatment i

$B_j = \sum_i \sum_k y_{ijk}$,   for each j, the total of observations in block j of the layout

$S_k = \sum_i \sum_j y_{ijk}$,   for each k, the total of observations on sub-plot treatment k

$M_{ij} = \sum_k y_{ijk}$,   for each main plot, the sum of all observations in it

and we also need a two-way table of the totals, over blocks, of the main-plot/sub-plot combinations as described in Example 7.1 below.

The correction term is

$$G^2/N = \frac{G^2}{rms} = S_0$$

and other summation terms are

$$S_B = \sum_j \frac{B_j^2}{ms}; \qquad S_T = \sum_i \frac{T_i^2}{rs}; \qquad S_S = \sum_k \frac{S_k^2}{rm}$$

$$S = \sum_i \sum_j \sum_k y_{ijk}^2; \qquad S_M = \sum_i \sum_j \frac{M_{ij}^2}{s}$$

The Analysis of Variance is set out as follows:

Source of variation	d.f.	Sum of squares	Mean square	F-test
Blocks	$r-1$	$S_B - S_0$	M_B	M_B/M_M
Main-plot treatments	$m-1$	$S_T - S_0$	M_T	M_T/M_M
Main-plot residual	$(r-1)(m-1)$	(by subtraction)	M_M	
Main plots	$rm-1$	$S_M - S_0$		
Sub-plot treatments	$s-1$	$S_S - S_0$	M_S	M_S/M_R
Interaction	$(m-1)(s-1)$	(from two-way table)	M_I	M_I/M_R
Sub-plot residual	$(r-1)m(s-1)$	(by subtraction)	M_R	
Total	$rms-1$	$S - S_0$		

By arguments similar to those in Chapter 4 (pages 36–7, 42) it is possible to show that the expected values of the mean squares M_B, M_T are the same as M_M on a Null Hypothesis of no block or main-plot treatment effects, and are greater on the Alternative Hypothesis [1], so that the usual F-tests are justified in the upper half of the analysis. Again, with some algebra, M_S, M_I are found to have the same expected value as M_R on a Null Hypothesis of no sub-plot treatment effects and no interaction, but a larger value otherwise; this leads to the F-tests which may be carried out in the lower part of the analysis.

Before considering the t-tests we may want to make among means, we give an example of the layout and Analysis of Variance.

Example 7.1. In an experiment on rice, there are 4 methods of irrigation to be compared, on main plots, with 3 fertilizer mixtures on sub-plots. Two complete replicates of these treatments are possible. Figure 7.1 shows the layout for the experiment; the 4 irrigation methods I1–I4 are randomized on to large strips, independently in each block,

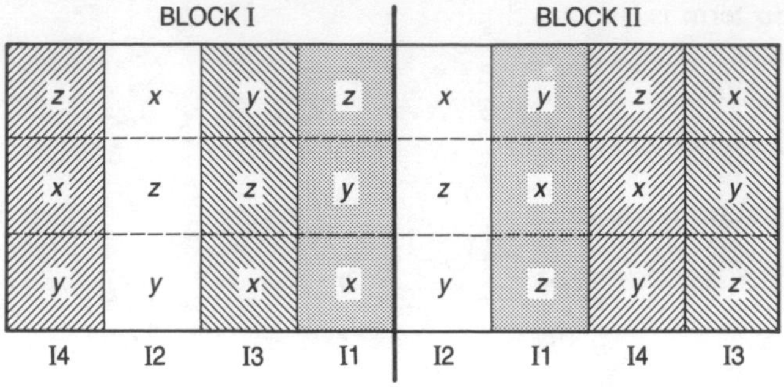

Fig. 7.1 Split-plot layout with I1–I4 on main plots and *X, Y, Z* on sub-plots

and within each of these main plots the 3 fertilizers are randomized on sub-plots, a fresh randomization being used for each main plot.

Some data on crop yield (tonnes/hectare) from this experiment are:

	Irrigation					Irrigation			
Block I	I1	I2	I3	I4	Block II	I1	I2	I3	I4
Fertilizer *x*	2.16	2.03	1.77	2.44	*x*	2.52	2.31	2.01	2.23
y	2.38	2.41	1.95	2.63	*y*	2.64	2.50	2.06	2.04
z	2.77	2.68	2.01	3.12	*z*	3.23	2.48	2.09	2.33

Block totals are (I) 28.35; (II) 28.44. $G = 56.79$, $N = 24$.
Main plots are the block/irrigation combinations, totalled over X, Y, Z:

		I1	I2	I3	I4
Block	I	7.31	7.12	5.73	8.19
	II	8.39	7.29	6.16	6.60

The two-way table of main plot/sub-plot treatment combination totals (summed over blocks) is:

	I1	I2	I3	I4	Total
X	4.68	4.34	3.78	4.67	17.47
Y	5.02	4.91	4.01	4.67	18.61
Z	6.00	5.16	4.10	5.45	20.71
	15.70	14.41	11.89	14.79	56.79

$$r=2, \; m=4, \; s=3. \; S=137.3993; \; S_B=\tfrac{1}{12}(28.35^2+28.44^2)=134.3797$$

$$S_0=\frac{56.79^2}{24}=134.3793; \; S_T=\tfrac{1}{6}(15.70^2+14.41^2+11.89^2+14.79^2)=135.7091$$

$$S_S=\tfrac{1}{8}(17.47^2+18.61^2+20.71^2)=135.0546$$

$$S_M=\tfrac{1}{3}(7.31^2+7.12^2+\cdots+6.16^2+6.60^2)=136.3604$$

From the two-way table, the sum of squares for main plots + sub-plots + interaction is

$$\tfrac{1}{2}(4.68^2+4.34^2+\cdots+4.10^2+5.45^2)-S_0=136.5855-134.3793=2.2062$$

The full Analysis of Variance is:

Source of variation	d.f.	Sum of squares	Mean square	
Blocks	1	0.0004		
Irrigations	3	1.3298	0.4433	$F_{(3,3)}=2.04$ n.s.
Main plot residual	3	0.6509	0.2170	
Main plots	7	1.9811		
Fertilizers (X, Y, Z)	2	0.6753	0.3377	$F_{(2,8)}=16.63$**
Irrigation × fertilizer interaction	6	0.2011	0.0335	$F_{(6,8)}=1.65$ n.s.
Sub-plot residual	8	0.1625	0.0203	
Total	23	3.0200		

n.s. = not significant.

There is no evidence of overall differences among the irrigation methods, but if any particular contrasts are to be examined they must be done in the top part of the analysis and be based on the main-plot residual, 0.2170, with 3 degrees of freedom. The standard error of the difference between two means is

$$\sqrt{\frac{2}{rs}\sigma^2}$$

estimated by

$$\sqrt{\frac{0.2170}{3}}=0.269$$

since $t_{(3)}=3.182$ at the 5% level, confidence intervals based on so few d.f. will be very wide.

Similarly, there is no evidence of any fertilizer differences; any contrasts for this factor must be based on the lower half of the analysis which gives the residual 0.0203 with 8 d.f. The standard error of the difference between the two sub-plot main effect means is

$$\sqrt{\frac{2}{rm}\,\sigma_S^2}$$

estimated by

$$\sqrt{\frac{0.0203}{4}} = 0.0712$$

7.3　Studying interactions

In Example 7.1, the only overall *F*-test that gives evidence of differences is that for interaction. Graphs of means, as described in Chapter 5, are again useful. But standard errors have to be used with care, because there are two different standard errors according to what type of comparison is being made. When two sub-plot treatments are compared at the *same* 'level' of the main-plot factor, the appropriate standard error of difference is

$$\sqrt{\frac{2}{r}\,\sigma_S^2}$$

whereas if two main-plot treatments are compared at the same 'level' of the sub-plot factor it is

$$\sqrt{\frac{2}{r}(\sigma_S^2 + \sigma_M^2)}$$

The analysis does not provide a direct estimate of σ_M^2, but of $\sigma^2 = \sigma_S^2 + s\sigma_M^2$, so that this second standard error is more conveniently written as

$$\sqrt{\frac{2}{rs}\{(s-1)\sigma_S^2 + \sigma^2\}}$$

For the same irrigation method, two different fertilizer means require

$$\sqrt{\tfrac{2}{2} \times 0.0203} = 0.142$$

to be used as the standard error in comparisons, while for different irrigation methods and the same fertilizer we need

$$\sqrt{\tfrac{2}{6}\{(2 \times 0.0203) + 0.2170\}} = 0.293$$

This second standard error is much larger than the first because the variability in the upper part of the analysis, for irrigation, was much greater than that in the lower part, for fertilizers. The relationship between the variabilities in a split-plot experiment is most commonly like this, so that if a split plot had not been used there would have been a corresponding loss of precision of the fertilizer (sub-plot) comparisons, which is unnecessary because there is no reason why large plots should have been used for the fertilizers.

In Example 7.1, the table of means for studying the interaction is:

Irrigation		I1	I2	I3	I4
Fertilizer	X	2.34	2.17	1.89	2.34
	Y	2.51	2.46	2.01	2.34
	Z	3.00	2.58	2.05	2.73

Horizontal comparisons (on the rows) use 0.293 as the standard error of difference between two means, and $t_{(8)}$ as an *approximate* basis for testing; in practice we require a weighted average of $t_{(8)}$ and $t_{(3)}$ with weights that lean heavily towards $t_{(8)}$ (Cochran and Cox (1992)). 'Vertical' comparisons (in columns) use 0.142 as standard error, and $t_{(8)}$ as the test statistic. These latter comparisons are the more helpful; $X < Y < Z$ on (almost) every irrigation, though the levels of significance vary quite a lot.

We could summarize the results by suggesting that irrigation method I3 is not worth further study, and also that any follow-up experiment should concentrate on fertilizers Y and Z.

7.4 Repeated splitting

There may be more than two levels of plot size: a third factor may be applied to split-split plots, where each sub-plot is subdivided into as many parts as necessary to contain each level of the third factor once. The analysis now has 3 sections, with 3 estimates of residual variation corresponding to the 3 levels of plot size. However, in order to have any reasonable replication of the factor on the largest plots there will need to be a considerable number of sub-sub plots, and the degrees of freedom in the lowest section of the analysis may even become unnecessarily large.

If factors A, B, C are on main plots, sub-plots and sub-sub plots, respectively, then A is compared on main plots, B and AB on sub-plots, C, AC, BC and ABC on sub-sub plots. Standard errors for studying interactions are quite complicated; Cochran and Cox (1992) gives some details.

7.5 Confounding in split-plot experiments

As is clear from Example 7.1, the main-plot and sub-plot 'factors' need not be 'factorial' in the sense of Chapter 5, but merely collections of treatments which are compared on plots of different sizes. However, when they *are* 'genuinely' factorial, the same problems of design, due to having a very large number of treatment combinations, may arise as we have discussed in Chapter 6.

Let us suppose that the main-plot factor is a 2-level factor A (levels a_0, a_1). We suppose also that a 2^3 set, made up from factors B, C and D, is to be accommodated on sub-plots. It is often undesirable to split a main plot into as many as 8 sub-plots, and we can restrict the number to 4 sub-plots per main plot by confounding one interaction. Since A is already on main plots, we must choose an interaction from the other three factors to be confounded, and the obvious one is BCD. The 8 treatment combinations are thus divided into the two sets $((1), bc, bd, cd)$ and (b, c, d, bcd), which are then combined, properly randomized, with a_0 and a_1 as shown in Fig. 7.2.

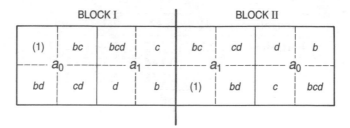

	BLOCK I				BLOCK II		
(1)	bc	bcd	c	bc	cd	d	b
a_0		a_1		a_1		a_0	
bd	cd	d	b	(1)	bd	c	bcd

Fig. 7.2 One complete replicate of a split-plot 2^4 scheme with A on main plots and the interaction *BCD* confounded on sub-plots

The effect of confounding *BCD* in this way is to transfer it from the lower to the upper part of the analysis; and *ABCD* will be completely confounded with main plots. We may see this by looking at the '±' table of contrasts for 4 factors (cf. page 111) and locating the + and the − combinations in the layout.

At least four repetitions, with new randomizations each time, of the scheme shown in Fig. 7.2 would be needed to give enough main plots for a satisfactory analysis. With 4 complete replicates i.e. 8 blocks, the items in an Analysis of Variance would be:

Source of variation	Degrees of freedom	
Blocks	7	(this includes *ABCD*)
A	1	
BCD	1	
Main plot residual	6	
	15	
B, C and D	3	
AB, AC, AD, BC, BD, CD	6	
ABC, ABD and ACD	3	
Sub plot residual	36	
Total	63	

Those main effects and interactions that are estimable, which in this example will be all except *ABCD*, are computed in the usual way as $(+, -)$ contrasts, and placed in the correct section of the analysis. We can easily see that with a moderate number of 2-level factors, the number of degrees of freedom for the sub-plot residual becomes larger than is necessary if we try to make the top part of the analysis reasonably precise. For example, only two complete replicates of the layout of Fig. 7.2 i.e. 32 unit plots, would leave us with an analysis in which only the lower part was worth doing. Of course, if A had been studied previously and *BCD* was not important, this would not matter.

7.6 Other designs for main plots

Laying main plots out as randomized complete blocks is probably the most common plan, but there is no reason why a completely randomized scheme should not be used, such as in the example of changing temperatures in an industrial experiment, or a Latin square when variation needs controlling in two directions. For example, the main plots of four replicates of the scheme discussed above might be set out as

$$a_0 \quad a_1 \quad a_1 \quad a_0$$
$$a_1 \quad a_0 \quad a_0 \quad a_1$$
$$a_1 \quad a_1 \quad a_0 \quad a_0$$
$$a_0 \quad a_0 \quad a_1 \quad a_1$$

each being split in a manner similar to that shown in Fig. 7.2.

7.7 Criss-cross design

This is a variant on the split-plot idea; it is also called 'strip-plots'. Although its uses are somewhat limited, it is interesting because it requires a complete partition of the residual sum of squares into components that can each be used for testing just one of the main effects or interactions in the analysis.

If two factors are used in an experiment and both of them must be applied to large-sized plots, then the main-plot/sub-plot approach is not sufficient. This situation is most likely to arise in a field experiment, particularly where mechanical cultivation has to be done and where chemical treatment of trees, plants or soil is also involved. If small plots are used, it is very likely that turning machinery will damage the crop; and there will have to be considerable use of 'guard' or shelter plants between the experimental plots to avoid carry over of chemical treatments on the ground or in the form of sprays, leading to much waste of land and material.

Large plots at right angles are used for the 2 sets of treatments, and the scheme is replicated a number of times in blocks. Figure 7.3 illustrates the scheme when 4 different times Q, R, S, T are possible for a cultivation process to be carried out, and there are also 3 pest control sprays A, B, C which can be used. Complete vertical strips in the diagram are treated with A, B or C and the horizontal strips receive the cultivation treatment at the 4 different possible times. At the end of each block, a substantial space will have to be left to allow machinery to get in safely, especially as this has to be done 4 times in the course of the experiment.

The smallest plots are the units where cultivation and spray factors cross one another, and the smallest variability is therefore to be expected in the interaction between the 2 sets (A, B, C) and (Q, R, S, T); in fact the interaction is estimated with greater precision than either of the main effects. By analogy with a randomized complete block experiment, the complete residual is the 'Blocks $\times$ Treatments' interaction; it has to be split into Blocks $\times (A, B, C)$, Blocks $\times (Q, R, S, T)$ and Blocks $\times$ Interaction. The main effect A versus B versus C is examined using Blocks $\times (A, B, C)$, Q versus R versus S versus T is examined using Blocks $\times (Q, R, S, T)$ and the interaction is examined using Blocks $\times$ Interaction.

The aim of an experiment like this is often to decide which combination of treatments, one from each set, is best; in these circumstances it is acceptable that the interaction has the best basis for comparison. But if there is no significant interaction, the results may be presented in terms of the main effects and the precision of these is not very good.

We will give the outline of the steps needed to analyse a design like that in Fig. 7.3, and further details of the model and the theory may be found in Cochran and Cox (1992). We may note that 48 unit plots have to be recorded to give enough degrees of freedom for all the comparisons to have a good chance of being reasonably precise; this is rather a large number. Three, rather than four, blocks (36 unit plots) might just have been adequate, particularly if there is an interaction so that main effects do not need to be studied.

Some new notation is needed for totals in the data table, and this is shown in the skeleton below.

Data table:						Sub-tables:			
Block I	II	III	IV	Total		Block I	II	III	IV
$A:Q$				(AQ)		A			
R				(AR)		B			
S				(AS)		C			
T				(AT)		(Table α)			
$B:Q$				(BQ)					
R				(BR)					
S				(BS)					
T				(BT)		Block I	II	III	IV
$C:Q$				(CQ)		Q			
R				(CR)		R			
S				(CS)		S			
T				(CT)		T			
						(Table β)			
Totals B_{I}	B_{II}	B_{III}	B_{IV}	G		$N =$ total number of unit plots			

	Q	R	S	T	Total
A	(AQ)	(AR)	(AS)	(AT):	T_A
B	(BQ)	(BR)	(BS)	(BT):	T_B
C	(CQ)	(CR)	(CS)	(CT):	T_C
Total	T_Q	T_R	T_S	T_T:	G

In the following Analysis of Variance table, we show which totals must be used to calculate the appropriate summation term for each item. The correction term for each

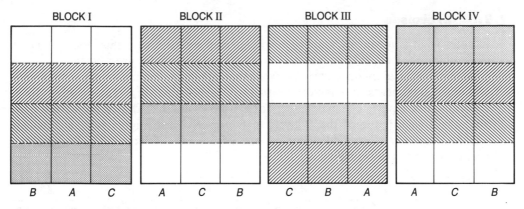

Fig. 7.3 Criss-cross design with one set of treatments A, B, C applied to strips in one direction and the second set Q, R, S, T to strips at right angles to the first. Key: ▨ Q, ▦ R, ☐ S, ▧ T

sum of squares is G^2/N.

Source of variation	d.f.	Sum of squares calculated from
A versus B versus C	2	T_A, T_B, T_C
Q versus R versus S versus T	3	T_Q, T_R, T_S, T_T
Interaction $(ABC) \times (QRST)$	6	(by subtraction)
	11	$(AQ), (AR), (AS), (AT), (BQ), (BR), (BS), (BT),$ $(CQ), (CR), (CS), (CT)$
Blocks	3	$B_I, B_{II}, B_{III}, B_{IV}$
Residual	33	(by subtraction)
Total	47	all individual observations on the 48 unit plots

The residual is now broken down as follows:

Blocks $\times (ABC)$ 6   The entries in Table α give
                             Blocks $+ (ABC) +$ Blocks $\times (ABC)$

Blocks $\times (QRST)$ 9   The entries in Table β give
                             Blocks $+ (QRST) +$ Blocks $\times (QRST)$

Blocks $\times$ Interaction   18   (by subtraction)

                                  33   from above

First the interaction must be tested as $F_{(6,18)}$. If there is evidence of interaction, or if particular comparisons have been nominated to be made, the 18 d.f. component of residual is required for use with the two-way table of means $(ABC) \times (QRST)$.

When there is no interaction, the two main sets of treatments may be compared, (ABC) as $F_{(2,6)}$ using the Blocks $\times (ABC)$ component of residual with 6 d.f., and $(QRST)$ as $F_{(3,9)}$. These components of residual will also be used in any comparisons of treatment means.

7.8 Exercises

1. Explain what is meant by a *split-plot* design and indicate why this design may be preferable to the alternative factorial arrangement for randomized blocks.

   Batches of metal ingots arrive in boxes at a workshop. The ingots are treated with one of 15 treatments in a factorial structure, then tested for breaking strength in Newtons. For 2 particular batches, 5 boxes were each selected at random and each box was assigned one of the 5 levels of factor *A*. Three ingots were selected at random from each box and allocated one of 3 levels of factor *B*. The breaking strengths were as given below.

	A1			A2			A3			A4			A5		
	B1	B2	B3	B1	B2	B3	B1	B2	B3	B1	B2	B3	B1	B2	B3
Batch 1	16	8	16	19	16	16	23	26	14	23	20	14	20	18	8
Batch 2	18	11	15	18	15	18	20	23	16	27	25	12	22	19	11

   Given that ingots in the same box are more likely to be similar than ingots in different boxes, analyse the data to investigate whether there are real differences between the ingots as a result of the treatments.

   Discuss how the investigation of the treatment differences should proceed and comment accordingly.

2. Discuss the advantages and disadvantages of split-plot experiments and explain the situations where a split-plot experiment may be appropriate.

   Three varieties of triticale were grown in a split-plot experiment in which 4 fertilizer levels formed the sub-plot treatments. The yields (t/ha) were as shown.

	Block 1				Block 2				
	F1	F2	F3	F4	F1	F2	F3	F4	
Salvo	6.1	6.3	6.5	6.5	5.8	5.7	6.6	6.4	F1: 80 kg/ha
Newton	4.9	5.6	5.8	5.4	5.3	5.2	5.4	5.5	F2: 100 kg/ha
Lasko	6.3	6.3	6.8	5.9	6.1	6.4	6.7	6.3	F3: 120 kg/ha
									F4: 140 kg/ha

   Analyse the data to investigate the differences between treatment means and examine where these differences occur. Comment on the suitability of the design.

3. Two varieties of barley were grown in a split-plot design which used nitrogen fertilizer as the sub-plot treatment. The yields (y t/ha) were as shown in the results table.

	V1			V2			Key
							V1 Ingri
	N1	N2	N3	N1	N2	N3	V2 Tipper
Block 1	7.7	8.3	8.6	7.2	7.7	7.7	N1 40 kg/ha
Block 2	7.4	8.0	8.2	7.0	7.1	7.8	N2 80 kg/ha
Block 3	6.8	6.7	7.6	6.9	7.3	7.2	N3 120 kg/ha
Block 4	6.9	6.7	7.1	6.8	7.3	7.5	
							$\Sigma y^2 = 1319.13$

Analyse the data and comment on any differences between treatment means. Describe the conditions under which this could have been considered a well-conducted experiment and the assumptions necessary for the validity of the analysis.

4. An experiment is to be designed to test the effects of 4 levels of nitrogen and 5 growth regulators on the yield of rapeseed. The field to be used for the experiment is known to be of variable fertility and 60 experimental units are available. Describe how a suitable split-plot experiment may be arranged and comment on its advantages over a comparable completely randomized factorial design, given that the effect of nitrogen has been examined in the past. Prove that the variances of the differences between means of 2 levels of nitrogen and 5 different growth regulators are given by expressions of the form $2(\sigma^2 + 5\sigma_m^2)/15$ and $\sigma^2/6$, respectively, and in the same notation obtain the variance of the difference between interaction means.

5. Four varieties of autumn kale were grown in a randomized-blocks experiment on 24 units arranged in split plots, the sub-plot treatments being applications of 40 and 80 kg/ha of nitrogen. The design and the amount of digestible organic matter (tonnes/ha) were as shown.

 (i) Compute the Analysis of Variance table and report on the significance of the results.
 (ii) Obtain the standard errors of the differences between nitrogen levels and between varieties, and their interactions.
 (iii) Criticize the suitability of the design.

	Block 1		Block 2		Block 3	
	A $N1$ $N2$		B $N2$ $N1$		C $N2$ $N1$	
	5.3 5.6		5.6 5.3		5.0 4.8	
	B $N1$ $N2$		D $N1$ $N2$		A $N2$ $N1$	
	5.8 5.8		5.0 5.4		5.3 5.5	
	C $N2$ $N1$		A $N2$ $N1$		D $N1$ $N2$	
	5.3 4.7		5.2 5.3		4.8 5.1	
	D $N2$ $N1$		C $N1$ $N2$		B $N2$ $N1$	
	5.2 5.0		4.6 4.7		5.4 5.3	

$N1$: 40 kg/ha nitrogen
$N2$: 80 kg/ha nitrogen

Varieties A: Maris Kestrel
  B: Proteor
  C: Midas
  D: Vulcan

$\Sigma x^2 = 653.62$

Nitrogen × Varieties totals table

	A	B	C	D	Total
$N1$	16.1	16.4	14.1	14.8	61.4
$N2$	16.1	16.8	15.0	15.7	63.6
Total	32.2	33.2	29.1	30.5	125.0

6. A report reads as follows: 'In this investigation into the effect of diet on blood sugar, it was decided to compare 3 diets and 4 strains of rat. Each diet was allotted to 4 rats, one chosen at random from each strain. The diets were fed for 4 weeks, and

at the end of each week duplicate determinations of blood sugar were made on each rat. The Analysis of Variance of the results was as follows.

	d.f.	Sum of squares	Mean square	F
Replicates	1	0.19		1.73
Diets	2	6.03	3.02	27.45
Strains	3	5.82	1.94	17.64
Diets × strains	6	3.18	0.53	4.82
Weeks	3	7.47	2.49	22.64
Diets × weeks	6	6.12	1.02	9.27
Strains × weeks	9	6.51	0.72	6.55
Diets × strains × weeks	18	10.02	0.56	5.09
Error (residual)	47	5.17	0.11	
Total	95	50.51		

The values of F for diets, strains, weeks and all their interactions are significant at the 0.1% level. We were surprised to find such very significant interactions between weeks and diets, and weeks and strains, and assumed at first that this was caused by some diets and some strains reaching saturation earlier than others: however, investigation of the means shows that this was not so. An explanation of these interactions must therefore await further study.'

Comment on the design of the experiment, the form of the analysis (making any modifications you think fit), and the conclusions drawn from it.

7. The 12 treatments in a factorial experiment on melons consisted of all combinations of 3 doses of a fungicide spray and 4 melon varieties. The experimental area was divided into 4 blocks each containing 12 melons. Since the spraying could only be done for a relatively large area for any one dose, the blocks were each divided into 3 large plots, each of which received one of the doses of fungicide and contained each of the 4 varieties of melon. The allocation of dose to, and position of variety within, each large plot was randomized.

In the report of the experiment the following Analysis of Variance table was presented, the observations being the melon weights in ounces. (F and V refer to fungicide and variety, respectively.)

Source	d.f.	Sum of squares	Mean square	F
Blocks	3	833.94		
V main effect	3	652.48	217.49	8.13***
F main effect	2	648.15	324.08	12.12***
VF interaction	6	135.12	22.52	0.84
Residual	34	909.03	26.74	
Total (corrected)	47	3077.38		

Explain why this Analysis of Variance table is not appropriate for this experiment, indicating the possible misleading effects that its use could have on the conclusions from the analysis.

Construct the appropriate Analysis of Variance table. (The following corrected sums of squares may be useful:

$$\begin{array}{ll}
\text{Block} \times V \text{ interaction} & 48.71 \\
\text{Block} \times F \text{ interaction} & 780.75 \\
\text{Block} \times V \times F \text{ interaction} & 79.57)
\end{array}$$

How do your conclusions from this table differ from those from the table given above?

8. A field experiment is carried out in a 'criss-cross' (strip-plot) design. Two cultivation methods, X and Y, are used to prepare the ground for planting; these methods are applied in strips in one direction. Four herbicides, a, b, c and d, for keeping ground cover clean of weeds, are applied in strips at right angles to the first set of strips. The complete experiment consists of 4 blocks, each block containing one replicate of the 8 treatment combinations. The data on yield of a corn crop are shown in the table below.

(i) Make a sketch plan to show clearly how this experiment would be laid out in the field, and how it should be randomized.
(ii) Analyse the data to examine the effects of the factors (X, Y) and (a, b, c, d) and their interaction. Provide confidence intervals for relevant differences between means.
(iii) Comment briefly on the strengths and weaknesses of such a design.

Block		I	II	III	IV		I	II	III	IV
Method X	a	43	27	37	41	a	97	67	83	99
	b	29	28	39	33	b	61	65	83	73
	c	35	34	40	26	c	75	75	75	65
	d	26	22	20	29	d	57	55	50	64
		133	111	136	129					

Table of crop yields (kg) — Summary totals

		X	Y		Total
	a	148	198	:	346
Method Y	a ...				

Method Y	a	54	40	46	58	a	148	198	:	346
	b	32	37	44	40	b	129	153	:	282
	c	40	41	35	39	c	135	155	:	290
	d	31	33	30	35	d	97	129	:	226
		157	151	155	172		509	635		1144
Block total		290	262	291	301					

The sum of squares of all 32 observations is 43048.

8
Industrial experimentation

8.1   Introduction

The fractional factorial experiments described in Chapter 6 were for many years used in industrial research and development. Often a very large number of factors were included in the same experiment, so fractional replication was the obvious technique for gaining useful information from experiments of a manageable size.

Recently, statistical quality control, product and process design in industry has taken up some methods of experimental design that can be very much less satisfactory than fractional replication. This arises from a change in the philosophy behind research and development, which introduced some new ideas about what measurements it is necessary to take when assessing whether a system is working well. Unfortunately, some potentially inefficient statistical methods came in the same package of ideas and the long experience of industrial statisticians was overlooked.

8.2   Taguchi methods in statistical quality control

There are a number of useful ideas in the package, which has come to be associated with the name of the Japanese engineer G. Taguchi, although other workers have also contributed. Provided that these good ideas are combined with efficient statistical methods in experimentation, they will make a useful addition to the range of methods available in industry and no doubt, in due course, become adapted in other studies as so many other statistical methods have done.

For a full discussion of the new methods in statistical quality control, Bissell (1994) should be consulted. In this chapter we will give an outline of the most important statistical points. To appreciate the new ideas, we must realize how important it is to control *variability*, both in a batch of the finished product and in its subsequent use by customers. Variability is important in other areas of experimental statistics too, not least in assessing new varieties or treatments in agricultural experiments whose results are going to determine future practice in a wide area or region where it is important to produce enough of a staple crop. We have tended to concentrate on *mean* yields; but if there is a high variability in yield, due perhaps to high susceptibility to relatively small climatic changes, there could well be a shortage of this staple crop in some parts of the region.

Some of the basic concepts behind what are called 'Taguchi methods' are:

(1) good quality *begins* with product design rather than with careful control of manufacturing processes;
(2) as remarked above, design must take account of variability in manufacture and also the environmental or usage conditions that the product is likely to meet during its working life;
(3) design principles can be applied to identify process settings that minimize variation in manufacture and minimize the effect of those variations on product performance.

In addition, as other commentators have also noted, customer satisfaction is inversely proportional to performance variation: what is required is something that behaves in a reliable and predictable way in all reasonable conditions of use. It is also said that the true criterion for judging quality should be the 'cost to society', whatever this means; often it has been interpreted as *loss* in a much more general sense than the financial cost of unsaleable production.

The Taguchi philosophy thus calls for experiments at the product or process design stage, whereas other common industrial experimental schemes – especially *response surfaces*, discussed in Chapter 9 – concentrate on optimizing the process when it is in operation. Because the study of methods of quality improvement takes place before bulk production begins, it is called *off-line* quality control.

One stage in Taguchi strategy is *system design*: developing a prototype product using existing scientific and technological knowledge; this does not usually involve experimental design methods. Then there is *parameter design*, to identify the settings that will lead to minimum variation in performance. First we shall look at *tolerance design*, which seeks to define tolerances, such as sizes of components, so as to minimize costs both during manufacture and in the subsequent lifetime of use of the product. *Loss functions* are an essential idea here.

8.3   Loss functions

We will suppose that there is a characteristic y being measured, and that it has some intended or ideal value; either there may be a *target* value m which it should attain as closely as possible, or it should have as *large* a value as possible (such as the measured strength of a material) or as *small* a value as possible (such as the amount of impurity in a chemical product). Loss is to be counted all down the line from producer to user, including any later manufacturing stages where failure to meet specification (e.g. of length or diameter in a component) may lead to a loss, financial or in terms of inconvenience.

(i) *'Nominal is best'* is the phrase often used to describe the situation where a target value is to be met. Some older and simpler methods of quality control do not distinguish between two items where, for example, one is exactly equal in size to its target m and the other is only just within some given tolerance limit: we wish to make decisions more subtle than just 'accept' or 'reject'. A *quadratic loss function* is useful because it measures the loss as becoming rapidly more important the further away from m the measurement gets, as shown in Fig. 8.1. A small departure from m may not affect manufacture or use seriously, but as the departure increases so the likelihood of failure increases; also there comes a point where the item cannot be used at all.

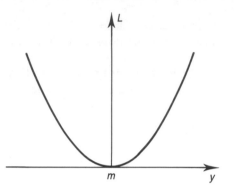

Fig. 8.1 Quadratic loss function: *y* is the observation, *m* is the nominal value, *L* is the loss

If the ith item measures y_i, then the loss for n items, using this *quadratic loss function*, is

$$L = k \sum_{i=1}^{n} (y_i - m)^2$$

which gives an *average loss*

$$\frac{L}{n} = k(\sigma^2 + \{\bar{y} - m\}^2)$$

The loss function can be considered as a measure of the proportion of dissatisfied customers. Any deviation from the nominal value will result in some dissatisfaction, so the aim is always to improve the product so that the nominal value is more nearly attained as the process improves. The loss function is more stringent than the control chart since the latter assumes that provided observations fall within the tolerance, then all is satisfactory. A simple example will show the fallacy of this. If two devices such as a nut and a bolt are to fit together, then both can individually fall within tolerance but together they may give a bad fit, as they may be too tight or too loose if departures are on opposite sides of the nominal values. For example, if the nut has a target diameter of 7.1 mm and tolerance 0.07 mm and the bolt has a target diameter of 7.0 mm and tolerance of 0.06 mm, then acceptable diameters for the parts are

nut: 7.03 to 7.17 mm

bolt: 6.94 to 7.06 mm

This means that it is 'acceptable' to have a nut of size 7.04 mm to fit on a bolt of size 7.05 mm, even though they will not fit. On the other hand Taguchi would suggest that both the nut and the bolt would individually be responsible for large losses because they are so far from target.

In medicine the LD50 is the term for the level of dose that will kill 50% of subjects (LD = Lethal Dose). The term is useful as a measure of toxicity of drugs and can be used to indicate the dose required to kill bacteria, insects or vermin. Taguchi applied the term to quality to indicate the value of the observation that would render 50% of customers satisfied. Suppose the LD50 is called D, the corresponding deviation of the observation from its target is called T, and the resulting loss is C; then

$$k = \frac{C}{D^2}$$

in the loss function.

(ii) *'Large is best'* will be appropriate where the measurement should be as large as possible, such as the strength of a material or the lifetime of an electronic component or machine part. Figure 8.2 shows a loss function that is suitable in these cases, namely $L = k/y^2$.

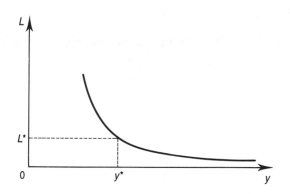

Fig. 8.2 Loss function $L = k/y^2$, where $k = L^*(y^*)^2$

For n items the loss will be

$$L = k \sum_{i=1}^{n} \frac{1}{y_i^2}$$

giving the average loss

$$\frac{L}{n} = \frac{k}{n} \sum_{i=1}^{n} \frac{1}{y_i^2}$$

Applying the approximate methods given on page 68, we find this is

$$\frac{L}{n} \doteq \frac{k}{\bar{y}^2} \left(1 + \frac{3\sigma^2}{\bar{y}^2} \right)$$

If we know a value for the loss at some given value of y, say loss $= L^*$ at $y = y^*$, then $k = L^*(y^*)^2$.

(iii) '*Small is best*' applies when we are studying undesirable types of measurement, such as chemical impurities or the amount of wear on a tyre or a turbine blade. An

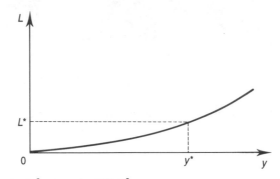

Fig. 8.3 Loss function $L = ky^2$, where $k = L^*/(y^*)^2$

appropriate loss function now is $L = ky^2$, shown in Fig. 8.3. When there are n items, the total loss is

$$L = k \sum_{i=1}^{n} y_i^2$$

and the average loss is

$$\frac{L}{n} = \frac{k}{n} \sum_{i=1}^{n} y_i^2 = k(\bar{y}^2 + \sigma^2)$$

Again, if we know that $L = L^*$ at $y = y^*$, we can obtain $k = L^*/(y^*)^2$.

8.4 Sources of variation

Since measurements are taken in order to assess variation, which in this context is often called by its engineering name *noise*, we need to identify the sources of variation. They are classified as *external* and *internal*. *External*, or *assignable*, sources of variation are systematic and are due to such things as environment, or human variation in operating the product. Statistical process control aims to control and reduce manufacturing variation by eliminating known assignable sources. *Internal* noise variation is sometimes related to these external sources, such as when a product shows greater variability if used in conditions of high humidity. Reducing inherent variation is not always easy; *parameter design* is devoted to minimizing this *internal noise*. These uncontrollable factors give an estimate of the residual error in an experiment, while the controllable factors are applied in a fractional factorial design using an orthogonal array. Robustness to internal variation often goes with robustness to external variation.

If we call the measured response the *signal* (also an engineering name), the variability of a product within each treatment or each treatment combination is the *signal-to-noise ratio*, and when this is low it identifies a factor that is susceptible to high variability and may therefore lead to unsatisfactory products.

One reason why modern statistical process control is often successful is that everyone is involved in quality improvement, whereas previously there was a temptation for the shop floor to continue routine production methods unless and until the quality control staff told them that the percentage of faulty products had grown too high. Now there is more realization that everyone should monitor their own performance and aim to improve it – and some realization also that the customers actually prefer 0% faulty! If this improvement in attitude can be combined with a proper use of experimental design, the Taguchi movement will prove beneficial.

Before an experimental programme can start there must be, as in any experiment, a clear definition of the problem and suitable measurements must be identified, as discussed in Chapter 3. Even if the aims at the outset are not at all obvious, the Taguchi approach of considering how much money would be wasted by making bad decisions often helps to identify the main points. The factors to be varied, and their experimental levels, must be decided; other factors under which we want the process to be robust will be part of 'noise', as will factors which it is realized we cannot control closely in an actual production run. Since we will need several 'replicates' of each treatment combination in order to compute a signal-to-noise ratio, these 'noise' factors are often regarded as providing the replication. We may have reservations about this if any of those 'factors' ought to have been made the basis of *blocks* – indeed would have been so made in other types of experiment.

The *signal-to-noise ratio* is an important part of analysis, and it is constructed so that large values are desirable. Therefore in the three situations described in Section 8.3, the following are used:

(i)   either $Z_i = -\log_{10}\left(\sum \dfrac{(y_i - m)^2}{n}\right)$ if absolute variation is important

or $Z_i = -\log_{10}\left(\sum \dfrac{(y_i - m)^2}{nm^2}\right)$ if relative variation is important

(these are related to $-10\log_{10} s^2$ and $10\log_{10} (\bar{y}/s)^2$, respectively);

(ii) $Z_i = -10\log_{10}\left(\sum \dfrac{1}{y_i^2}\middle/ n\right)$

(iii) $Z_i = -10\log_{10}\left(\sum y_i^2 / n\right)$

The values of these functions Z_i, for each treatment combination in the experiment, may be analysed in an Analysis of Variance. But there are statistical problems in using these $\{Z_i\}$ in this way; they do not satisfy the assumptions needed for the usual tests in the analysis and it is difficult to interpret the results. Bissell recommends carrying out the usual analysis of the treatment responses (in this case the means of the internal 'replicates') and making in addition an Analysis of Variance of the variances in each treatment combination. For this we use $\log_e s^2$ or $\log_e s$. The natural logarithms have the advantage that the standard error of $\log_e s^2$ is $1/(n-1)$ and of $\log_e s$ is $1/[2(n-1)]$, so that (in most experiments) these will be constant as required for the Analysis of Variance.

The best way of obtaining useful information is to combine the usual analysis of treatment response with an analysis of the internal variances. Carrying out these analyses, and studying the results, separately extracts the most information about the behaviour of the experimental material.

8.5 Orthogonal arrays

A class of design which has appeared in a number of other forms is often used as part of the package of 'Taguchi Methods'. Orthogonal arrays are fractional factorial designs, and therefore do not study all interactions as we saw in Chapter 6; so-called 'saturated' arrays provide no residual degrees of freedom. The array style of presenting a design is due to Rao (1946): the columns of a matrix are related to factors, often being numbered, although we should prefer to denote them by capital letters as in Chapter 5. The rows represent the units or plots to be used in the design. Using this method, the array

Factor		1	2	3
Run	1	1	1	1
	2	1	2	2
	3	2	1	2
	4	2	2	1

is called an L_4-array, and is equivalent to a 2^3 system with confounding, if a new factor is associated with each column. The four runs or plots in that situation would be (1), bc, ac, ab (and their order should of course be randomized for experimental use). We have in fact a $\frac{1}{2}$-replicate of a 2^3 system, with ABC confounded. A better scheme would be to allocate factors A, B to columns 1, 2; then column 3 is the AB interaction. Even so, there are no residual degrees of freedom; more than four runs are needed to achieve that.

The L_8-array is:

Column		1	2	3	4	5	6	7
Run	1	1	1	1	1	1	1	1
	2	1	1	1	2	2	2	2
	3	1	2	2	1	1	2	2
	4	1	2	2	2	2	1	1
	5	2	1	2	1	2	1	2
	6	2	1	2	2	1	2	1
	7	2	2	1	1	2	2	1
	8	2	2	1	2	1	1	2

In the notation of arrays it would be called 2^7, but again it is not a 7-factor scheme, only a 3-factor one if all the interactions are preserved. One mathematical way of showing this is to use *linear graphs* (Taguchi jargon), and the two graphs for L_8 are shown in Fig. 8.4. Graph I shows that if columns 1, 2, 4 are chosen for main effects, A, B, C say, then columns 3, 5, 6 give the interactions AB, AC, BC. Column 7 in a 2^3 factorial design is the interaction ABC. However, in Taguchi methods it could often be allocated to another factor, D; this would result in aliasing D with ABC and all the corresponding aliases being made. Graph II shows that the interaction between columns 1 and 2 is found in column 3; 1 and 4 in 5; 1 and 7 in 6.

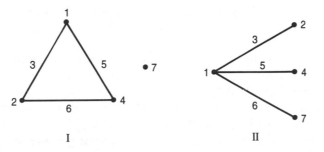

Fig. 8.4 Linear graphs for an L_8 array

A common, and very valid, objection to Taguchi's method is that the designs offer little protection against the presence of interactions. The idea is often to explore as many main factors as possible in a 'saturated' design, on the grounds of economy. But if a fractional experiment is conducted it is inevitable that information is going to be sacrificed in return for a reduction in the size of the experiment. We cannot just ignore interactions; if they cannot be estimated it is because they are confounded with something else, possibly with a main effect. Sometimes these interactions will be ignored when in fact they are important, and this is a weakness in the approach. Analysis is wrong if it does not recognize this.

The Taguchi reply to this is that interactions are rarely important in industrial experimentation, and when they *are* likely to be important the linear graph(s) for the design array help the experimenter to assign factors to columns in such a way that it may still be possible to include some interactions in the design scheme. But at the end of an experimental programme a *confirmation run* of the combination of factor levels thought to be best may give an unexpected result, which could indicate that an important interaction has been omitted.

Residual 'error' terms and degrees of freedom

There is often a lack of suitable residual 'error' terms in Taguchi designs, because of the small number of runs (plots) and the large number of effects the design tries to estimate. In factorial experiments, we have seen that it is common to treat high-order interactions as residual when there is a lack of replicates, but this can be more difficult to carry out in fractionally replicated experiments.

The Taguchi approach here ignores the problem of Type I error. It is subjective, even rather cavalier: to pool several of the smaller mean squares in the Analysis of Variance, say the smaller 50% or so, or those with an F-value less than 2 when tested against the highest-order terms available. The resulting pseudo-residual mean square is then used as the basis for carrying out F tests on the other factors. Also, when there really are very few degrees of freedom available for the residual, F-tests are seriously lacking in power, and again some doubtful solutions are suggested, either to treat any F-value that is greater than 2 as 'indicative', or to use levels such as 10%, 15%, much less strict than the usual conventional level of 5% for statistical tests so that the number of false 'significances' arising by pure chance is greatly increased.

Some alternative methods have been suggested for evaluating mean squares in saturated designs. Daniel (1959) gave a simple graphical method for detecting factors that may be significant in 2^p designs, and for deciding which interactions may reasonably be pooled for residual in a large design. Bissell (1989) gives a graphical method for 3-level factors.

A third method is a numerical test of homogeneity for k mean squares, each having the *same* degrees of freedom, v; therefore it is suitable only for experiments with all factors at the same number of levels. The mean m and the variance s^2 of these mean squares is calculated, and the test statistic $\frac{1}{2}(k-1)vs^2/m^2$ is found. If there are really no differences among the k means, this statistic should follow (approximately) $\chi^2_{(k-1)}$. When this test gives a significant result, *and* there is some good reason to nominate one or more of the larger mean squares for omission, the test may be repeated with less than k items. Some experimenters would repeat this process until they found a non-significant result and then use what was left for the residual.

Some more examples of orthogonal arrays follow, for 2-level, 3-level and mixed-level designs. Taguchi and Konishi (1987) give many others.

The analysis of data from orthogonal-array designs is carried out as for any factorial system in any number of levels: the total for the 1's in a column is found, as for the 2's (and 3's if any), and also the usual correction term. For two levels, a sum of squares

$$\frac{T_0^2 + T_1^2}{r} - \frac{G^2}{2r}$$

is allocated to that column or factor (or interaction as appropriate). Such d.f. as are not used in this way will form residual 'error'. The extension to 3-level factors is obvious.

An example of an L_{12} array ('2^{11}') is:

		1	2	3	4	5	6	7	8	9	10	11
Runs	1	1	1	1	1	1	1	1	1	1	1	1
	2	1	1	1	1	1	2	2	2	2	2	2
	3	1	1	2	2	2	1	1	1	2	2	2
	4	1	2	1	2	2	1	2	2	1	1	2
	5	1	2	2	1	2	2	1	2	1	2	1
	6	1	2	2	2	1	2	2	1	2	1	1
	7	2	1	2	2	1	1	2	2	1	2	1
	8	2	1	2	1	2	2	2	1	1	1	2
	9	2	1	1	2	2	2	1	2	2	1	1
	10	2	2	2	1	1	1	1	2	2	1	2
	11	2	2	1	2	1	2	1	1	1	2	2
	12	2	2	1	1	2	1	2	1	2	2	1

An L_9 array ('3^4') is:

		1	2	3	4
Runs	1	1	1	1	1
	2	1	2	2	2
	3	1	3	3	3
	4	2	1	2	3
	5	2	2	3	1
	6	2	3	1	2
	7	3	1	3	2
	8	3	2	1	3
	9	3	3	2	1

An L_{18} array ('$2^1 \times 3^7$') is:

		1	2	3	4	5	6	7	8
Runs	1	1	1	1	1	1	1	1	1
	2	1	1	2	2	2	2	2	2
	3	1	1	3	3	3	3	3	3
	4	1	2	1	1	2	2	3	3
	5	1	2	2	2	3	3	1	1
	6	1	2	3	3	1	1	2	2
	7	1	3	1	2	1	3	2	3
	8	1	3	2	3	2	1	3	1
	9	1	3	3	1	3	2	1	2
	10	2	1	1	3	3	2	2	1
	11	2	1	2	1	1	3	3	2
	12	2	1	3	2	2	1	1	3
	13	2	2	1	2	3	1	3	2
	14	2	2	2	3	1	2	1	3
	15	2	2	3	1	2	3	2	1
	16	2	3	1	3	2	3	1	2
	17	2	3	2	1	3	1	2	3
	18	2	3	3	2	1	2	3	1

8.6   Choice of design

Suppose that it is required to fit a particular experiment to a Taguchi design. It is very likely that the number of factors is not the same as the number of columns in relevant orthogonal arrays. However, that need not render the problem insoluble. For example, when 12 runs are possible and 9 two-level factors are to be investigated the $L_{12}(2^{11})$ array is an obvious possibility. It has 11 columns, but this does not matter because only 9 of them need be used: the other two can be discarded. If interactions are to be estimated, the choice of columns affects the capacity of the experiment to investigate particular interactions. For this example, with 9 factors, the L_{12} array has no linear graph because it does not allow any interactions to be estimated. Those whose experience is in agriculture may disagree with the claim that interactions are seldom important, even when an experimenter is able to start from a position of some knowledge about the materials that are to be used. Whenever there seems likely to be an interaction of some importance, a *linear graph* may be used to identify the allocation of the factors so that that interaction can still be estimated.

The basic property of orthogonal arrays is that for each factor at one particular level all the other factors are represented at all their levels. This is always so for fractional replicate designs, but there are other designs which also have this property. For example, in the $L_8(2^7)$ array, column 1 is at level 1 in the first four runs and at level 2 in the second four. In each of these groups of four, column 2 has two 1's and two 2's. These columns are thus orthogonal, and in fact an orthogonal array consists of a set of columns, each pair of which is orthogonal.

Two examples

We now give in detail two examples from the American Supplier Institute (ASI) (1987). Example 8.1 shows the Taguchi approach to a problem which would have been handled differently by other workers and Example 8.2 uses an L_8 array in 8 replicates.

Example 8.1. Manufacture of tiles. ASI describe an application of Taguchi methods to a tile manufacturing process in a Japanese factory. The company purchased a tunnel kiln and produced the tiles by baking them in the kiln as they rested on a truck which moved slowly along a track through the length of the kiln. Although tiles in the centre of the stack were attaining the specification standards, the tiles on the outside were very variable and over half of them fell outside the specification limits.

Now most engineering managers would try to solve such a problem by trying to find the source of variation and then to remove it. That procedure would be of no use here since the cause of the variation was well known, namely the variability of temperature within the kiln. However, it would have been extremely expensive to try to equalize the variation within the kiln by engineering methods. Instead, the composition of the tiles was changed so that they became less susceptible to temperature variation, and in this way the proportion of tiles lying within the specification limits was dramatically increased.

The essential principles followed were:

1. for each factor maximize the signal-to-noise ratio SN and thus find the optimum level;
2. control the nominal value using factors that affect the mean but do not affect SN.

It was decided to use an L_8 orthogonal array for the experiment as follows:

A	B	C	D	E	F	G	A	B	C	D	E	F	G
1	1	1	1	1	1	1	2	1	2	1	2	1	2
1	1	1	2	2	2	2	2	1	2	2	1	2	1
1	2	2	1	1	2	2	2	2	1	1	2	2	1
1	2	2	2	2	1	1	2	2	1	2	1	1	2

The array related to the following control factors in the experiment:

		Level 1	Level 2
A	Limestone content	5%	1%
B	Limestone texture	Coarse	Fine
C	Agalmatolite content	43%	53%
D	Agalmatolite type	Old	New
E	Changing quantity	1300 kg	1200 kg
F	Proportion of reuse	0	4%
G	Feldspar content	0	5%

These factors, *A–G*, can all be controlled because they are affected by the *composition* of the material used for the production of tiles. Now the variability of the response was affected by the positions of the tiles within the kiln, and since it was too expensive to control the positional effects, these were treated as noise factors and described as the middle, bottom, side, top and top corner of the stack. Since the locations of the points of the noise factor are systematic, these really form a block effect in the design.

Block 1   Middle
Block 2   Bottom
Block 3   Side
Block 4   Top
Block 5   Top corner

The result of the experiment gave the tile sizes in mm as follows:

		A	B	C	D	E	F	G	Block 1	2	3	4	5
Run	1	1	1	1	1	1	1	1	151.9	151.4	150.4	150.2	149.6
	2	1	1	1	2	2	2	2	151.5	150.8	150.0	149.4	149.1
	3	1	2	2	1	1	2	2	153.1	151.8	151.8	151.4	150.6
	4	1	2	2	2	2	1	1	152.2	151.3	151.1	150.6	150.0
	5	2	1	2	1	2	1	2	151.5	150.8	150.6	150.2	149.7
	6	2	1	2	2	1	2	1	156.5	152.1	150.3	148.5	144.6
	7	2	2	1	1	2	2	1	154.5	153.3	151.8	150.4	149.6
	8	2	2	1	2	1	1	2	153.0	152.0	151.3	150.0	149.5

Now if 150 mm is the target size, it is clear that some of the runs are much more reliable than others. Runs 1 and 2 are both fairly good since their means are close to the target and their ranges are small. The mean of Run 6 is also good but its range is very wide, so that factor combination is likely to lead to highly variable tiles. An Analysis of Variance of the data gives the following results for the tile sizes:

Source	d.f.	Sum of squares	Mean square	F
Blocks	4	72.506	18.126	(11.40)
A	1	0.100	0.100	0.06
B	1	10.201	10.201	6.42**
C	1	0.025	0.025	0.02
D	1	2.916	2.916	1.83
E	1	0.064	0.064	0.04
F	1	0.361	0.361	0.23
G	1	0.121	0.121	0.08
Residual	28	44.522	1.590	
Total	39	130.816		

From this analysis it is clear that the only factor that really affects the mean is *B*. The between-block variance is over 11 times as great as the residual, which confirms that there is strong evidence of positional variation between tiles. In fact, more than half the overall variation between the tiles in this experiment is explained by the blocks. If the positional effects in the experiment had been ignored, then it is possible that the difference in means of factor *B* could have remained undetected. If blocks are treated as residual variation, the Analysis of Variance is much less informative.

The analysis without blocks is:

Source	d.f.	Sum of squares	Mean square	F
A	1	0.100	0.100	<1
B	1	10.201	10.201	2.79 n.s.
C	1	0.025	0.025	<1
D	1	2.916	2.916	<1
E	1	0.064	0.064	<1
F	1	0.361	0.361	<1
G	1	0.121	0.121	<1
Residual	32	117.028	3.657	
Total	39	130.816		

As can be seen from this Analysis of Variance table, if the blocks variation is merged with the residual, then the unexplained variation dwarfs the explained variation. Now in 'classical' experimental work the first analysis would normally be taken as correct, since variation is attributed to assignable causes. However, in Taguchi methodology the second analysis is taken to be correct, even though it may appear wrong to anyone used to other experimental work. The difference is that the factors A, B, C, D, E, F, G may be considered as controllable factors, while the positional effects cannot be controlled. The manufacturer has the power to alter the composition of his tiles, but once the tiles have been composed he wants them all to have the same measurements at the end of the process. Unfortunately, there are positional differences throughout the kiln and it would not be practical to remove them. It would be possible to compensate for positional effects by making tiles of different sizes for different parts of the kiln, but quite impractical to include this feature in the manufacturing process.

   Taguchi's solution to the problem is as follows:

   (i) choose the levels of the factors so that the signal-to-noise ratio SN is large,
   (ii) control the target mean by adjusting the factors that account for most variability in the parameters.

The treatment means were as shown below:

Factor	Level 1	Level 2	Difference
A	150.91	151.01	$+0.10$
B	150.46	151.47	$+1.01$
C	150.99	150.94	-0.05
D	151.23	150.69	-0.54
E	151.00	150.92	-0.08
F	150.87	151.06	$+0.19$
G	151.01	150.91	-0.01

This table confirms that the largest difference in means is between the levels of factor B.

   Now it is important to investigate the noise factor in greater detail to improve the constancy of the product. The block variability for each factor combination is a measure of the noise that is to be removed or at least reduced. The objective in this case is to try to reach a constant size for the 150 mm tiles, so this comes under the category 'nominal is best'. For this criterion SN is measured by

$$SN = 10 \log \left[\left(\frac{\bar{y}}{s} \right)^2 - \frac{1}{n} \right]$$

which is generally well approximated by the alternative formula

$$SN = 20 \log \left(\frac{\bar{y}}{s} \right)$$

These are calculated for each run.

		A	B	C	D	E	F	G	$\bar{y}$	s	SN
Run	1	1	1	1	1	1	1	1	150.70	0.933	44.17
	2	1	1	1	2	2	2	2	150.16	0.991	43.61
	3	1	2	2	1	1	2	2	151.74	0.904	44.49
	4	1	2	2	2	2	1	1	151.04	0.820	45.30
	5	2	1	2	1	2	1	2	150.56	0.673	46.99
	6	2	1	2	2	1	2	1	150.40	4.398	30.68
	7	2	2	1	1	2	2	1	151.92	2.017	37.54
	8	2	2	1	2	1	1	2	151.16	1.433	40.46

The means are all effectively 150 but the proportionate variation in s is very marked, the last three runs giving the highest standard deviations, especially Run 6. These results are reflected in the SN ratios where Run 6 gives the lowest value and Runs 7 and 8 are also low.

The SN ratios are now averaged with respect to the factor levels. As with the analysis of tile sizes, the SN means can be averaged over levels 1 and 2 and the difference gives a measure of the difference in SN response for the levels of that factor. This is because of the orthogonality of the design, which assumes that comparisons for each factor are independent of other comparisons.

	Level			
Factor	1	2	SN mean difference	Preference
A	44.39	38.92	-5.47	$A1$
B	41.36	41.95	0.59	$B2$
C	41.44	41.87	0.43	$C2$
D	43.30	40.01	-3.29	$D1$
E	39.95	43.36	3.41	$E2$
F	44.23	39.08	-5.15	$F1$
G	39.42	43.89	4.47	$G2$

The preference column is based on the level of each factor that has the higher SN ratio. However, the preference is much stronger for factors A, D, E, F, G than for B, C which are not so crucial. Each factor is now examined in turn.

A. Since limestone was inexpensive, the choice of $A1$ was not hard to implement.
B. Tile variability is little affected by B, so it does not really matter which level is used; however, the texture was found to affect the size of the tile.
C. There is a preference for $C2$ because of the SN ratio, but the difference is slight and $C2$ was much more expensive than $C1$ so $C1$ was chosen.
D. It was better to use the existing agalmatolite type $D1$ even though it was more expensive, because of the SN ratio.

E. *E*2 gives a better SN ratio at the expense of productivity; however, the lower changing rate may provide a higher production rate since the wastage rate may be lower.

F. It was better to use 0% waste instead of 4% as reduced variability resulted. However, the level of waste may be reduced if more tiles conform to standard.

G. Adding Feldspar appears to reduce variation.

The choices were between

$$A1, B1, C1, D1, E2, F1, G2 \quad \text{and} \quad A1, B2, C1, D1, E2, F1, G2$$

When the new formula was introduced, the tiles produced showed a much greater tendency to conform to standard. In addition, it was established that the amount of Agalmatolite was not as important as was thought, so a considerable saving in the ingredients was obtained.

Once again we mention that the *cause* of the variation between tiles has not been removed, but the *tendency* for tiles to finish with different sizes has been tackled through a strategic choice of their composition.

Example 8.2. Brackets. In an experiment reported to the ASI by Flex Technologies Inc. in 1987 the problem faced was the development of a thermoplastic fuel rail bracket which had proved susceptible to breakage in practice. The factors most likely to affect stress had been identified. It was decided to run an experiment with the following factors:

		Level 1	Level 2
A	Rear side radius	None	5 mm
B	Upper start of outboard webs	Rail edge	C/L
C	Lower start of outboard webs	Bottom	C/L
D	Radial rib on bracket face	None	2 mm
E	Bracket wall thickness	2 mm	4 mm
F	Bolt pad on bracket	2 mm	None
G	Centre rib on bracket	Yes	No

An $L_8(2^7)$ orthogonal array was used:

		A	B	C	D	E	F	G
Run	1	1	1	1	1	1	1	1
	2	1	1	1	2	2	2	2
	3	1	2	2	1	1	2	2
	4	1	2	2	2	2	1	1
	5	2	1	2	1	2	1	2
	6	2	1	2	2	1	2	1
	7	2	2	1	1	2	2	1
	8	2	2	1	2	1	1	2

Eight schematic models were created according to the above design characteristics and tested to find the greatest stress through finite element analysis. The results were as follows:

Run	Stress						
1	88.98	88.87	87.09	84.36	82.63	81.35	76.58
2	72.17	70.58	70.40	68.05	67.31	62.31	61.13
3	334.49	311.81	310.38	267.60	258.89	256.26	244.26
4	128.94	128.22	125.41	119.61	119.57	118.16	117.83
5	117.26	116.87	115.06	112.16	110.84	105.60	104.47
6	152.65	149.12	142.12	139.02	138.87	135.16	133.62
7	122.20	113.29	111.86	111.68	110.65	110.05	103.78
8	170.05	151.74	143.01	133.35	130.58	122.79	119.18

Since stress should be minimized in order to improve the product, the 'small is best' criterion applies, and so an appropriate SN transformation is

$$SN = -10 \log \left(\frac{1}{n} \sum y^2 \right)$$

The transformation gives SN values:

Run	A	B	C	D	E	F	G	SN
1	1	1	1	1	1	1	1	-38.28
2	1	1	1	2	2	2	2	-36.32
3	1	2	2	1	1	2	2	-48.63
4	1	2	2	2	2	1	1	-41.19
5	2	1	2	1	2	1	2	-40.76
6	2	1	2	2	1	2	1	-42.69
7	2	2	1	1	2	2	1	-40.71
8	2	2	1	2	1	1	2	-42.44

The basic Analysis of Variance is:

Source	d.f.	Sum of squares	Mean square
A	1	0.40	0.40
B	1	29.34	29.34
C	1	31.38	31.38
D	1	3.56	3.56
E	1	20.03	20.03
F	1	3.48	3.48
G	1	2.98	2.98
Total	7	91.17	

An adjusted analysis is:

Source	d.f.	Sum of squares	Mean square
B	1	29.34	29.34
C	1	31.38	31.38
E	1	20.03	20.03
Error	4	10.42	2.61
Total	7	91.17	

Factors responsible for variability appear to be B, C, E. In Taguchi philosophy, the others can be used as 'residual'. These and other engineering considerations led to the optimal design

$$A2, B1, C1, D1, E1. F2, G2$$

The predicted response to this treatment from the previous analysis was estimated to have an SN ratio of -39.11. A confirmation run for 10 replications gave $SN = -38.35$ which was even better than the prediction. The overall improvement in quality was estimated as tenfold, and losses per piece reduced from 4.91 cents to 0.39 cents.

9
Response surfaces and mixture designs

9.1 Introduction

Many of the methods introduced so far can be used whether or not the treatments included in an experiment have any particular relationship to one another. However, we have already mentioned in Chapter 5 some special ways of dealing with quantitative treatments, which are different levels of the same experimental factor; these ideas can be extended considerably and we now need to study *response curves*. A quadratic component of the responses to varying levels of a factor (page 90) is the simplest example of a response curve, and we remarked that cubic, quartic and higher-order components can also be found when a factor is included in an experiment at several levels.

Looking at linear, quadratic and perhaps higher-order components is a great deal more useful than simply comparing some of the mean responses at different levels of the factor: when a factor is included in an experiment at a number of levels, the aim is generally to fit some sort of a curve which summarizes the pattern of all the responses, and any other method of analysis will throw away information. There is a close parallel with regression analysis, where all the data are used together, to fit an appropriate statistical model.

Appropriate models for response curves vary considerably. In the operation of an industrial process, the *yield* (a general term which may have a variety of meanings, as for any experimental measurements) could be linearly related to the factor level; or the relationship could be quadratic if there is an optimum level of the factor, within the range of experimental values used, which corresponds to a maximum or minimum yield. Fitting a curve by a higher-order polynomial than a quadratic may give a better fit, but it can be hard to explain the results in any practical way. A quadratic model $y = a + bx + cx^2$, where y is response or yield and x is the level of the quantitative factor applied, gives a response curve that is symmetric about the value of x where the maximum or minimum occurs. For many industrial experiments that seek optimum values this is acceptable, but biological response curves are often not symmetrical and alternative models are needed (Section 9.9).

9.2 Are experimental conditions 'constant'?

Let us take a very simple example where a factor is applied at different levels x. Figure 9.1 indicates what may happen in an experiment carried out in two stages in very closely controlled environmental conditions. Three values x_0, x_1, x_2 in the first stage give the corresponding mean responses y_0, y_1, y_2. An initial analysis of these would suggest a strong linear component together with a quite substantial quadratic one. The pattern of the response between these three levels of x must be explored further.

In the second stage, various possibilities are illustrated in Fig. 9.1. The points x_a, x_b, x_c may give the mean responses ($\bullet$), and these would suggest a mainly quadratic response over the whole experimental range (or *region*) of x-values. If, alternatively, x_a, x_b, x_c gave the mean responses ($\times$), we may be willing to propose a linear relationship overall. Finally ($\circ$) would indicate a curved relationship, but we are not yet clear whether a maximum may have been reached within the experimental region or whether there may be a *plateau* (Section 9.9).

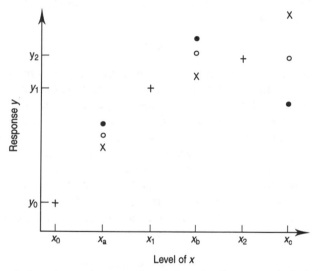

Fig. 9.1 Responses in two stages of an experiment. Key: $+$ = Responses in first stage; second stage responses denoted by *either* $\bullet$, *or* $\times$, *or* $\circ$

However, we can only combine the two stages if they are giving responses y that can strictly be compared with one another. Suppose that the environmental temperature and humidity are critical to the size of the response y. If the second stage had been carried out in conditions that were, even only slightly, different from the first, then every second-stage response would be systematically larger (or systematically smaller) than every first-stage response. In such a situation, we could not combine them into one equation, because at best the values of a would be different and possibly b and c could also be affected. Industrial experiments usually assume that stages *can* be combined, biological ones that they cannot.

9.3 Response surfaces

When two quantitative factors were included in a 3-level experiment in Chapter 5 we were able to divide main effects and interactions into single-degree-of-freedom components involving linear and quadratic parts (page 91). In general, as soon as there are two or more factors we are looking not just for a response curve but for a *response surface*. The simplest of these 'surfaces' is a general first-order model that includes a *surface.* The simplest of these 'surfaces' is a general first-order model that includes a term for interaction. From now on we shall label the factors $x_1, x_2, x_3, \ldots$; x_1 will be included at levels $x_{1(i)}$, where i goes through all the separate levels actually present in the experiment, but we shall try to simplify notation by not specifying every individual level in detail. As the examples will show, this does not usually lead to any confusion.

When two factors interact in a linear model we require a term x_1x_2. So if y is a measurement depending on two quantitative factors x_1 and x_2, the general first-order model is

$$y = a + b_1x_1 + b_2x_2 + b_{12}x_1x_2 + e$$

where e is a residual, natural variation, term with all the usual properties. For computation this can be treated as

$$y = b_0u_0 + b_1u_1 + b_2u_2 + b_3u_3$$

in which $u_0 = 1$, $u_1 = x_1$, $u_2 = x_2$ and $u_3 = x_1x_2$. Multiple regression methods can therefore be used. Usually in industrial work the aim is to find the values of x_1 and x_2 that give maximum y. A linear surface is therefore not adequate. But we do not usually know exactly where to choose experimental values of x_1, x_2 in order to search for a maximum, and so a first experiment in a series of experiments fits a linear surface and uses the gradients $\partial y/\partial x_1$ and $\partial y/\partial x_2$ to indicate the direction in which we should move to choose values of x_1 and x_2 for the next experiment.

From existing knowledge about the situation being studied, it should be possible to decide on an experimental region for the first experiment, and 3 equally spaced points are chosen in this region for x_1 and 3 for x_2. Each of these can be coded as $(-1, 0, 1)$, as shown in Fig. 9.2.

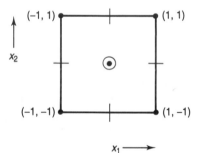

Fig. 9.2 Points ● that may be used in a first-order design with two factors

Designs for response surface analyses are usually based on 2- or 3-level factorials according to whether a first-order or second-order surface is to be fitted. With only 2 factors, a 2^2 design is very small and there is the possibility of replicating it in full. It can instead (or in addition) be augmented by including a few 'centre-points' at $(0, 0)$. Unless at least one of these things is done, there will be no estimate of residual variation against which to test the goodness of fit of a linear model. In the case of 3 or more factors, complete replication is not usually easy in industrial experiments, because it is difficult to carry out very many treatment combinations in one experiment. For 4 or more factors, first experiments are very often fractional replicates.

9.4   Experiments with three factors, x_1, x_2 and x_3

An experiment may begin by using a full 2^3 set of points $(-1, -1, -1)$ up to $(1, 1, 1)$, at the corners of a 'cube' (Fig. 9.3). A simple linear equation

$$y = a + b_1 x_1 + b_2 x_2 + b_3 x_3 + e$$

could be fitted, and this would provide 4 d.f. for the residual to use in testing the goodness of fit. However, this model does not allow for any interactions between the factors. We often prefer (using an obvious and convenient notation for subscripts)

$$y = a + b_1 x_1 + b_2 x_2 + b_3 x_3 + b_{12} x_1 x_2 + b_{13} x_1 x_3 + b_{23} x_2 x_3 + e$$

but now we have only 1 d.f. for estimating the goodness of fit. A moderate number of points at the centre $(0, 0, 0)$ of the design would provide such an estimate; two complete replicates of the design, perhaps with a few centre-points in addition, would give a more reasonable number of residual d.f.

To apply standard multiple regression analysis, writing the model as

$$Y = X\beta + \varepsilon$$

$$\beta = (a, b_1, b_2, b_3, b_{12}, b_{13}, b_{23})$$

and X is made up from the columns

$$(1, x_1, x_2, x_3, x_1 x_2, x_1 x_3, x_2 x_3)$$

If the design used is

x_1	x_2	x_3	Observation
−1	−1	−1	y_1
1	−1	−1	y_2
−1	1	−1	y_3
1	1	−1	y_4
−1	−1	1	y_5
1	−1	1	y_6
−1	1	1	y_7
1	1	1	y_8
0	0	0	y_9

with one centre-point y_9 then

$$y = \begin{pmatrix} y_1 \\ \vdots \\ y_9 \end{pmatrix}, \qquad \varepsilon = \begin{pmatrix} e_1 \\ \vdots \\ e_9 \end{pmatrix}$$

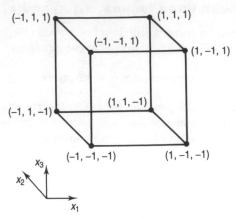

Fig. 9.3 A 2^3 set of points for use in a design with three factors

and

$$X = \begin{bmatrix} 1 & -1 & -1 & -1 & 1 & 1 & 1 \\ 1 & 1 & -1 & -1 & -1 & -1 & 1 \\ 1 & -1 & 1 & -1 & -1 & 1 & -1 \\ 1 & 1 & 1 & -1 & 1 & -1 & -1 \\ 1 & -1 & -1 & 1 & 1 & -1 & -1 \\ 1 & 1 & -1 & 1 & -1 & 1 & -1 \\ 1 & -1 & 1 & 1 & -1 & -1 & 1 \\ 1 & 1 & 1 & 1 & 1 & 1 & 1 \\ 1 & 0 & 0 & 0 & 0 & 0 & 0 \end{bmatrix}$$

The solution, in standard notation, is

$$\hat{\beta} = (X'X)^{-1}X'y$$

Here

$$X'X = \begin{bmatrix} 9 & & & & & & \\ & 8 & & & & \bigcirc & \\ & & 8 & & & & \\ & & & 8 & & & \\ & & & & 8 & & \\ & \bigcirc & & & & 8 & \\ & & & & & & 8 \end{bmatrix}$$

so

$$(X'X)^{-1} = \begin{bmatrix} 1/9 & & & & & & \\ & 1/8 & & & & \bigcirc & \\ & & 1/8 & & & & \\ & & & 1/8 & & & \\ & & & & 1/8 & & \\ & \bigcirc & & & & 1/8 & \\ & & & & & & 1/8 \end{bmatrix}$$

We have

$$X'y = \begin{pmatrix} y_1+y_2+\cdots+y_9 \\ y_2+y_4+y_6+y_8-y_1-y_3-y_5-y_7 \\ \vdots \\ y_1+y_2+y_7+y_8-y_3-y_4-y_5-y_6 \end{pmatrix}$$

therefore

$$\hat{\beta}_1 \equiv \hat{a} = \bar{y}, \ldots$$

and so on down to

$$\hat{\beta}_7 \equiv \hat{b}_{23} = \tfrac{1}{8}(y_1+y_2+y_7+y_8-y_3-y_4-y_5-y_6)$$

The sum of squares for regression is $y'X(X'X)^{-1}X'y - G^2/N$, where G^2/N is the correction term in the usual notation; the total sum of squares is $\sum_{i=1}^{N} y_i^2 - G^2/N$ (and in this example $N=9$). The residual mean square gives the estimate of σ^2.

Degrees of freedom are:

Regression	6
Residual	2
Total	8

In practice we would regard 2 d.f. as insufficient for the residual. The variances of the parameters are calculated from $\sigma^2(X'X)^{-1}$, and since it is a diagonal matrix we obtain simply

$$\text{Var}\,[\hat{a}] = \frac{\sigma^2}{9}; \qquad \text{Var}\,[\hat{b}_1] = \cdots = \text{Var}\,[\hat{b}_{23}] = \frac{\sigma^2}{8}$$

With this design, the off-diagonal elements in $(X'X)^{-1}$ are 0 and so the estimates of parameters are uncorrelated. In such a case the estimates divided by their standard errors provide a test of the significance of each estimate. But we do not usually wish to omit individual non-significant terms, because the model stands as a whole. The regression mean square can therefore be tested against the residual mean square. Two possible outcomes follow:

(i) if the linear surface fits satisfactorily, find the direction of 'steepest ascent' on the surface and proceed in this direction to choose factor levels for the next experiment;

(ii) if the linear surface does not fit, a quadratic surface is required. If the experiment was designed so as to fit a linear surface only, some more points will be needed in a further experiment before a quadratic can be fitted. (As has already been remarked, in industrial work we usually assume that background conditions do not change from one experiment to another, so that the results can be combined.)

It is not appropriate to look at individual terms in a model, because there are some restrictions on how terms may be omitted from a quadratic polynomial:

(i) if a quadratic term (e.g. b_{11}) is in the model, the corresponding linear term (e.g. b_1) must also be present;

(ii) if a product term is included (e.g. b_{12}) then both corresponding linear terms (e.g. b_1 and b_2) must also be present.

Therefore in equations with many variables we can only omit the linear term if (i) the quadratic term for that variable is not required *and* (ii) no cross-product terms involving that variable are required.

Conditions (i) and (ii) can be checked by the 'extra sum of squares' principle. To examine the effect of omitting one or more terms, we apply the *extra sum of squares* principle: compute the equations for two models, one with and one without the term(s) in question, and compute the sum of squares for regression removed by each model. The difference between these two sums of squares is the amount due to the term(s) in question and has the same number of degrees of freedom as there are terms to be omitted. It is the extra sum of squares accounted for by these terms, and if its mean square is not significantly greater than the residual mean square on the *full* model with all linear and quadratic and cross-product terms included then we may omit these terms without seriously affecting the fit. The fewer terms in a model, the easier it is to explain the situation practically.

Example 9.1. Using the design described above, three factors were examined in an industrial process, so as to maximize the yield y of a chemical product: x_1 was the time of running a process and x_2 the operating temperature, while x_3 was the percentage of one of the raw materials included in the mixture being processed. The vector of resulting yields was

$$y = \begin{pmatrix} 16.75 \\ 19.75 \\ 21.90 \\ 23.60 \\ 19.75 \\ 21.75 \\ 22.85 \\ 25.40 \\ 20.40 \end{pmatrix}$$

the values of x_1 were 50 min and 70 min, of x_2 120°C and 130°C, of x_3 25% and 30%. The order of running the 9 treatment combinations was randomized and no blocking was needed in this small experiment.

In the usual notation, $N=9$ and $G=192.15$. Following the calculations described earlier,

$$X'y = \begin{pmatrix} 192.15 \\ 9.25 \\ 15.75 \\ 7.75 \\ -0.75 \\ -0.15 \\ -2.25 \end{pmatrix}$$

so that

$$\hat{\boldsymbol{\beta}} = \begin{pmatrix} 1/9 & & & \text{O} \\ & 1/8 & & \\ & & \ddots & \\ \text{O} & & & 1/8 \end{pmatrix} X'y = \begin{pmatrix} 21.35 \\ 1.156 \\ 1.969 \\ 0.969 \\ -0.094 \\ -0.019 \\ -0.281 \end{pmatrix}$$

The first item is $\hat{a}$, and the next three entries are the 'main-effect' coefficients $\hat{b}_1$, $\hat{b}_2$, $\hat{b}_3$; finally there are the 'interaction' coefficients, all of which are quite small. As explained above, the total sum of squares is

$$4153.7625 - 192.15^2/9 = 51.3600$$

and the sum of squares for regression is

$$\hat{\boldsymbol{\beta}}'X'y - G^2/N = 4152.3194 - 4102.4025 = 49.9169$$

the residual sum of squares is therefore 1.4431. The results may be set out in the following table:

Source of variation	d.f.	Sum of squares	Mean square	
Regression	6	49.9169	8.3195	$F_{(6,2)} = 11.53$
Residual	2	1.4431	0.7216	
Total	8	51.3600		

Although there is no significant F-value, because the residual d.f. are so inadequate, there is no evidence that the model fits badly. The variance of each coefficient is estimated as $0.7216/8 = 0.0902$, giving the standard deviation 0.300. This strongly suggests that the 'interaction' terms (x_1x_2, x_1x_3, x_2x_3) can be omitted, and we may use the model

$$\hat{y} = 21.35 + 1.156x_1 + 1.969x_2 + 0.969x_3$$

to summarize this first experiment. Clearly, as each of x_1, x_2, x_3 is increased so y increases, and we therefore wish to examine higher values of each x than those used so far.

If we use this very simple model, we can apply the method of *steepest ascent* in its most basic form. The gradients b_1, b_2, b_3 tell us that $\hat{y}$ changes by $+1.156$ units when x_1 increases by 1 and the values of x_2 and x_3 remain the same; by $+1.969$ units when x_2 similarly increases by 1; and by $+0.969$ units when x_3 similarly increases by 1. The path of steepest ascent, along which y increases as rapidly as possible, is therefore found by moving b_2 units in x_2 and b_3 units in x_3 for every b_1 units in x_1; i.e. increasing x_1 by 1 unit should be accompanied by an increase of $b_2/b_1 = 1.969/1.156 = 1.703$ units in x_2, and of $b_3/b_1 = 0.969/1.156 = 1.521$ units in x_3.

The values of x_1 were 50, coded -1, and 70, coded $+1$, so that $x_1 = 0$ corresponds to 60 min and one unit of x_1 is 10 min. Likewise, $x_2 = 0$ corresponds to 125°C and one unit of x_2 is 5°C; also $x_3 = 0$ corresponds to $27\frac{1}{2}\%$ and one unit of x_3 is $2\frac{1}{2}\%$. At this stage we must use any physical knowledge we have of the system to decide how far from 0 it is reasonable or possible to go in each x-variable. We should also check the residuals at each of the 9 experimental points already used, to look for any suggestion of a systematic pattern of good or bad fit over any parts of the experimental region. If there is, for example, already some suggestion of curvature at the upper end of the first experimental region we shall not want to go too much further for the follow-up study, whereas a completely adequate linear fit might persuade us to go further up the path of steepest ascent.

Suppose, in this example, that we know the process cannot reasonably be run for more than 80 min. The maximum of x_1 is therefore $+2$ coded units. On the line of steepest ascent, $x_1 = 2$ gives $x_2 = 2 \times 1.703 = 3.406$, and $x_3 = 2 \times 1.521 = 3.042$. Since the zero of x_2 was at 125°C and one unit is 5°C, we should take $x_2 = 125 + 5 \times 3.406 = 142$°C at this uppermost point. Similarly, the zero of x_3 was $27\frac{1}{2}\%$ and the unit is $2\frac{1}{2}\%$, so we should take $x_3 = 27.5 + 2.5 \times 3.042 = 35.1\%$. These values may have to be modified to give an experimental region in which the process will actually work!

There are various possibilities. We could run another 9 points of similar pattern to the first, using as $(-1, 0, 1)$ the values $x_1 = 60, 70, 80$; $x_2 = 125, 133\frac{1}{2}, 142$; $x_3 = 27\frac{1}{2}, 31\frac{1}{4}, 35$. Sensible practical approximations to these would be $(125, 134, 143)$ for x_2 and $(27, 31, 35)$ for x_3.

9.5 Second-order surfaces

Alternatively in Example 9.1 we might concentrate on the higher values and take $x_1 = 70, 75, 80$; $x_2 = 133, 138, 143$; $x_3 = 31, 33, 35$. Coding each of these as $(-1, 0, 1)$ we will probably try to find a maximum for y in this region, since we cannot increase the x-values (x_1, at least) any further. This requires a *second-order surface*. For three factors, this is

$$y = a + b_1 x_1 + b_2 x_2 + b_3 x_3 + b_{12} x_1 x_2 + b_{13} x_1 x_3 + b_{23} x_2 x_3 + b_{11} x_1^2 + b_{22} x_2^2 + b_{33} x_3^2$$

which involves 10 coefficients that require estimating. Of course any set of more than 10 points in the experimental region will allow this fitting, but there are design considerations which improve the ease and the quality of the estimation. Since quadratic terms are to be fitted, each x must be used at 3 (or more) levels. A 3^p factorial design is very large for most values of p, and also gives poor precision (i.e. large standard errors) for b_{11}, b_{22} and b_{33}. Even though modern computing facilities make ease of fitting unimportant, the simpler and more symmetrical designs have good properties, and we consider three of them.

A *central composite design* can be used to locate a maximum when we appear to have climbed the path of steepest ascent far enough to use the centre $(0, 0, 0)$ of the most recent experiment as the centre of the final one which allows a second-order surface to be fitted. If there are p factors, the most recent experiment will have contained $(0, 0, \ldots, 0)$ and all 2^p combinations $(\pm 1, \pm 1, \ldots, \pm 1)$; to this we add another $2p$ points of the form $(\pm a, 0, \ldots, 0)$, $(0, \pm a, 0, \ldots, 0)$, and so on. Using just one point at the centre (though there would be no reason to limit it to one if we needed more residual d.f.) the final central composite design will have $2^p + 2p + 1$ points from which we can estimate the required coefficients; for $p = 2, 3, 4$, respectively, this is 9, 15 and

25 points, whereas 3^p is 9, 27 and 81, so there is considerable economy of experimental work.

The value of α may be chosen *either* to ensure that the regression coefficients are mutually orthogonal, which requires $\alpha = \sqrt{2} = 1.414$ for $p = 2$, or $\alpha^2 = 2\sqrt{2}$ so that $\alpha = 1.682$ for $p = 3$, and so on, the normal equations being very easy to solve, *or* to make them *rotatable*, which we discuss below. Figure 9.4 shows how the design of Fig. 9.3 is modified by taking the extra 6 points distant α from the centre $(0, 0, 0)$ to make a central composite design.

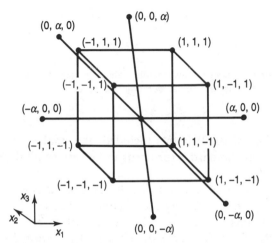

Fig. 9.4 Central composite design for three factors

When a new centre appears to be needed for the final experiment we shall need a *non-central composite design*. The maximum may seem to be near to one of the treatment combinations other than $(0, 0, \ldots, 0)$; for example with 3 factors we may want to work in the region of the highest combination $(1, 1, 1)$. This time we add p points to the previous experiment, either above or below according to where the maximum seems to be. From $(1, 1, 1)$, the 3 additional points would be $(1 + \alpha, 1, 1)$, $(1, 1 + \alpha, 1)$ and $(1, 1, 1 + \alpha)$. If some other corner of the design of Fig. 9.3 is to be used, the rule is always to go α units *outwards* from the cube; thus from $(1, -1, 1)$ the extra points would be $(1 + \alpha, -1, 1)$, $(1, -1 - \alpha, 1)$ and $(1, -1, 1 + \alpha)$.

The idea of a *rotatable design* arises from considering the standard error of the fitted $\hat{y}$ values. This standard error depends on the coordinates of the point on the response surface at which y is evaluated and on the coefficients b. In a rotatable design, this standard error is the same for all points that are the same distance from the central point $(0, 0, \ldots, 0)$. Thus with two variables x_1, x_2 we will be using a selection of points in the (x_1, x_2)-plane whose origin is $(0, 0)$, and it turns out that for a rotatable design we require points equally spaced round the circle, centre $(0, 0)$, together with some at $(0, 0)$ itself. There are six coefficients in the 2-variable second-order equation:

$$y = a + b_1 x_1 + b_2 x_2 + b_{12} x_1 x_2 + b_{11} x_1^2 + b_{22} x_2^2$$

and the absolute minimum for estimation is a pentagon with one centre-point, although in practice this would not be acceptable since it gives no residual degrees of freedom.

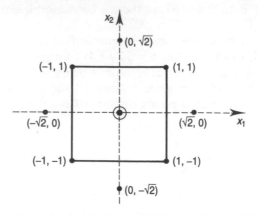

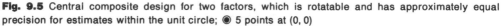

Fig. 9.5 Central composite design for two factors, which is rotatable and has approximately equal precision for estimates within the unit circle; ⊚ 5 points at (0, 0)

The larger the number of $(0, 0)$ points included, the smaller the standard error of $\hat{y}$ near $(0, 0)$; whether this standard error needs to be very small depends on whether the maximum is near to $(0, 0)$.

The best design for general use is probably the *central composite rotatable design*. For two variables (Fig. 9.5), there are the 2^2 combinations $(-1, -1)$, $(1, -1)$, $(-1, 1)$, $(1, 1)$, plus the 2×2 points $(0, \sqrt{2})$, $(0, -\sqrt{2})$, $(-\sqrt{2}, 0)$, $(\sqrt{2}, 0)$ for rotatability; finally 5 centre-points $(0, 0)$ will give approximately equal precision for $\hat{y}$ within the circle centre $(0, 0)$ and radius 1 as recommended by Box and Hunter (1957). For rotatability,

$$\alpha = 2^{p/4}$$

when a complete replicate of the 2^p set is used, but for larger values of p a half-replicate is adequate to estimate all necessary coefficients, and then $\alpha = 2^{(p-1)/4}$; the following table shows how the most useful central composite rotatable designs are made up for estimating second-order surfaces:

p	2^p	Star $(2p)$	Centre†	α	Total number of points
2	4	4	5	1.414	13
3	8	6	6	1.682	20
4	16	8	7	2.000	31
5	16*	10	6	2.000	32
6	32*	12	9	2.378	53

* Half-replicate.
†For approximately equal precision in unit sphere.

Cochran and Cox (1992) and Box *et al.* (1978) give full discussions of the design and analysis of experiments for fitting response surfaces.

9.6   Contour diagrams in analysis

In two variables, the maximum is estimated by solving the two equations

$$\frac{\partial y}{\partial x_1} = 0 \quad \text{and} \quad \frac{\partial y}{\partial x_2} = 0$$

with a second-order surface these are

$$b_1 + 2b_{11}x_1 + b_{12}x_2 = 0$$

and

$$b_2 + 2b_{22}x_2 + b_{12}x_1 = 0$$

Remembering the 'steepest-ascent' approach, we may think of this as being the summit of a hill, which we tried to reach as quickly as possible by climbing up the steepest available path from the experimental region of values of x_1 and x_2 where we began. A *contour diagram*, like a geographical map, shows the characteristics of the surface that has been climbed.

Suppose that the maximum value of y is 120. If we now write $y = 110$ on the left-hand side of the fitted equation, we shall have, in a simple case, the equation of an ellipse, every point on which gives a yield of 110 units (like a contour on a map, which shows points the same distance above sea level). Repeating this at convenient spacing of y will give a great deal of information about a surface, more clearly than carrying out a mathematical study of the equation; most good graph-plotting packages, especially those intended for statisticians, will do this. Figure 9.6(a) shows a situation where there is relatively rapid change in y as the levels of x_1 and x_2 are changed; and in Fig. 9.6(b) the value of y changes much more slowly so that it is less critical to the optimum operation of the system to have x_1 and x_2 controlled exactly. (Occasionally the elliptical contours may become circles.)

Other less simple pictures are possible. Figure 9.7 shows a *stationary ridge* where, so long as (x_1, x_2) lies on the line at the centre of the diagram the value of y is equally

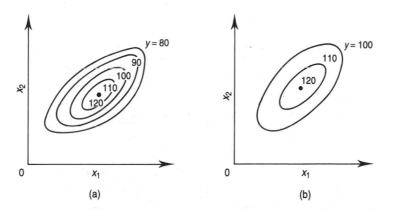

(a)   (b)

Fig. 9.6 Contour diagrams showing change in yield y as values of x_1, x_2, are varied: $y = 120$ is the maximum yield

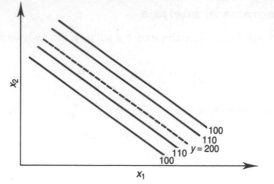

Fig. 9.7 A stationary-ridge contour diagram

high. Figure 9.8 will occur when there is a steady rise within the experimental region and beyond it, so that eventually a maximum may be reached by proceeding along the *rising ridge*. A more difficult situation to interpret is that of Fig. 9.9, where there is a saddle from which the surface rises in one direction and falls in another.

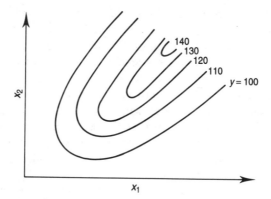

Fig. 9.8 A rising-ridge contour diagram

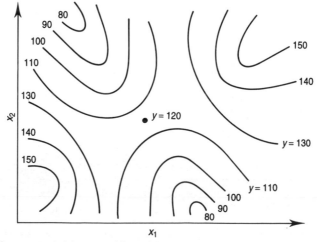

Fig. 9.9 Contour diagram containing a saddle point

When there are several variables x_1, x_2, x_3, ... contour diagrams may be obtained for two variables at a time, having fixed the values of all others; for example, when dealing with 3 variables we may fix x_3 at -1 and draw a contour diagram for x_1 and x_2, then fix x_3 at 0 and repeat the drawing, and do the same again with x_3 fixed at $+1$. Interpreting the results of this procedure will be greatly helped by any prior knowledge of the system being studied. In fact, modern computer graphics equipment will produce 'three-dimensional' looking contours, but for 4 or more x-variables we must study subsets of the total number.

9.7   Transformations

When a suitable model appears to have been found, it is very useful to look at *residuals* and – as in any regression model – to consider whether the pattern of residuals indicates skewness so that a transformation of y, or of the x-variables, may give better results. Ridges, such as those mentioned above, can arise when the x-variables actually measured are not the fundamental ones that show how a system is working but simply those that it was possible, or easiest, to measure. Products or ratios of measured x-variables may be more informative. Ridges indicate that, at least, some form of *interaction* is present among the x's.

Furthermore, there is no reason why the coding $(-1, 0, 1)$ should always mean a set of equal steps on an additive linear scale (as it did in Example 9.1); sometimes it is much better to work on a logarithmic (i.e. ratio) scale, for example $\frac{1}{2}$, 1 and 2 units can correspond to $(-1, 0, 1)$ on the coded scale.

Example 9.2. The use of residuals may be illustrated by the following set of data:

x_1	x_2	x_3	y
-1	-1	-1	4.4
1	-1	-1	22.2
-1	1	-1	15.3
1	1	-1	20.3
-1	-1	1	6.5
1	-1	1	25.1
-1	1	1	23.0
1	1	1	29.0

A linear surface fitted to the data gives

$$\hat{y} = 18.225 + 5.925x_1 + 3.675x_2 + 2.675x_3$$

The Analysis of Variance is

	d.f.	Sum of squares	Mean square	
Regression	3	446.135	148.712	$F_{(3,4)} = 6.11$
Residual	4	97.300	24.325	
Total	7	543.435		

The fit may be acceptable, although not very good (the critical value of $F_{(3,4)}$ at 5% is 6.59) and it gives the residuals:

x_1	x_2	x_3	Residual
−1	−1	−1	−1.55
1	−1	−1	4.40
−1	1	−1	2.00
1	1	−1	−4.85
−1	−1	1	−4.80
1	−1	1	1.95
−1	1	1	4.35
1	1	1	−1.50

The two large positive residuals are at $(1, -1, -1)$ and $(-1, 1, 1)$ which are opposite corners of the design; also the two large negative residuals are at $(1, 1, -1)$ and $(-1, -1, 1)$, which again are opposite corners. We cannot say that the residuals are showing a purely random pattern, and it seems very likely that the surface needs more terms added to the model.

The results suggest that x_1, x_2, x_3 should all be increased, and in a *steepest-ascent* analysis a new centre could be at $(1, 1, 1)$ or, probably better, at $(2, 2, 2)$ with a second-order design and analysis being used.

9.8 Mixture designs

There is a special class of response surface design considered by Scheffé (1958, 1963) and many other workers: we will indicate the main outlines of the idea in this section and refer the reader to Khuri and Cornell (1987) for a full description and examples.

Some types of work give rise to restrictions on the levels of factors included. When mixtures of ingredients are used, the sum of their levels (%) included in any treatment must be fixed (100%; or sometimes 100% minus one fixed component). Examples are making detergents from mixtures of basic components, varying the proportions of different nutrients in animal feeds or in agricultural fertilizers, making up different blends of fuels or altering the relative quantities of ingredients in the recipe of a food product.

Design considerations for mixtures are therefore different from standard factorials, and the types of response surface that may be fitted are also somewhat more restricted to allow for the relationships between the factor levels $\{x_i\}$.

In an experiment with q factors, write

$$\sum_{i=1}^{q} x_i = 1 \qquad (\text{all } x_i \geqslant 0)$$

The experimental region is what may be called a *simplex*: with only two components this degenerates to a straight line, $x_1 + x_2 = 1$, and all experimental points must lie on that line: see Fig. 9.10(a).

With three factors, the restriction is $x_1 + x_2 + x_3 = 1$, an equilateral triangle and all chosen treatments (factor level combinations) must lie on that plane: see Fig. 9.10(b). The actual designs proposed for this type of experiment would make it possible to 'predict' a response for any combination $(x_1, x_2, \ldots)$ in the experimental region and to examine the effects of components singly and in combination.

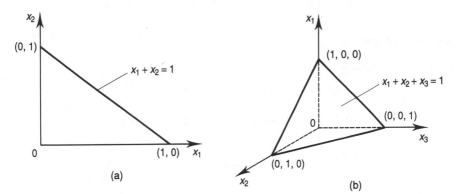

Fig. 9.10 Experimental regions for (a) two, (b) three factors

Simplex lattice designs

These designs cover the simplex factor space uniformly, thus providing an obvious class of design for this class of experiment. A $\{q, m\}$ design has q components (factors) each spaced as $x_i = 0, 1/m, 2/m, \ldots, 1$, and all possible combinations of these are used in an experiment.

For a $\{3, 2\}$ lattice the levels are $0, \frac{1}{2}, 1$, as in Fig. 9.11(a), while for a $\{3, 3\}$ lattice Fig. 9.11(b) shows that a centroid point $(\frac{1}{3}, \frac{1}{3}, \frac{1}{3})$ is used as well as the points along the sides of the triangle. When $q = 4$, there are 4 points not on the sides, typically $(0, \frac{1}{3}, \frac{1}{3}, \frac{1}{3})$; others are $(1, 0, 0, 0)$, etc., $(\frac{2}{3}, \frac{1}{3}, 0, 0)$ etc., giving 20 points in all. A $\{q, m\}$ design in general has

$$\frac{(q+m-1)!}{m!\,(q-1)!}$$

points. These designs were proposed by Scheffé, who later gave the modification called 'simplex centroid'.

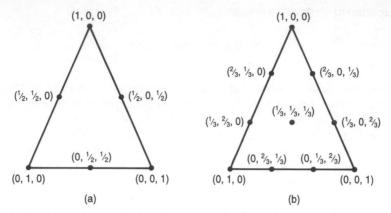

Fig. 9.11 Simplex lattice designs for three factors with (a) three, (b) four levels

Simplex centroid designs

These have $2^q - 1$ points:

$$q \qquad \text{permutations like} \quad (1, 0, 0, \ldots, 0)$$

$$\binom{q}{2} \qquad \text{permutations like} \quad (\tfrac{1}{2}, \tfrac{1}{2}, 0, \ldots, 0)$$

$$\binom{q}{3} \qquad \text{permutations like} \quad (\tfrac{1}{3}, \tfrac{1}{3}, \tfrac{1}{3}, 0, \ldots, 0)$$

etc., finishing with the centroid point

$$\left(\frac{1}{q}, \frac{1}{q}, \ldots, \frac{1}{q} \right)$$

Whether the experimenter will prefer lattice or centroid designs, and whether it is reasonable to have zeros (several zeros in some combinations) in the mixture will depend on the type of study being carried out.

Models to be fitted

A linear model in 2 components is usually $y = a + b_1 x_1 + b_2 x_2$. But in a mixture we have the constraint $x_1 + x_2 = 1$, so that we can write

$$y = a(x_1 + x_2) + b_1 x_1 + b_2 x_2$$

$$= c_1 x_1 + c_2 x_2$$

with no constant term. When looking at polynomial models we can, for example, write

$$x_1^2 = x_1(1 - x_2)$$

and we arrive at a model that contains linear and product (i.e. $x_1 x_2$) terms, but no squares, as the standard quadratic.

A summary of appropriate models is

Linear: $\qquad y = \sum\limits_{i=1}^{q} b_i x_i$

Quadratic: $\qquad y = \sum\limits_{i=1}^{q} b_i x_i + \sum\sum\limits_{i<j} b_{ij} x_i x_j$

Cubic: $\qquad y = \sum\limits_{i=1}^{q} b_i x_i + \sum\sum\limits_{i<j} b_{ij} x_i x_j + \sum\sum\sum\limits_{i<j<k}^{q} b_{ijk} x_i x_j x_k$

$$+ \sum\sum\limits_{i<j}^{q} c_{ij} x_i x_j (x_i - x_j)$$

In this 'cubic' model there are squared terms, but they appear in a special way: the final term in the cubic model may not always be needed. The meanings of the various terms are:

b_i is the response at $(x_i = 1,$ all other $x_j = 0)$; thus if the blending of components is strictly additive the purely linear model should fit the data;

b_{ij} arise when there is curvature in the mixture surface, for example synergistic or antagonistic blending, i.e. if the effects of the components are enhanced when they are present together or if they work against one another;

b_{ijk} and also possibly c_{ij} will be needed when the deviations from a mixture surface are more complex.

Given adequate data and replication to obtain residual 'error' variance, the models will be fitted in order of complexity. Upon reaching the cubic, the model without c_{ij} is fitted first. It will usually be helpful to plot contours, as for any other response surface fitting. Without these it can be very hard to see the meaning of more complicated models or to locate any region of acceptable mixtures.

Restrictions on $\{x_i\}$

Individual x's may not be allowed to go down to 0 or up to 1. The design must then be modified, keeping to similar principles as before. One or more components of the mixture may have a lower bound: $x_i \geqslant L_i$. Suppose there are three components, $x_1 \geqslant L_1$, $x_2 \geqslant L_2$, $x_3 \geqslant L_3$; it must follow that, for example, x_3 has an upper bound $1 - L_1 - L_2$. Thus the upper bounds are implied whether or not they are stated. Consider first the case where they are not stated, but only implied. Then in general

$$x_i \geqslant L_i \geqslant 0 \qquad i = 1, 2, \ldots, q$$

and

$$\sum\limits_{i=1}^{q} L_i = L < 1$$

For $q = 3$, Fig. 9.12(a) shows that when the condition $x_1 \geqslant L_1$ is imposed, the other boundary of the region satisfies $x_2 = x_3 = 1 - L_1$. Adding a lower bound for x_2 reduces the experimental region further, as in Fig. 9.12(b), and including a lower bound for x_3, as in Fig. 9.12(c), leaves the shaded region as the experimental region.

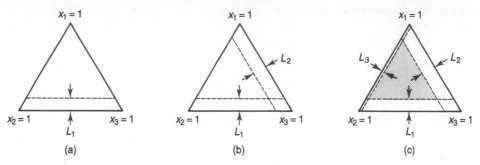

Fig. 9.12 Experimental regions when there are restrictions on $\{x_i\}$

Pseudocomponents can be defined as

$$x'_i = \frac{x_i - L_i}{1 - L}$$

and will then satisfy

$$\sum_{i=1}^{q} x'_i = 1$$

In the pseudocomponent system $(x'_1, x'_2, x'_3) = (1, 0, 0)$ corresponds to $(x_1, x_2, x_3) = (1 - L_2 - L_3, L_2, L_3)$, and so on. The subregion defined in pseudocomponents may have the same designs placed on it as have already been used; these will correspond to original component settings of $x_i = L_i + (1 - L)x'_i$.

For example, if $L_1 = 0.2$, $L_2 = 0.4$, $L_3 = 0.2$, so that $L = 0.8$,

$$x'_i = \frac{x_1 - 0.2}{0.2}, \qquad x'_2 = \frac{x_2 - 0.4}{0.2}, \qquad x'_3 = \frac{x_3 - 0.2}{0.2}$$

and the vertices of the pseudocomponent triangle are

$$(1, 0, 0) \rightarrow (0.4, 0.4, 0.2) \qquad \text{for } (x_1, x_2, x_3)$$

$$(0, 1, 0) \rightarrow (0.2, 0.6, 0.2)$$

$$(0, 0, 1) \rightarrow (0.2, 0.4, 0.4)$$

which define the effective experimental region (or subregion) for $\{x_i\}$. The corners are no longer pure components, even though derived from pure pseudo-components. Other experimental points are affected in the same way:

$$(x'_1, x'_2, x'_3) = (0, \tfrac{1}{2}, \tfrac{1}{2}) \rightarrow (x_1, x_2, x_3) = (0.2, 0.5, 0.3)$$

$$(\tfrac{1}{3}, \tfrac{1}{3}, \tfrac{1}{3}) \rightarrow \qquad\qquad = (0.27, 0.46, 0.27)$$

$$(\tfrac{1}{6}, \tfrac{1}{6}, \tfrac{2}{3}) \rightarrow \qquad\qquad = (0.23, 0.43, 0.33)$$

Extreme vertices designs

The situation where both lower and upper bounds are specified is more complex and requires further types of design. Conditions are

$$0 < L_i \leqslant x_i \leqslant U_i < 1, \qquad i = 1, 2, \ldots, q$$

The resulting experimental region is a more complicated shape, the number of vertices actually depending on the values of $\{L_i, U_i\}$.

Example 9.3. $q=3$;

$$0.2 \leqslant x_1 \leqslant 0.6$$

$$0.1 \leqslant x_2 \leqslant 0.6$$

$$0.1 \leqslant x_3 \leqslant 0.5$$

Figure 9.13 illustrates this.

$$R_L = 1 - \sum_{i=1}^{q} L_i, \quad R_U = \sum_{i=1}^{q} U_i - 1, \quad R_P = \min\,(R_L, R_U)$$

where $R_i = U_i - L_i$ is the range of x_i. Hence $R_1 = 0.4$, $R_2 = 0.5$, $R_3 = 0.4$, and also $R_L = 0.6$, $R_U = 0.7$, $R_P = 0.6$.

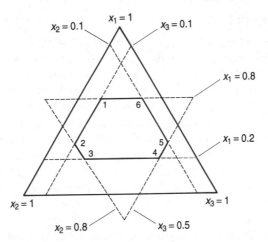

Fig. 9.13 Lower and upper bounds in a three-factor design

There is a combinatorial method for determining the number of vertices, which we see from Fig. 9.13 is 6 in this example; $^qC_r = C(q, r)$ is the number of combinations of q components taken r at a time: $C(3, 1) = C(3, 2) = 3$ and $C(3, 3) = 1$. These are divided into three groups, containing

  (I) $L(q, r)$ combinations of r whose ranges sum to less than R_P
  (II) $E(q, r)$ combinations of r whose ranges sum to exactly R_P
  (III) $G(q, r)$ combinations of r whose ranges sum to more than R_P

and the number of vertices of the constrained region is

$$N_V = q + \sum_{r=1}^{q} L(q, r)(q - 2r) + \sum_{r=1}^{q} E(q, r)(1 - r)$$

$C(3, 1)$: x_1, $R_1 = 0.4$; x_2, $R_2 = 0.5$;

      x_3, $R_3 = 0.4$. $L(3, 1) = 3$

$C(3, 2):$ $x_1, x_2, R_1 + R_2 = 0.9; x_1, x_3, R_1 + R_3 = 0.8$

$x_2, x_3, R_2 + R_3 = 0.9$ $G(3, 2) = 3$

$C(3, 3):$ $x_1, x_2, x_3, R_1 + R_2 + R_3 = 1.30$ $G(3, 3) = 1$

So

$$N_v = 3 + 3(3 - 2) + 0 = 6$$

Experimental points are determined by writing down all 2^q combinations of L_i, U_i, and finding all combinations that satisfy $x_1 + x_2 + x_3 + x_4 = 1$, where $x_i = L_i$ or U_i, $i = 1, 2, 3, 4$, *or* which can be made to satisfy this by adjusting one coordinate only.

Example 9.3: continued. The $2^3 = 8$ combinations of L_i, U_i are as follows:

(1) $(L_1, L_2, L_3) = (0.2, 0.1, 0.1)$ (5) $(U_1, L_2, L_3) = (0.6, 0.1, 0.1)$

(2) $(L_1, L_2, U_3) = (0.2, 0.1, 0.5)$ (6) $(U_1, L_2, U_3) = (0.6, 0.1, 0.5)$

(3) $(L_1, U_2, L_3) = (0.2, 0.6, 0.1)$ (7) $(U_1, U_2, L_3) = (0.6, 0.6, 0.1)$

(4) $(L_1, U_2, U_3) = (0.2, 0.6, 0.5)$ (8) $(U_1, U_2, U_3) = (0.6, 0.6, 0.5)$

In fact none of these satisfies $\sum_{i=1}^{q} x_i = 1$. We are allowed to adjust *one* coordinate. Some adjustments give impossible results, e.g. (1) cannot become (0.8, 0.1, 0.1) because $x_1 \leqslant 0.6$, nor (0.2, 0.7, 0.1) because $x_2 \leqslant 0.6$, and similarly for x_3. However, (2) gives two possible results: (0.4, 0.1, 0.5), (0.2, 0.3, 0.5), which are the points marked 5 and 4, respectively, in Fig. 9.13. From (3) we may obtain (0.3, 0.6, 0.1) and (0.6, 0.1, 0.3), points 2 and 3; from (4) (0.2, 0.3, 0.5) and (0.2, 0.6, 0.2), points 4 (again) and 3 (again); from (5) (0.6, 0.3, 0.1) and (0.6, 0.1, 0.3), points 1 and 6. We now have all six vertices, but if we had not made the calculation of N_v we might prefer to enumerate all the remaining possibilities and check that there was nothing new ((6) and (7) give repeats of points 5, 6, 1 and 2, while (8) yields no possible adjustments).

Depending on the number of factors q, and the type of model it is required to fit, experimental points will consist of the vertices of the experimental region together with points on the edges or faces of the region, especially centroids, the overall centroid and centroids of subregions in fewer dimensions. For large q, a computer algorithm for generating design points is useful. As a general rule, centroids are important when models are non-linear.

Including process variables in mixture experiments

Not all variables need to be subject to mixture constraints. There may be in the same experiment the usual types of factor such as operating temperature of a system and process time. It is possible to set up a simplex for mixture variables x at each point of a factorial arrangement for z, the process variables, e.g. 2^n for n process variables.

For three x's and two z's, the separate models would be

$$Y = \sum_{i=1}^{3} \beta_i x_i + \sum_{i,j=1,2,3} \beta_{ij} x_i x_j \qquad \text{for the mixture}$$

and

$$Y = \alpha_0 + \alpha_1 z_1 + \alpha_2 z_2 + \alpha_{12} z_1 z_2 \qquad \text{for the } 2^2 \text{ factorial}$$

The full model for the combined design consists of 24 terms, in the product of these two models.

Sometimes it may not be thought necessary to have all terms in the model, or all points in the design, and the 2^n factorial may be reduced to a suitable fraction.

9.9   Other types of response surface

The 'classical' types of response function so far described only explain symmetrical forms of curve or surface; the only way of removing any asymmetry is to use a transformation of y or of the x-variables. Many biological experiments, in particular, will yield curves like those illustrated for one x-variable in Fig. 9.14. Either (a) there is a maximum response which is approached asymptotically, y_{max}, or (b) the descent from y_{max} is at a different rate from the ascent to it.

Nelder (1966, 1968) introduced a family of curves that include both of these types of response, and which is therefore very useful and flexible for agricultural and biological work. These are the *inverse polynomials*, one of the general class of responses discussed by McCullagh and Nelder (1989). The equation contains $1/y$ and $1/x$ (hence the name).

The *plateau* type of response (Fig. 9.14(a)) is fitted by the linear inverse polynomial

$$\frac{1}{y}=\frac{a}{x}+b$$

where the constants a and b must be estimated. The gradient on the (x, y) graph near the origin is $1/a$, and the asymptote y_{max} is $1/b$. The *asymmetric* response of Fig. 9.14(b) is fitted by

$$\frac{1}{y}=\frac{a}{x}+b+cx$$

with one extra parameter c. Again the gradient on the (x, y) graph near the origin is $1/a$; the size of y_{max} is $1/(b+2\sqrt{ac})$ and this maximum occurs at $x=\sqrt{a/c}$. This is a quadratic inverse polynomial, and the rate of fall to the right of the maximum is determined by c.

It is convenient to rewrite these two equations as:

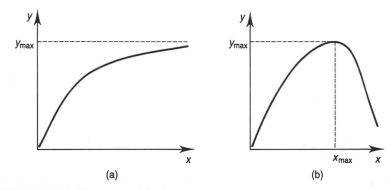

Fig. 9.14 (a) 'Plateau', (b) unsymmetrical responses

$$\text{Linear:} \quad \frac{x}{y} = a + bx$$

$$\text{Quadratic:} \quad \frac{x}{y} = a + bx + cx^2$$

because these forms make it easier to construct surfaces involving two or more x-variables.

When a factor is present at only two levels in an experiment, we can only fit the linear inverse polynomial; in order to study the general shape of a response at least 3 levels are needed, and although this is not usually enough to give a proper idea of the form of the curve, it is only a, b, c in the quadratic inverse polynomial that give the most important information. Adding a cubic term dx^3 on the right-hand side would allow the behaviour of the curve beyond the maximum, in the right-hand tail, to be studied; often this is not of any real interest, and the corresponding values of x can also be beyond the range of physically useful, even possible values.

Several factors

Models for any number of factors, x_1, x_2, x_3, ... are built up by multiplying the right-hand sides of linear or quadratic inverse polynomials in the following ways. When there are just two factors at two levels, we can only have

$$\frac{x_1 x_2}{y} = (a_1 + b_1 x_1)(a_2 + b_2 x_2)$$

which reduces to

$$\frac{x_1 x_2}{y} = \alpha + \beta_1 x_1 + \beta_2 x_2 + \beta_{12} x_1 x_2$$

As usual, the term in $x_1 x_2$ is present when the 2 factors interact. Similarly, for 3 factors a model allowing for all possible interactions would be

$$\frac{x_1 x_2 x_3}{y} = \alpha + \beta_1 x_1 + \beta_2 x_2 + \beta_3 x_3 + \beta_{12} x_1 x_2 + \beta_{13} x_1 x_3 + \beta_{23} x_2 x_3 + \beta_{123} x_1 x_2 x_3$$

Two factors, each present at 3 or more levels, would allow the model

$$\frac{x_1 x_2}{y} = (a_1 + b_1 x_1 + c_1 x_1^2)(a_2 + b_2 x_2 + c_2 x_2^2)$$

to be fitted; this reduces to

$$\frac{x_1 x_2}{y} = \alpha + \beta_1 x_1 + \beta_2 x_2 + \beta_{12} x_1 x_2 + \gamma_1 x_1^2 + \gamma_2 x_2^2 + \gamma_{12} x_1 x_2^2 + \gamma_{21} x_1^2 x_2 + \gamma_{22} x_1^2 x_2^2$$

As with any other response surface, it is best not to look at each individual term to see if it should be present in the model, but instead to build-up and test models hierarchically; we begin, for example, with this latter model which can be called 'quadratic × quadratic' as the full model for an experiment, and then examine the 'quadratic × linear' model

$$\frac{x_1 x_2}{y} = (a_1 + b_1 x_1 + c_1 x_1^2)(a_2 + b_2 x_2)$$

This contains three fewer terms (not just one fewer), and is a simpler model that has practical meaning, namely that the response to x_2 is predominantly linear while that to x_1 is quadratic. Another candidate for study is 'linear × quadratic'

$$\frac{x_1 x_2}{y} = (a_1 + b_1 x_1)(a_2 + b_2 x_2 + c_2 x_2^2)$$

and if either or both of these simpler models seems as good as the full model, we may finally go on to the 'linear × linear'

$$\frac{x_1 x_2}{y} = a + \beta_1 x_1 + \beta_2 x_2 + \beta_{12} x_1 x_2$$

discussed at the beginning of this section.

An alternative family of models

Inverse polynomials use an iterative computing routine (due to Nelder) for estimating parameters; these estimators do not have such good distributional properties as ordinary regression estimators. The assumption involved in the analysis, namely that logs of the residual terms follow a normal distribution, suggests another form of equation (Ali (1983), Ali *et al.* (1986)) which replaces $1/y$ by $\log y$ and is otherwise like the inverse polynomials already discussed. Thus the linear form in one variable is

$$\log y = \frac{a}{x} + b$$

which is $x(\log y) = a + bx$; in two variables the linear × linear form will become

$$x_1 x_2 (\log y) = a + \beta_1 x_1 + \beta_2 x_2 + \beta_{12} x_1 x_2$$

These may be called *loglinear response surfaces*; they can be fitted using standard regression routines, with an appropriate addition to the program so that the values of $x_1 x_2 \log y$ are calculated for each point as the 'dependent variable'. The 'independent' or predictor variables must be made to include $1/x$. The estimated parameters will have distributional properties like regression coefficients.

Combinations of linear and quadratic parts can be built into these models as with ordinary inverse polynomials; for example x_1 linear and x_2 quadratic is found from

$$x_1 x_2 (\log y) = (a_1 + b_1 x_1)(a_2 + b_2 x_2 + c_2 x_2^2)$$

$$= a + \beta_1 x_1 + \beta_2 x_2 + \beta_{12} x_1 x_2 + \gamma_2 x_2^2 + \gamma_{12} x_1 x_2^2$$

It is interesting to note, if we write this in the more basic form

$$\log y = \frac{\alpha}{x_1 x_2} + \frac{\beta_1}{x_2} + \frac{\beta_2}{x_1} + \beta_{12} + \gamma_2 \frac{x_2}{x_1} + \gamma_{12} x_2$$

that we now have a ratio term x_2/x_1, and (the inverse of) a product term $x_1 x_2$. More complicated models will contain further ratio terms; and we have already noted that product terms are needed when interactions exist among the factors. In biological situations it is quite possible for ratio terms to have very real meanings; for example, the amount of one x-variable alone may not be the critical determinant of a response but instead it is the ratio of two of the x's that is most important. In a full quadratic model, written in the form with $\log y$ as the left-hand side, there will be a complete set of ratio and product terms, and these will be important in interpreting the results of an experiment analysed using this class of models.

As in any response surface, contour diagrams are a valuable aid to interpreting the results, and these can be computed in the same way as for 'classical' response surfaces (page 177).

It has been found that for many sets of data there is little to choose between inverse polynomials in $1/y$ and the logarithmic form where $1/y$ is replaced by $\log y$ when carrying out an analysis.

The problem of 'origins'

Although inverse and loglinear functions have advantages for suitable types of data, there is one big disadvantage. If the origin from which x_1 is measured is changed by an amount α_1, so that $(x_1 + \alpha_1)$ is the variable to be used in the model, ordinary polynomials are not affected because only the values of the constants change and the form of the model remains the same. But the whole model is affected with inverse and loglinear functions. Also, and importantly, it is not possible with these functions to use $x=0$ as an experimental point (because there are terms in $1/x$).

The values of $\alpha_1, \alpha_2 \ldots$ in agricultural experiments on (for example) the addition of fertilizers to the soil where a crop is growing, are the amounts already present in the soil: x_1 of nitrogen is applied, α_1 is already available to the plant from the soil, so the plant actually responds to $(x_1 + \alpha_1)$. The values of $\alpha_1, \alpha_2, \ldots$ need to be estimated in addition to the other parameters in the model: for example, we could have

$$\frac{(x+\alpha)}{y} = a + b(x+\alpha) + c(x+\alpha)^2$$

which is equivalent to

$$\frac{1}{y} = \frac{a}{x+\alpha} + b + c(x+\alpha)$$

with only one x-variable being studied, and corresponding results in more complex models. A computer search method may be applied to find α_1. First guess a likely range of values for α_1 from existing knowledge of the system, and then divide this range up into a 'grid' of steps (e.g. if $0 < \alpha_1 \leqslant 1$, then we might use the grid $\alpha = 0.1, 0.2, 0.3, \ldots, 0.9, 1.0$) and compute the fit of a suitable model for each value of α. This involves a considerable amount of computing, especially when there are several factors x, and the grid for α should therefore not be made too fine in the first run, as it can be refined in later stages of the computing. There is of course the added complication that the choice of α can affect the choice of model; provided there are not too many x-variables it may be best to use fully quadratic models initially, and to try simplifying them when suitable orgins $\alpha_1, \alpha_2, \ldots$ seem to have been located.

Size of a model

If an inverse polynomial model is fitted to data from a 4-factor experiment in which 2 factors X, Z are at 2 levels, factor V is at 3 levels and factor W at 4 levels, a suitable model would be linear in x and z, and quadratic in v and w. This could be used as the basic model against which others would be tested. No cubic term in w is usually necessary, as this would influence only the right-hand tail of the response surface. The resulting model is

$$\frac{xzvw}{y} = (a_1 + b_1 x)(a_2 + b_2 z)(a_3 + b_3 w + c_3 w^2)(a_4 + b_4 v + c_4 v^2)$$

where a_i, b_i, c_i are constants. This expands to give a model which contains 36 terms altogether, and so demands considerable computing power to fit it.

If the effect of V is linear only, with no quadratic component (either because V is used only at 2 levels or because we wish to test the hypothesis that V only has a linear effect), there will be no term $c_4 v^2$ in the final bracket in the original model. This will reduce the number of terms in the expanded model to 24, removing from the list above any that contain v^2; the computation will be correspondingly less demanding.

When a log-linear model is fitted instead of an inverse polynomial, the term $xzvw/y$ on the left is replaced by $(xzvw) \log y$.

Clearly, when origins are also to be fitted to a model in several variables, it is vital to have a reasonable idea where to search for the values of $\alpha_1, \alpha_2, \ldots$, as otherwise the whole enterprise can become impossible.

9.10 Exercises

1. Discuss briefly the use of (i) first-order designs and models, and (ii) second-order designs and models, when searching for an optimum value of a response variable.

   A programme of experiments is to be carried out to find the reaction temperature R (°F) and reaction time T (h), which will produce a binder of highest viscosity in developing asphalt. A first-order design used two levels of R and of T, and the resulting viscosities y were measured twice for each combination, as follows:

R	T	Viscosity
300	$\frac{1}{2}$	y_{11}, y_{12}
300	2	y_{21}, y_{22}
340	$\frac{1}{2}$	y_{31}, y_{32}
340	2	y_{41}, y_{42}

After a suitable coding of the levels of R and T, use the method of least squares to fit an appropriate first-order model based on the values of y.

Show that this is a rotatable design, and indicate how to augment it to produce

(i) a central composite design with 3 levels of each factor, and
(ii) a rotatable central composite design.

In each case a maximum of 22 observations can be used.

2. (i) The design matrix and vector of responses for a 2^n experiment are given by

$$
D = \begin{array}{ccc} x_1 & x_2 & x_3 \end{array}
\begin{bmatrix}
1 & -1 & -1 \\
1 & -1 & -1 \\
-1 & 1 & -1 \\
-1 & 1 & -1 \\
-1 & -1 & 1 \\
-1 & -1 & 1 \\
1 & 1 & 1 \\
1 & 1 & 1 \\
0 & 0 & 0 \\
0 & 0 & 0 \\
0 & 0 & 0 \\
0 & 0 & 0
\end{bmatrix}
\qquad
Y = \begin{bmatrix}
55.7 \\
54.4 \\
60.2 \\
60.8 \\
64.3 \\
64.8 \\
68.3 \\
68.9 \\
61.4 \\
61.9 \\
61.8 \\
62.3
\end{bmatrix}
$$

Analyse the data and determine whether or not a first-order design is appropriate. Estimate the coefficients in the response function.

(ii) What are the advantages of augmenting the basic design by the addition of points at the centre? Comment on the design in part (i) and suggest an improvement. What conclusions can be reached about the response function on the basis of the set of data given above?

3. Plants were grown in culture solutions A, B, C which contained zinc at 0.12, 0.16, 0.20 mg/L, respectively. They were also subjected to longer or shorter periods of artificial lighting during growth. Only 12 plants completed their development and produced seed; the seed dry weights per plant were recorded for these. Taking x_1 as the variable for zinc, x_2 for the amount of light and y for the seed dry weight, propose a suitable model and fit it to the data:

Shorter light period:	A	2.4, 2.9, 2.5	(from 3 plants)
	B	3.7, 3.9	(from 2 plants)
	C	3.0	(1 plant);
Longer light period:	A	2.7	(1 plant)
	B	3.2, 3.8, 4.3	(from 3 plants)
	C	3.2, 3.5	(from 2 plants).

4. Write down the design matrix for a $\frac{1}{2}$-replicate of a 2^3 design based on the defining contrast ABC, and augmented by adding 4 observations at the centre, for investigating the relationship of a response y to 3 variables x_1, x_2, x_3. Indicate what models could be studied using this design. How might the number of residual degrees of freedom be increased?

5. A response surface study was carried out to find the conditions on variables x_1, x_2, x_3, x_4 which maximized the response y. A $\frac{1}{2}$-replicate of a 2^4 design was used with the interaction $ABCD$ as the defining contrast. The region of the variables was

	-1	$+1$
x_1:	5	10
x_2:	2	4
x_3:	30	35
x_4:	50	60

Using the usual notation the observations were

$$\begin{bmatrix} (1) & : & 30.5 \\ ab & : & 34.0 \\ cd & : & 28.0 \\ ac & : & 31.8 \\ ad & : & 30.4 \\ bc & : & 32.8 \\ bd & : & 30.6 \\ abcd & : & 32.6 \end{bmatrix}$$

From the data determine the first two steps along the path of steepest ascent using an increment of 2 units of x_1. State briefly what the next stage in the experimental procedure would be. What is the main drawback in using the method of steepest ascent?

10
The analysis of covariance

10.1 Introduction

The observation y at the end of an experiment may depend not only on the treatment applied to the experimental unit, and the block or row or column of the design in which the unit lies, but also on a measurement x recorded on that unit before the experimental treatment was applied. The idea is that if x_{ij} is large on unit (ij) then y_{ij} will also tend to be large; more specifically, a linear regression of y_{ij} on x_{ij} will help to explain the final value of y_{ij}. Combining this with the appropriate linear model for the experimental design being used (completely randomized, randomized block, Latin square, etc.) we have the *Analysis of Covariance*, and x is called the *covariate* or *calibrating variate*. The method can be extended to various other situations in which not all systematic unit-to-unit variation can be removed by blocking, but some of it persists on individual units.

First we consider the case where a value x_{ij} is available for each of the units (i, j); often x is a continuous variate but it does not have to be. In a medical trial, a measurement y which indicates the rate of recovery from an illness may be taken after a standard course of treatment; but the rate of recovery is quite likely to depend on the initial state of the patient, perhaps the severity of the illness or the previous general condition of the patient. We may be able to measure an x which helps to indicate this initial state (sometimes there are several such x's), or we may rely on an assessment by a doctor made on an integer scale of, say, 1 to 5.

In agricultural trials, the rate (%) of germination of seeds will often be a guide to the final crop yield on each plot; in a small trial this may be estimated fairly accurately, while in a larger one a visual estimate will give information accurate enough to be well worth using. Another example from horticulture is in trials with strawberry plants, where a plantation is established for about 3 years of work, during which some plants will die and their neighbours will to some extent expand to fill the spaces. A typical unit plot might consist of 4 rows of 12 plants each, giving 48 plants to receive the experimental treatment. The record x will be the number of plants lost from each plot, and a regression of y, which is the total yield from the surviving plants in the plot, on x is likely to be a better way of correcting for losses than simply multiplying y up by a factor to 'correct' it to 48 plants. Covariance allows for the empty spaces being partly filled in, whereas multiplying up does not; hence the regression coefficient found in the covariance calculation is unlikely to be 1.

For animals taking part in dietary trials, their initial weight when they enter the trial is likely to be a useful covariate, even when they are also grouped into blocks of similar animals such as littermates of the same sex. Variation in animal experiments is often very high, and any way of accounting for more of it in the experimental analysis is valuable.

10.2 Analysis for a design in randomized blocks: general theory

The model behind this analysis will be the usual randomized block model (page 42) with a regression term added:

$$y_{ij} = \mu + \tau_i + \beta_j + \gamma(x_{ij} - x..) + \varepsilon_{ij} \qquad [10.1]$$

where γ is constant and $x..$ is the mean of all the $\{x_{ij}\}$. All other terms have their usual properties.

For the least-squares solution, we have to minimize

$$\sum \sum \{y_{ij} - \mu - \tau_i - \beta_j - \gamma(x_{ij} - x..)\}^2$$

This is simplified by writing

$$\xi_{ij} = x_{ij} - x_i. - x._j + x..$$

now we find that

$$\sum_j \xi_{ij} = 0 = \sum_i \xi_{ij}$$

for all i and for all j, respectively. Also $\sum \sum \xi_{ij}^2$ will be the residual sum of squares in the Analysis of Variance of x (page 43).

In addition to Analyses of Variance for y and x, we require an analysis using cross-product terms xy. The standard notation (page 43) for analysis of a randomized complete block design needs to be extended when two variables x and y are involved. In the Analysis of Variance of y, let us write $G_{(y)}$ instead of G, $T_{i(y)}$ instead of T_i and $B_{j(y)}$ instead of B_j, so that the sum of squares for treatments is

$$\sum_i \frac{T_{i(y)}^2}{r} - \frac{G_{(y)}^2}{N}$$

There will be a similar sum of squares for blocks, and the total sum of squares is

$$\sum_i \sum_j y_{ij}^2 - \frac{G_{(y)}^2}{N}$$

For ease of reference, we shall call these sums of squares T_{yy}, B_{yy} and S_{yy}, respectively, with the residual sum of squares $(=S_{yy} - T_{yy} - B_{yy})$ labelled R_{yy}. There will be exactly similar terms in the variable x.

For an analysis of cross-product terms, we replace squares of y or of x by the corresponding products in xy. Extending the notation in an obvious way,

$$S_{xy} = \sum_i \sum_j x_{ij} y_{ij} - \frac{G_{(x)} G_{(y)}}{N}$$

Likewise

$$T_{xy} = \sum_i \frac{T_{i(x)} T_{i(y)}}{r} - \frac{G_{(x)} G_{(y)}}{N}$$

and

$$B_{xy} = \sum_j \frac{B_{j(x)} B_{j(y)}}{v} - \frac{G_{(x)} G_{(y)}}{N}$$

The sums of squares and products are collected together in the following table (S.S. stands for 'sum of squares' and S.P. for 'sum of products'):

Source	d.f.	S.S. for y	S.P. for xy	S.S. for x
Blocks	$r-1$	B_{yy}	B_{xy}	B_{xx}
Treatments	$v-1$	T_{yy}	T_{xy}	T_{xx}
Residual	$(r-1)(v-1)$	R_{yy}	R_{xy}	R_{xx}
Total	$rv-1$	S_{yy}	S_{xy}	S_{xx}

We noted above that $\sum_i \sum_j \xi_{ij}^2$ is in fact R_{xx}. In the same way, R_{xy} will be $\sum_i \sum_j \xi_{ij}(y_{ij} - y_{i.} - y_{.j} + y_{..})$; however, this reduces to $\sum_i \sum_j \xi_{ij} y_{ij}$ because the other 3 terms each sum to zero. For example, $\sum_i \sum_j \xi_{ij} y_{i.} = \sum_i y_{i.}(\sum_j \xi_{ij})$, which is zero because $\sum_j \xi_{ij} = 0$ for each i.

It is useful to introduce 'adjusted' treatment and block effects, $\tau_i' = \tau_i + \gamma(x_{i.} - x_{..})$ and $\beta_j' = \beta_j + \gamma(x_{.j} - x_{..})$, because this simplifies the linear model [10.1], which becomes

$$y_{ij} = \mu + \tau_i' + \beta_j' + \gamma \xi_{ij} + \varepsilon_{ij} \qquad [10.2]$$

and

$$\sum_i \tau_i' = \sum_i \tau_i + \gamma \sum_i (x_{i.} - x_{..}) = 0$$

since each of the summations on the right-hand side is zero. Similarly, $\sum_j \beta_j' = 0$.

Least-squares minimization is also much simplified by these adjustments; we require $\hat{\mu}$, $\{\hat{\tau}_i'\}$, $\{\hat{\beta}_j'\}$ and $\hat{\gamma}$ found by minimizing

$$\Sigma = \sum_i \sum_j (y_{ij} - \mu - \tau_i' - \beta_j' - \gamma \xi_{ij})^2$$

$$= \sum_i \sum_j (y_{ij} - \mu - \tau_i' - \beta_j')^2 - 2\gamma R_{xy} + \gamma^2 R_{xx} \qquad [10.3]$$

after removing those terms which are zero by virtue of the results developed above. Minimizing the first term leads to the same standard results as in an ordinary randomized block model (page 42), except that it is τ_i' and β_j' that are being estimated. Hence:

$$\hat{\mu} = \frac{G_{(y)}}{N}; \qquad \hat{\tau}_i' = \frac{T_{i(y)}}{r} - \frac{G_{(y)}}{N}; \qquad \hat{\beta}_j' = \frac{B_{j(y)}}{v} - \frac{G_{(y)}}{N}$$

which give

$$\hat{\tau}_i = \hat{\tau}'_i - \hat{\gamma}(x_i. - x..)$$

$$= \frac{T_{i(y)}}{r} - \frac{G_{(y)}}{N} - \hat{\gamma}(x_i. - x..)$$

and

$$\hat{\beta}_j = \hat{\beta}'_j - \hat{\gamma}(x._j - x..)$$

$$= \frac{B_{j(y)}}{v} - \frac{G_{(y)}}{N} - \hat{\gamma}(x._j - x..)$$

Finally,

$$\frac{\partial \Sigma}{\partial \gamma} = 0$$

gives

$$\hat{\gamma} = \frac{R_{xy}}{R_{xx}}$$

The treatment means in the adjusted model including the covariance term are

$$\hat{\mu} + \hat{\tau}_i = \frac{T_{i(y)}}{r} - \hat{\gamma}(x_i. - x..)$$

where $\hat{\gamma} = R_{xy}/R_{xx}$; these adjusted means are often called $y'_i.$. The adjustment may be positive or negative; if $\hat{\gamma}$ is positive, the adjustment will reduce the 'raw' means in those cases where $x_i. > x..$, i.e. for treatments applied to units for which the mean x-value $x_i.$ is greater than the overall mean $x..$, and will increase the 'raw' means for treatments having smaller-than-average $x_i.$. If $\hat{\gamma}$ is negative, so that a large x goes with a small y and vice versa, the adjustments go the opposite way.

From the form of Σ in [10.3], we see that the residual sum of squares, after removing the contributions from all the parameters including γ, is $R_{yy} - 2\hat{\gamma}R_{xy} + \hat{\gamma}^2 R_{xx}$. This will have $(r-1)(v-1) - 1$ degrees of freedom, 1 d.f. having been taken up in estimating γ; and by the usual rules for a simple linear regression, this single degree of freedom accounts for the sum of squares R^2_{xy}/R_{xx}.

Steps in the analysis

(i) First we wish to know whether the covariance term is really needed in the model. This may be checked by an *F*-test to compare R^2_{xy}/R_{xx} with the residual mean square, which is (again using standard regression results) $R_{yy} - (R_{xy}^2/R_{xx})$ divided by $\{(r-1)(v-1)-1\}$; this residual mean square is also the estimate of σ^2, the variance of ε_{ij} in [10.1]. This first step may be summarized in the table:

	d.f.	Sum of squares	Mean square	*F*
Covariance	1	R_{xy}^2/R_{xx}		Tests need of γ
Residual	$\{(r-1)(v-1)-1\}$	(by subtraction)	Estimates σ^2	
	$(r-1)(v-1)$	R_{yy}		

When the *F*-test, with 1 and $\{(r-1)(v-1)-1\}$ degrees of freedom, gives a value that is not significant, the covariance is not making any worthwhile reduction in variability; therefore it may be abandoned and the analysis carried out as an ordinary Analysis of Variance for *y*.

(ii) The second step, when the *F*-test for covariance is significant, uses the 'extra sum of squares' principle to compare models with and without $\{\tau_i\}$. When $\{\tau_i\}$ is omitted from the model, the sums of squares and products for *T* have to be transferred to *R*, to give $(T+R)$ as the residual. A 'sum-of-squares for covariance' can be extracted from $T_{yy} + R_{yy}$ as $(T_{xy} + R_{xy})^2/(T_{xx} + R_{xx})$, and the remainder after extracting it may be called 'adjusted treatments-plus-residual'. The residual from step (i) above is the 'adjusted residual' and the difference is 'adjusted treatments'. Example 10.1 below shows details of this analysis.

(iii) The third step is to test differences between means. In order to calculate adjusted means $y'_i.$, we need $\hat{\gamma} = R_{xy}/R_{xx}$ and $x.. = G_{(x)}/N$. The adjustment from the 'raw' means $y_i.$ to the adjusted means $y'_i.$ is best set out as a table, as in Example 10.1. We apply standard regression results again to find the standard error of the difference between two adjusted treatment means; the standard errors will be different for each pair of means because $x_i.$ changes from one treatment to another. For comparing treatments *A* and *B*, the standard error is

$$\sqrt{\hat{\sigma}^2\left(\frac{2}{r} + \frac{(x_A. - x_B.)^2}{R_{xx}}\right)}$$

The adjusted residual mean square gives $\hat{\sigma}^2$, and we assume that each treatment is replicated *r* times.

Example 10.1. An experiment on maize was laid out as 5 randomized complete blocks, with 5 treatments *A–E*, and the measurement *y* to be analysed is the crop weight. There was a maize crop the previous year on the same field, split into exactly the same units, and the crop weights *x* are available from that planting. No different treatments were used, so *x* may be taken as indicating the basic fertility of the different units in the field.

The complete table of records is:

Treatment	A		B		C		D		E		Block totals	
	y	*x*	*y*	*x*	*y*	*x*	*y*	*x*	*y*	*x*	*y*	*x*
Block I	33.7	53.0	37.1	51.2	33.7	47.9	34.3	48.0	32.6	44.8	171.4	244.9
II	36.5	62.8	40.7	60.9	42.7	56.7	42.7	52.9	51.3	59.4	213.9	292.7
III	37.1	62.0	44.4	59.9	51.0	62.8	45.8	61.5	56.5	59.2	234.8	305.4
IV	37.1	63.5	40.4	54.7	50.2	66.2	44.0	62.2	50.4	56.2	222.1	302.8
V	37.3	64.0	46.2	69.5	49.0	61.2	25.8	47.9	47.0	54.1	205.3	296.7
Treatment totals	181.7	305.3	208.8	296.2	226.6	294.8	192.6	272.5	237.8	273.7	1047.5 $(G_{(y)})$	1442.5 $(G_{(x)})$
N=25												

For the Analysis of Variance of y,

$$S_0 = \frac{1047.5^2}{25} = 43890.25; \quad S = 45198.85; \quad S_B = \frac{221738.71}{5} = 44347.742$$

$$S_T = \frac{221603.49}{5} = 44320.698$$

These values lead to the first column of figures in the table below.
    Similarly for x,

$$S_0 = \frac{1442.5^2}{25} = 83232.25; \quad S = 84219.91; \quad S_B = \frac{418637.19}{5} = 83727.438$$

$$S_T = \frac{417017.51}{5} = 83403.502$$

which lead to the third column.
    The sum of products for treatments is

$$\tfrac{1}{5}\{(181.7 \times 305.3) + (208.8 \times 296.2) + (226.6 \times 294.8) + (192.6 \times 272.5)$$
$$+ (237.8 \times 273.7)\} - \tfrac{1}{25}(1047.5 \times 1442.5)$$

$$= \frac{301690.61}{5} - \frac{1511018.75}{25} = 60338.122 - 60440.75 = -102.628$$

Although a sum of squares is bound to be positive, this is not so for a sum of products (as we know from calculating correlation and regression coefficients).
    For blocks, we have

$$\tfrac{1}{5}\{(171.4 \times 244.9) + (213.9 \times 292.7) + (234.8 \times 305.4) + (222.1 \times 302.8) + (205.3 \times 296.7)\}$$
$$- 60440.75 = 60891.34 - 60440.75 = 450.590$$

The total sum of products is

$$\sum_i \sum_j y_{ij} x_{ij} - G_{(y)} G_{(x)} / N = 61062.43 - 60440.75 = 621.680$$

The results may now be set out in a table, the row for 'residual' being completed by taking (total − blocks − treatments) for each column.

Source	d.f.	S.S. for y	S.P. (xy)	S.S. for x
Blocks	4	$B_{yy} = 457.492$	$B_{xy} = 450.590$	$B_{xx} = 495.188$
Treatments	4	$T_{yy} = 430.448$	$T_{xy} = -102.628$	$T_{xx} = 171.252$
Residual	16	$R_{yy} = 420.660$	$R_{xy} = 273.718$	$R_{xx} = 321.220$
Total	24	$S_{yy} = 1308.600$	$S_{xy} = 621.680$	$S_{xx} = 987.660$

$$\frac{R_{xy}^2}{R_{xx}} = \frac{273.718^2}{321.220} = 233.241$$

which we use to test whether a covariance adjustment will reduce the estimate of residual variation.

	d.f.	Sum of squares	Mean square	
Covariance	1	$\dfrac{R_{xy}^2}{R_{xx}} = 233.241$	233.241	$F_{(1,15)} = 18.67$***
Adjusted residual	15	187.419	$12.495 = \hat{\sigma}^2$	
Residual	16	$R_{yy} = 420.660$		

The *F*-test indicates that the covariance adjustment is very useful. The sums of squares and products for $(T + R)$ are

$$T_{yy} + R_{yy} = 851.108, \qquad T_{xy} + R_{xy} = 171.090, \qquad T_{xx} + R_{xx} = 492.472$$

From these, $(T_{xy} + R_{xy})^2 / (T_{xx} + R_{xx}) = 59.439$. The 'adjusted treatments-plus-residual' may again be shown in a table.

	d.f.	Sum of squares
Covariance adjustment	1	59.439
Adjusted treatments-plus-residual	19	791.669
Treatments-plus-residual	20	851.108

Having now 'adjusted' the sums of squares for treatments plus residual and for residual, the adjusted sum of squares for treatments can be found, and an *F*-test for treatments against adjusted residual may be carried out if required.

Adjusted analysis

Source	d.f.	Sum of squares	Mean squares	
Treatments	4	604.250	151.063	$F_{(4,15)} = 12.09$***
Residual	15	187.419	12.495	
Treatments-plus-residual	19	791.669		

If we are interested in the Null Hypothesis that all treatments have a similar effect, the $F_{(4,15)}$ test provides very strong evidence against it; we are, however, more likely to want to compare particular means or to extract certain contrasts out of the complete sum of squares for treatments as discussed further below.

In comparing means, we use $\hat{\sigma}^2 = 12.495$ from the adjusted analysis; covariance has substantially reduced this, because without the adjustment we should be using $R_{yy}/16 = 26.291$ as the estimate. The information on cropping potential given by x has removed a systematic unit-to-unit component of variation, and what now remains as $\hat{\sigma}^2$ should be a genuine estimate of random natural variation.

Adjusted means
We require

$$x.. = \frac{G_{(x)}}{N} = \frac{1442.5}{25} = 57.70$$

and also the estimate

$$\hat{\gamma} = \frac{R_{xy}}{R_{xx}} = \frac{273.718}{321.220} = 0.852$$

The calculations of $y'_i. = y_i. - \hat{\gamma}(x_i. - x..)$ may again be set out as a table.

Treatment	$y_i.$	$x_i.$	$-\hat{\gamma}(x_i. - x..)$ $= -0.852(x_i. - 57.70)$	$y'_i.$
A	36.34	61.06	−2.86	33.48
B	41.76	59.24	−1.31	40.45
C	45.32	58.96	−1.07	44.25
D	38.52	54.50	+2.73	41.25
E	47.56	54.74	+2.52	50.08
			+0.01	

As a check, the third column of figures should add up to zero (save for a very small rounding error).

It seems that, by an accident of the particular randomization used for the design, treatments D and E were both put on to plots where x was below average, with the other treatments – A in particular – being on better-than-average plots as measured by x. The means for D and E are adjusted upwards to compensate.

Suppose that we have some valid reason for comparing treatments A and D; they may have been hand-weeded at different stages in the season, to remove competition for nutrients by keeping the ground clean. The standard error for the difference between these means is

$$\sqrt{\hat{\sigma}^2\left(\frac{2}{5} + \frac{(x_A. - x_D.)^2}{R_{xx}}\right)} = \sqrt{12.495\left(0.4 + \frac{6.56^2}{321.22}\right)} = \sqrt{6.6719} = 2.583$$

The difference between their adjusted means is 7.77 (D being the larger) and so 7.77/ 2.583 = 3.01 follows $t_{(15)}$, the d.f. being the same as those for $\hat{\sigma}^2$. The value 3.01 is clearly significant and gives evidence that A and D do not have the same mean. Before adjustment, they were close to one another and would have appeared to have effectively the same mean.

In order to see how precisely estimates can be made, a confidence interval will be better than a significance test. The adjusted difference between D and A is 7.77 and its standard error is 2.583, as we have just found. The rule 'observed difference $\pm t$ times its standard error' thus gives 7.77 $\pm$ 2.131 $\times$ 2.583, using the 5% critical value of $t_{(15)}$. The 95% confidence interval is 7.77 $\pm$ 5.50, or (2.27 to 13.27), which is very wide and shows that even after adjustment there is a relatively large residual variation.

10.3   Individual contrasts

The adjustment made by considering $(T + R)$ compared linear models with and without the treatment term τ_i. Rather than remove the whole treatment effect, we may wish to consider one contrast from it (or more than one, though examining each separately). For example, a contrast of the type discussed in Chapter 4 may be important, or a main effect in a factorial experiment.

Let us call the contrast C. An analysis has to be carried out in which C is calculated for both sums of squares, C_{yy} and C_{xx}, and for C_{xy}, by exactly the same method as described above. To illustrate this, suppose that a 3-level factor is to have its linear and quadratic components of response studied. The 2 d.f. for this factor will be the contrasts $L(-1, 0, 1)$ and $Q(1, -2, 1)$, and *values* of L and Q can be calculated for the covariate x just as for the response variable y, leading to L_{xx}, L_{yy}, Q_{xx} and Q_{yy} by the usual calculation (page 91). Remembering that for the products we must replace squares in one variable by the corresponding products for the two variables, L_{xy} will be (value of L for x) $\times$ (value of L for y) divided by the appropriate divisor, which is $2r$ when each factor level is replicated r times. Similarly Q_{xy} is the product of the two 'values' divided by $6r$.

Example 10.2. As a very simple illustration (such an experiment would be too small for most purposes) suppose that factor A is at 3 levels, a_0, a_1 and a_2, each replicated 4 times. A covariate x is recorded as well as the response variable y, and the results are summarized in the following table, which also shows 'values'.

Treatment	a_0	a_1	a_2	Value	Divisor
Total y	28	40	56		
L_y	−1	0	1	28	8
Q_y	1	−2	1	4	24
Total x	30	35	45		
L_x	−1	0	1	15	8
Q_x	1	−2	1	5	24

$G_{(y)}=124$, so the sum of squares for treatments for y is

$$\tfrac{1}{4}(28^2+40^2+56^2)-124^2/12=98.667$$

$L_{(y)}=28^2/8=98.000$ and $Q_y=4^2/24=0.667$, which add to 98.667. $G_{(x)}=110$, so the sum of squares for treatments for x is

$$\tfrac{1}{4}(30^2+35^2+45^2)-\frac{110^2}{12}=29.167$$

$L_{(x)}=15^2/8=28.125$ and $Q_x=5^2/24=1.042$, which add to 29.167. The sum of products for treatments is

$$\frac{(28\times30)+(40\times35)+(56\times45)}{4}-\frac{124\times110}{12}=1190-1136.667=53.333$$

To divide this into linear and quadratic components, 'values' are multiplied: $L_{(xy)}=(28\times15)/8=52.500$ and $Q_{(xy)}=(4\times5)/24=0.833$, which add to 53.333.
   Either L or Q can now be taken (or both in turn) as C.
   Once the necessary sums of squares and products have been found for C, the analysis previously described for $(T+R)$ can be carried out for $(C+R)$. For a single-degree-of-freedom contrast, of course, only an F-test is required to examine its importance; no adjustment of means need be done.

10.4 Dummy covariates

If some accident affects one or more plots, and it bears no relationship to the experimental treatments, a dummy $(0,1)$ x-variable may be set up which takes the value 1 on affected plots and 0 elsewhere. When more than one unit plot is affected, we must be satisfied that there is not likely to be any substantial difference in the effect of the accident.

Example 10.3. The length of time (s) taken for 4 patients I, II, III, IV to react to 3 different formulations of a sedative drug A, B, C has been measured. Samples of the drug have been analysed, and those marked * contained a similar impurity. Is there evidence that this affects the reaction time? Compare the reaction times to formulations A, B and C.

Patient	I	II	III	IV
Drug A	11	56	15	6
B	26	83	34	13*
C	20*	71*	41	32

We will use a dummy covariate with values 1 where a result is starred and 0 elsewhere.

Patient	I		II		III		IV		Drug totals	
	y	x	y	x	y	x	y	x	y	x
Drug A	11	0	56	0	15	0	6	0	88	0
B	26	0	83	0	34	0	13	1	156	1
C	20	1	71	1	41	0	32	0	164	2
Patient totals	57	1	210	1	90	0	51	1	408	3

The Analysis of Variance for y is carried out in the usual way, that for x is particularly simple, and the sums of products (xy) are found as described above (pages 195–6).

Source	d.f.	SS. (y)	S.P. (xy)	S.S. (x)
Patients	3	5478	4	0.25
Drugs	2	872	19	0.50
Residual	6	332	−21	1.50
Total	11	6682	2	2.25

$R_{xy}^2/R_{xx} = (-21)^2/1.5 = 294$. Also $R_{xy}/R_{xx} = -14$.

(i) *Test for evidence of effect on reaction time*

Source	d.f.	Sum of squares	Mean square	
Covariance	1	294	294	$F_{(1,5)} = 38.68^{**}$
Adjusted residual	5	38	7.6	
Residual	6	332		

There is evidence of an effect on reaction time, since covariance using the dummy x significantly reduces variability.

(ii) *Adjusted analysis*
$D_{xy} + R_{xy} = -2$, $D_{xx} + R_{xx} = 2$, so the adjustment to $D_{yy} + R_{yy}$ is $(-2)^2/2 = 2$. The unadjusted value of $D_{yy} + R_{yy}$ is 1204.

Source	d.f.	Sum of squares
Covariance	1	2
Adjusted drugs + residual	7	1202
Drugs + residual	8	1204

We can now complete the adjusted analysis.

Source	d.f.	Sum of squares	Mean square	
Adjusted drugs	2	1164	582	$F_{(2,5)} = 76.6$***
Adjusted residual	5	38	$7.6 = \hat{\sigma}^2$	
Adjusted drugs + residual	7	1202		

Clearly there are large differences among the drugs.
To compare means, $\hat{\gamma} = -14$ and $x.. = 3/12 = 0.25$.

Drug	$y_i.$	$x_i. - x..$	$+14(x_i. - x..)$	$y'_i.$
A	22.0	−0.25	−3.5	18.5
B	39.0	0	0	39.0
C	41.0	+0.25	+3.5	44.5

No specific comparisons are asked for, but because A is so different from B and C it may be useful to compare B and C first. The standard error of the difference between their means is

$$\sqrt{7.6\left(\frac{2}{4} + \frac{(x_B. - x_C.)^2}{1.5}\right)} = \sqrt{7.6\left(0.5 + \frac{0.25^2}{1.5}\right)} = 2.029$$

The difference in means is 5.5; so $5.5/2.029 = 2.71$, which we test as $t_{(5)}$. There is some evidence of a real difference. If B and C are different, clearly A is different from both. However, a 95% confidence interval for the difference of means $(C - B)$ is very wide: $t_{(5)} = 2.571$ at the 5% critical point, so the limits are $5.5 \pm 2.571 \times 2.029$, or 5.50 ± 5.22, i.e. (0.28 to 10.72).

10.5 Systematic trend not removed by blocking

Sometimes the design of an experiment overlooks a source of systematic variation which should have been removed. Fertility trends in field experiments are an obvious example: a completely randomized layout which really needed blocks, or a randomized block which required rows and columns. For example, Fig. 10.1 shows a field layout in 5

BLOCK	I		II		III		IV		V	
	A	−2	C	0	B	−3	D	1	B	−2
	D	−3	A	−1	C	1	B	0	D	1
	C	1	D	0	A	1	A	−2	C	−3
	B	4	B	1	D	1	C	1	A	4

Fig. 10.1 Residuals in a randomized block layout

randomized blocks, with the residuals calculated for each plot. Because the columns of the layout formed blocks, the residuals sum to zero in columns as they do for treatments. In rows, however, the sums of residuals are respectively -6, -2, -3, $+11$.

There is a strong suspicion that a design that removed row-to-row variability ought to have been used. We may examine the need for this by using a covariate x such that $x = 6$ for every plot in the first row, $x = 2$ everywhere in the second row, $x = 3$ in the third row, and $x = -11$ in the fourth row.

10.6   Accidents in recording

Accurate recording of the responses from experimental units is a vital part of any study, but occasionally accidents occur which can be partly redeemed by tricks in analysis.

Suppose that *data from two units have been combined*, so that the total value $y_1 + y_2$ is the only information available from these units. This total value is ascribed to one of the units, 0 to the other, and a dummy covariate x is defined which is set equal to $+1$ and -1 on these units, respectively, and to 0 on all other units. Thus $x..$ is 0 in this method.

There is a related problem which can be handled more simply. Two values of y may not be properly labelled or identified, so that either may have come from either of two units, although we do still have two separate values. In this situation, two analyses may be performed, with the allocation of the y values done in both the possible ways. We will accept the analysis that leads to smaller residuals. If the result of doing these two analyses is inconclusive, the best procedure is to omit the two y-values that are in question and analyse the remaining data from the experiment; this will require a 'general block design' analysis (Chapter 11) if there are blocks, but will still be straightforward in a completely randomized scheme.

Especially in experiments with long-term crops and trees, covariance can be extremely useful in accounting for variation that occurs in an irregular way, and Pearce *et al.* (1988) and Dyke (1988) discuss applications of covariance thoroughly.

10.7   Assumptions in covariance analysis

The model [10.1] contains only one regression coefficient, γ; the relationship between y_{ij} and x_{ij} is independent of treatments and of blocks (or rows and columns). This has to be so, because the aim of covariance is to control unit-to-unit variation which would otherwise remain part of the term ε_{ij} in [10.1]. In most situations, it is probably reasonable to assume independence, but some graphical assistance can be useful. Cox (1992) gives a thorough discussion of the assumptions implicit in a covariance analysis; we illustrate a few important points in Fig. 10.2.

In Fig. 10.2(a) the relationship between y_{ij} and x_{ij} does indeed seem to be the same for both treatments, where there is also evidence of a treatment difference; Fig. 10.2(b) shows no treatment difference but a relationship where γ is the same for both treatments. Figure 10.2(c) illustrates a position where the values of $y_i.$ would be different, but where after correction for covariance those of $y_i'.$ would not. We have to keep a look out for the possibility shown in Fig. 10.2(d), where γ may be different for the two treatments, because in that case model [10.1] does not apply.

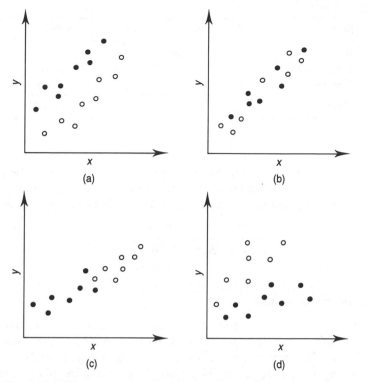

Fig. 10.2 (a) Treatment difference; same value of regression coefficient. (b) and (c) No treatment difference; same value of regression coefficient. (d) Treatment difference; different values of regression coefficient. Key: ● Treatment *A*, ○ Treatment *B*

10.8   Missing values

One of the most important uses of dummy variables in covariance is in 'estimating' missing observations (page 69). A computer-based non-orthogonal block analysis will in practice often be carried out when an observation is lost from a randomized block, but the older method using covariance is a good one.

We considered an example on page 69, and will use the same data again. The empty space in y is filled by 0; in the corresponding position in x we place -1, and everywhere else in x we place 0. By carrying through the usual Analysis of Covariance, $\hat{y}$ gives the 'estimate' m, which is the same as we found by minimizing the residual sum of squares.

y Block	I	II	III	IV	Total	x	I	II	III	IV	Total
Treatment *A*	0	21	21	18	60	*A*	−1	0	0	0	−1
B	13	15	19	13	60	*B*	0	0	0	0	0
C	12	14	17	13	56	*C*	0	0	0	0	0
D	11	18	15	20	64	*D*	0	0	0	0	0
	36	68	72	64	240		−1	0	0	0	−1

Analysis of Covariance:

Source	d.f.	S.S. (y)	S.P. (xy)	S.S. (x)
Blocks	3	200	6	3/16
Treatments	3	8	0	3/16
Residual	9	190	9	9/16
Total	15	398	15	15/16

$$\hat{\gamma} = R_{xy}/R_{xx} = 16$$

Although we have the same estimate m, the covariance method does better than the earlier approximate analysis; we leave the reader to check that completing the covariance analysis gives the *exact* correction (page 70). This is shown by the fact that

$$\frac{T_{xy} + R_{xy}}{T_{xx} + R_{xx}} = \frac{9}{12/16} = 12$$

10.9   Double covariance

Sometimes two (or more) x's suggest themselves; in medical work, in particular, this may be the best way of controlling variability even when a blocking system has been used. If two covariates, say x and w, are included in a linear model, we have the equivalent of fitting a multiple (2-variable) regression and the calculations for this part of the analysis are exactly the same as for multiple regression. By hand, we would require sums of squares for y, x and w, and products xy, wy and wx. Clearly this is tedious, though less so if at least one of x, w is a dummy covariate; a computer program for multiple covariance will save much time and effort, and so encourage the method to be used whenever it might be helpful. Two (or more) missing values can be handled by taking x as a $(-1, 0)$ variate corresponding to one of the missing values and w as a $(-1, 0)$ variate corresponding to the other.

10.10   Exercises

1. An experiment was carried out to investigate the effect of 2 methods of pruning and the use of 3 different insecticides on the crop of apples from a particular variety of apple tree. The treatments consisted of the 6 combinations of pruning method and insecticide and these were assigned to trees in 4 randomized complete blocks. The observation from each tree consisted of the total crop, in pounds, from 4 consecutive years. An additional measurement on each tree was also available: the total crop, in bushels, from the 4 years preceding the experiment, when no differential treatments were applied.

   Some summary statistics from the results of the experiment are presented below. The labels y and x refer to the experimental and preceding crop, respectively. Using an analysis that you consider appropriate, draw and present your conclusions on the comparative effect of the different treatments. (Any assumptions needed for your analysis to be valid should be stated clearly.)

Totals over all blocks

Pruning method (P)		1			2	
Insecticide type (I)	1	2	3	1	2	3
Crop (pounds) (y)	1069.5	1159.5	1165.4	990.5	920.8	1167.7
Preceding crop (x) (bushels)	33.8	33.4	33.4	31.7	29.9	37.2

Corrected sums of squares and cross products

	y^2	x^2	xy
Blocks	23071.14	25.632	1033.22
P main effect	3911.96	0.135	22.98
I main effect	6190.50	3.506	137.75
PI interaction	4006.29	3.752	107.56
Total	97450.35	57.258	2073.60

2. Carry out an Analysis of Covariance on the yield data, y (kg) of a crop, adjusted to take out the different effects of shelter as assessed by a score x taken before the treatments A–D were applied to the crop.

Layout				Yields, y				Scores, x			
A	B	C	D	15.0	15.5	18.3	18.8	10	11	10	9
D	C	A	B	19.1	16.6	12.6	12.9	9	12	10	8
B	A	D	C	14.8	13.3	19.8	15.0	7	12	11	8
C	D	B	A	18.5	18.0	15.0	13.0	9	8	8	10

3. Twelve cages each containing 5 male and 5 female mice are used in an experiment to test 4 treatments. The amount of food eaten by the mice in a cage in 4 weeks is the response to be analysed, and it is known that this is likely to be greater for males than for females. Unfortunately, some of the mice die at an early stage in the experiment (in the first 2 days) and cannot be replaced. The figures below give the total food eaten per cage suitably coded, together with the surviving numbers of males and females, respectively, shown in brackets. Test whether the treatments have a significant effect.

Treatment						
A	20	(5, 5)	24	(5, 5)	21	(5, 5)
B	14	(5, 4)	8	(4, 5)	18	(5, 5)
C	11	(5, 4)	14	(5, 5)	4	(3, 5)
D	13	(4, 4)	20	(5, 5)	16	(5, 5)

4. (i) Describe the uses of the Analysis of Covariance in designed experiments.
   (ii) In a randomized blocks experiment an observation was lost by accident and the remaining values were as given below:

	Treatment				
	A	B	C	D	E
Block 1:	43	53	38	47	39
Block 2:	48	50	36	*	36
Block 3:	50	57	43	50	38
Block 4:	42	52	39	49	35

Discuss alternative ways in which the data may be analysed. Estimate the missing value and obtain the Analysis of Variance through the covariance technique. Calculate the adjusted means for the 5 treatments.

If a further observation had been lost in block 3 describe how you would analyse the data if the second missing value were for

(a) treatment C,
(b) treatment D.

5. (i) Discuss methods that are available for the analysis of data from randomized complete block experiments that have missing observations.
   (ii) Show that the estimated value of a single missing observation in a randomized complete block experiment with t treatments and b blocks, which is chosen to minimize the residual sum of squares, is equal to

$$\frac{tT' + bB' - G'}{(b-1)(t-1)}$$

where T', B' and G' are totals for the treatment with an observation missing, for the block with an observation missing and for the whole experiment, respectively.

   (iii) The data below are from a randomized complete block experiment with 4 blocks (I, II, III, IV) and 5 treatments (A, B, C, D, E). The '*' represents a missing observation.

		Treatments					
		A	B	C	D	E	Total
Blocks	I	14	*	19	18	19	70
	II	15	14	19	16	18	82
	III	14	12	16	15	18	75
	IV	16	17	20	16	21	90
Total		59	43	74	65	76	317

Estimate the missing value and construct the corresponding approximate Analysis of Variance table.

(iv) Apply an Analysis of Covariance method to obtain a more accurate analysis. Use this to set 95% confidence limits to the difference in means of *B* and *E*.

6. Four treatments *A–D* were used in 6 blocks for an experiment on a wheat crop. The crop weight (*y* kg) was analysed, and the previous year's crop (*x*) was used as an indication of soil fertility on each plot. Analyse the results and report on them, examining in particular

(i) the difference between the means for *C* and *D*,
(ii) the comparison of *A* with (*B*, *C* and *D*).

Treatment	*A*		*B*		*C*		*D*	
Block	*y*	*x*	*y*	*x*	*y*	*x*	*y*	*x*
I	24.4	33.0	26.0	29.5	26.5	20.3	21.6	17.5
II	25.5	28.0	30.7	19.9	29.0	18.6	31.3	19.0
III	27.0	24.7	32.2	21.3	40.6	29.9	39.4	22.3
IV	27.1	26.1	30.4	22.4	37.7	32.0	32.8	22.4
V	27.4	20.3	35.3	26.2	38.5	27.5	36.6	21.0
VI	26.8	22.5	31.6	21.1	35.3	20.7	27.3	20.0

7. An experiment was conducted in each of 4 workshops, I–IV, using the same batch of raw material supplied at the beginning of the day. Five different ways, *A–E*, of processing the material were used. Initially it had not been supposed that the raw material itself would deteriorate if kept for a longer time prior to use; however, the person in charge of the experiment became convinced during the day that this *was* happening.

The results below show the processing treatments in the order they were applied, together with a measure of the quality of each final product on a scale of 1–100. Examine whether a linear time trend will improve the analysis of these data. Complete the analysis and write a report.

Workshop	I		II		III		IV	
Time 1	*A*	75	*B*	78	*B*	80	*A*	84
2	*D*	72	*E*	73	*D*	75	*C*	78
3	*B*	64	*A*	67	*C*	71	*E*	72
4	*E*	59	*D*	61	*A*	63	*D*	64
5	*C*	52	*C*	54	*E*	55	*B*	58

11
Balanced incomplete blocks and general non-orthogonal block designs

11.1 Introduction

Experiments often have to use material that is not homogeneous, but can be grouped into blocks of similar units so that much of the heterogeneity is taken out between blocks. So far we have used randomized complete blocks, and their extensions such as Latin squares, to help control heterogeneity whatever structure may be present in the treatments, and in addition we have used confounding when there is a factorial structure.

Sometimes blocks have to be very small, 2 units if twin animals of the same sex are used in a dietary trial or if a person's two arms are used to compare treatments for a skin complaint; not more than 4 or 5 if members of a tasting panel are scoring different foods for general appeal and quality at a single session. In research that needs complicated laboratory work, only very small numbers of treatments can be dealt with at one time; different times would be the blocks. There may well be more treatments to include in an experiment than there are units in the blocks available. One of the earliest designs proposed for solving this problem was the *balanced incomplete block*.

There are now available general computer programs for analysing any design with 'incomplete' blocks, i.e. where the block size is less than the number of treatments, and these will also deal with experiments that include blocks of different sizes. It may therefore seem unnecessary to single out balanced incomplete blocks for special mention before moving on to consider general block designs later in the chapter. However, the property of *balance*, more specifically called *total balance*, which they possess is a useful one to remember when designing experiments, because it is the next most efficient simple form of design after the orthogonal randomized complete block.

Any block design in which some treatments are missing from some blocks is *non-orthogonal*, because the treatment parameters can no longer be estimated independently of the block parameters. Obviously we must adjust the estimates of treatment effects to allow for the possibility that some treatments are missing from better blocks and others are missing from poorer blocks. This is shown in the analysis, when we find that T_i (in the usual notation) has to be replaced by a value Q_i for each treatment, which is T_i adjusted to reflect which blocks treatment i actually appeared in.

11.2 Definition and existence of a balanced incomplete block

A balanced incomplete block is a design in which the standard error of the difference between their means is the same for *any* pair of treatments; this is achieved by making every pair of treatments occur together in blocks the same number of times (the number of *concurrences* of the pair); in the standard notation this is called λ. Blocks in a simple balanced incomplete block are the same size, k, although this condition can be relaxed without losing total balance (page 230).

We will use the notation

v = number of treatments
b = number of blocks
r = number of replicates of each treatment
k = number of units in each block
N = total number of units in the experiment
λ = number of times each pair of treatments occurs together in
     a block (number of *concurrences*)

There are restrictions on the existence of balanced incomplete block designs, because the numbers r, v, b, k, N, λ have to satisfy some conditions for a design to be possible. By considering the total number of units in the design, clearly

$$rv = bk = N \qquad [11.1]$$

Now consider one particular treatment. It occurs r times and therefore must appear together with others in a block $r(k-1)$ times in total. But it also appears λ times with each of the other $(v-1)$ treatments. Hence the total number of concurrences of all pairs of treatments is $\lambda(v-1) = r(k-1)$ so that

$$\lambda = \frac{r(k-1)}{v-1}$$

Obviously by its definition λ must be an integer, so we have

$$\lambda = \frac{r(k-1)}{v-1} \quad \text{is an integer} \qquad [11.2]$$

There is no guarantee that a design can actually be found even when these conditions are satisfied, and although many different methods of constructing balanced incomplete blocks have been used, there are still some 'unsolved problems' – in theory, with a given set of numbers, a design can exist but one has not been found. The table below shows the most useful designs with N not excessively large. Cochran & Cox (1992) provide plans for the most common designs, and Cox (1992) gives information about the existence of designs for various values of v, b etc.

Useful small sizes of balanced incomplete block design

v	k	Number of units	
4	2	12	(two replicates needed)
4	3	12	(two replicates needed)
5	2	20	
5	3	30	
5	4	20	
6	2	30	
6	3	30	

v	k	Number of units
6	5	30
6	3	60
6	4	60
7	3	21
7	4	28
7	2	42
7	6	42

11.3 Methods of construction

(i) A design will always exist when $k = (v-1)$, provided we use v blocks; an example is

$$\text{I } ABC; \quad \text{II } ABD; \quad \text{III } ACD; \quad \text{IV } BCD$$

It has $k = 3$, $v = b = 4$, so that $r = (v-1) = 3$, $N = 12$ and $\lambda = (v-2) = 2$. This layout illustrates a good way of seeing total balance: each treatment has just one block that is in a special relationship with it, namely it does not appear in that block – D is related to I, C to II, B to III, A to IV. N is only 12 here, so two complete replicates of this scheme would be needed for a satisfactory experiment; and of course the actual layout within each block must be properly randomized before it is used in an experiment.

Any design in which $b = v$ and $r = k$ is called *symmetric*

(ii) A balanced incomplete block design can also be made up by using all the possible pairs out of the v treatments; this number is $\binom{v}{2}$, so $b = \binom{v}{2}$, which will be quite a large number if v is at all large. By definition, $\lambda = 1$, and also $k = 2$, so that we require $N = v(v-1)$ units altogether and $r = (v-1)$. This type of design, using all possible pairs, is called *unreduced*.

An example for $v = 5$ (treatments A, B, C, D, E) is to use as the 10 blocks $AB, AC, AD, AE, BC, BD, BE, CD, CE, DE$. Allocation of treatments between the two members of each block will be made at random, for example by tossing a coin.

(iii) A design *complementary* to a known balanced incomplete block will also be a balanced incomplete block; its blocks contain all the treatments that are not in the corresponding blocks of the known design. For example, the design complementary to the example given in (ii) above has blocks of size 3, and these are $CDE, BDE, BCE, BCD, ADE, ACE, ACD, ABE, ABD, ABC$. Given the values of r, v, b, k, N, λ in the known design, the new ones are

$$r^* = b - r, \quad v^* = v, \quad b^* = b, \quad k^* = v - k, \quad \lambda^* = \frac{(b-r)(v-k-1)}{(v-1)}$$

(iv) In a *resolvable* design, the blocks may be arranged in groups, each group consisting of one complete replicate of the whole set of treatments; therefore there must be r groups in all. For all resolvable designs, $b \geqslant v + r - 1$; when there is equality they are called *affine resolvable*.

A construction method for resolvable designs is as follows. By the nature of the design, k^2/v is an integer, and in the case $k^2 = v$ we can use a completely orthogonalized $k \times k$ square which will lead to a design with $\lambda = 1$, $r = k+1$, $b = k(k+1)$. If we take $k = 3$, $v = 9$, $\lambda = 1$, $r = 4$, $b = 12$, $N = 36$, we will associate the treatments A – J in turn with (1) the Latin letters, (2) the Greek letters, (3) the rows, (4) the columns of the completely orthogonalized 3×3 square

$$
\begin{array}{lll}
A: a\alpha & B: b\beta & C: c\gamma \\
D: c\beta & E: a\gamma & F: b\alpha \\
G: b\gamma & H: c\alpha & J: a\beta
\end{array}
$$

This gives ABC, DEF, GHJ; AEJ, BFG, CDH; ADG, BEH, CFJ; CEG, AFH, BDJ as the 12 blocks.

There will also be a complementary design, with $k^* = v - k = 6$, $v^* = v = 9$, $b^* = b = 12$, $r^* = b - r = 8$, $\lambda^* = 5$. The first block is $DEFGHJ$, and so on until the last one is $ACEFGH$.

(v) From a resolvable design, a design with $v^* = v + k + 1$ can be found by adding a new treatment to each group of blocks and also adding a block that contains all the extra treatments; for example, we can add $KLMN$ to the design given in (iv) and obtain:

$$ABCK, DEFK, GHJK; AEJL, BFGL, CDHL; ADGM, BEHM, CFJM; CEGN,$$
$$AFHN, BDJN; KLMN$$

Here $k^* = k + 1$, $r^* = r$, $b^* = b + 1$ and $\lambda^* = rk/(v+k)$ (which in this example is 1).

Conversely, removing any one block of this design, together with all the treatments it contains, will leave a balanced incomplete block.

(vi) *Cyclic* methods of construction exist for some designs: when one block has been given (for example, in tables) all the others can be generated from it by moving on one letter.

A generator for the design with $v = 11$, $k = 5$ when treatments are labelled $ABCDEFGHIJK$ is $AEFGI$, and so the remaining blocks will be $BFGHJ$, $CGHIK, \ldots, KDEFH$. This method only works when $b = v$ and $r = k$.

11.4   Linear model and analysis

The linear model for any block design, orthogonal or otherwise, is

$$y_{ij} = \mu + \tau_i + \beta_j + \varepsilon_{ij} \qquad (i = 1 \text{ to } v, j = 1 \text{ to } b)$$

and for a balanced incomplete block the conditions [11.1] and [11.2] above also apply. In order to estimate $\{\tau_i\}$, the same minimization is required as for a randomized complete block, and the normal equations are:

$$\sum_i \sum_j (y_{ij} - \mu - \tau_i - \beta_j) = 0$$

$$\sum_j (y_{ij} - \mu - \tau_i - \beta_j) = 0$$

$\sum$ being over all the blocks in which τ_i appears; there is one such equation for each $i=1$ to v, and

$$\sum_i (y_{ij}-\mu-\tau_i-\beta_j)=0$$

$\sum$ being over all the treatments that occur in block j; there is one such equation for each $j=1$ to b.

These give

$$G=N\hat{\mu}+r\sum_i \hat{\tau}_i+k\sum_j \hat{\beta}_j$$

or

$$G=N\hat{\mu}$$

since

$$\sum_i \hat{\tau}_i=0=\sum_j \hat{\beta}_j$$

(as on page 42); also

$$T_i=r\hat{\mu}+r\hat{\tau}_i+{\sum_j}' \hat{\beta}_j \qquad \text{for each } i=1 \text{ to } v \qquad \textbf{(a)}$$

where ${\sum_j}'$ indicates the sum over those blocks containing τ_i; and

$$B_j=k\hat{\mu}+k\hat{\beta}_j+{\sum_i}' \hat{\tau}_i \qquad \text{for each } j=1 \text{ to } b \qquad \textbf{(b)}$$

where ${\sum_i}'$ indicates the sum over those treatments present in block j.

In solving these equations, we use conditions [11.1] and [11.2]. Multiplying **(a)** through by k we have

$$kT_i=rk\hat{\mu}+rk\hat{\tau}_i+k{\sum_j}' \hat{\beta}_j \qquad \textbf{(a*)}$$

and carrying out the summation relevant to treatment i in **(b)** gives

$${\sum_j}' B_j=rk\hat{\mu}+k{\sum_j}' \hat{\beta}_j+r\hat{\tau}_i+\lambda\left(\sum_{\neq i} \tau_i\right) \qquad \textbf{(b*)}$$

because in these blocks treatment i will occur λ times with each of the other treatments. The symbol $(\sum_{\neq i} \tau_i)$ stands for the sum of all treatment parameters *except* that for treatment i; but since the sum over all i, $\sum_i \tau_i=0$, we can replace that symbol by $(-\tau_i)$.

Subtracting **(b*)** from **(a*)** to eliminate $\hat{\mu}$ gives

$$kT_i-{\sum_j}' B_j=rk\hat{\tau}_i-r\hat{\tau}_i+\lambda\hat{\tau}_i$$

$$=\{r(k-1)+\lambda\}\hat{\tau}_i$$

$$=\lambda v\hat{\tau}_i$$

using [11.2].

There is an alternative notation for ${\sum_j}' B_j$, which looks simpler, and we shall use it from now on, giving a new definition: $B^{(i)}=$ total yield of all plots in all those blocks containing treatment *i*. Finally,

$$Q_i = kT_i - B^{(i)}$$

(Note that $\sum_{i=1}^{v} Q_i = 0$.) The estimates of parameters are then

$$\hat{\mu} = \frac{G}{N}; \qquad \hat{\tau}_i = \frac{Q_i}{\lambda v}$$

We will not give the algebraic detail of the remainder of the analysis because it is a relatively simple special case of the general block method (page 228). The results are as follows:

(1) the variance of the difference $(\hat{\tau}_i - \hat{\tau}_j) = 2k\sigma^2/v\lambda$;
(2) the sum of squares for treatments after adjustment for block differences = $\sum_{i=1}^{v} Q_i^2/vk\lambda$;
(3) an Analysis of Variance is completed with a blocks term of the same form as in an orthogonal design, which is $\sum_{j=1}^{b} B_j^2/k - G^2/N$;
(4) the residual mean square, with $(rv - v - b + 1)$ degrees of freedom, estimates σ^2.

Note that we can make the usual *F*-test for the null hypothesis 'all treatment effects are the same' (i.e. all $\{\tau_i\}$ are zero) if we wish; but a different form of analysis would be needed for block comparisons.

We have only given above what is called the 'within blocks' or 'intrablock' analysis. There is some additional information for comparing treatments from the 'between blocks' or 'interblock' analysis described by Yates (1940), Cochran and Cox (1992) and others. However, the assumptions necessary, and the theory, are less straightforward, and the method is not helpful in the relatively small designs that are commonly used, so we refer readers elsewhere for this topic.

Example 11.1. Seven different confectionery products *A*–*G*, made from the same ingredients but using slightly different recipes, were examined by consumer tasting panels. Each session of a panel only tested three recipes; the total score given for a variety of characteristics was recorded, as follows:

Session (Block)	I	II	III	IV	V	VI	VII	
	A: 20	*B*: 25	*C*: 24	*A*: 16	*B*: 20	*C*: 19	*A*: 19	
	B: 23	*D*: 21	*E*: 18	*D*: 14	*E*: 16	*D*: 17	*F*: 20	
	C: 16	*F*: 20	*F*: 19	*E*: 15	*G*: 25	*G*: 22	*G*: 24	Total
Block totals	59	66	61	45	61	58	63	413

$S_0 = G^2/N = 413^2/21 = 8122.333$; $S = 8341$; $S_B = \frac{1}{3}(59^2 + 66^2 + \cdots + 63^2) = 8212.333$; $r = 3, v = 7, b = 7, k = 3, N = 21, \lambda = 1$.

Treatments (recipes)	A	B	C	D	E	F	G		
Totals	55	68	59	52	49	59	71	:	413
$B^{(i)}$	167	186	178	169	167	190	182		
$Q_i = kT_i - B^{(i)}$	-2	18	-1	-13	-20	-13	31		

The second row sums to $kG = 1239$, and $\sum Q_i = 0$, as they should.
$\sum Q_i^2 / vk\lambda = 2028/21 = 96.571$.

Analysis of Variance

Source	d.f.	Sum of squares	Mean square	
Blocks (ignoring treatments)	6	90.000	—	
Treatments adjusted for blocks	6	96.571	16.095	$F_{(6,9)} = 4.01^*$
Residual	8	32.096	$4.012 = \hat{\sigma}^2$	
Total	20	218.667		

The sum of squares for treatments, adjusted for blocks, has made allowance for the fact that not every treatment appeared in every block, and this sum of squares is therefore not the same as it would be in a randomized complete block. The 'blocks sum of squares' in this analysis is *not* a valid basis for comparing blocks because it does not make a similar allowance; it is, however, that part of the total sum of squares that has to be removed before a valid comparison of treatment and residual mean squares can be made.

There is evidence from the *F*-test of differences among means, and we may test any contrasts that are specified. Differences between pairs of means have variance $2k\hat{\sigma}^2/v\lambda = 6\hat{\sigma}^2/7 = 3.439$.

The values of the adjusted treatment effects $\hat{\tau}_i$ are given by $Q_i/\lambda v$:

A	B	C	D	E	F	G
-0.286	2.571	-0.143	-1.857	-2.857	-1.857	4.429

To obtain treatment means we add $\hat{\mu} = 413/21 = 19.667$ to each. The standard error of the difference between any pair of means, or, equally, between any two $\{\tau_i\}$, is $\sqrt{3.057} = 1.854$, and in *t*-tests or confidence interval calculations $t_{(8)}$ must be used.

11.5   Row and column design: the Youden square

Such a design is not a square, but a rectangle! The idea was introduced by Youden (1937, 1940) for work in a chemical laboratory where block size had to be small and the order of testing within a block was also important. A row-and-column design is required, but blocks have to be 'incomplete'. Every balanced incomplete block with $b = v$ can be written as a layout where every treatment appears just once on each row and the columns are the blocks of a balanced incomplete block. In fact every Latin square with just one row omitted is a design of this type, and explains why the word 'square' came to be used – Youden squares are 'incomplete Latin squares'. Recently they are being described as 'Youden Rectangles'.

A typical layout is

$$
\begin{array}{ccccc}
E & A & C & B & D \\
B & C & E & D & A \\
A & B & D & C & E \\
C & D & A & E & B \\
\end{array}
$$

and a design of useful size ($v=7, k=4$) is

$$
\begin{array}{ccccccc}
F & C & A & G & E & D & B \\
D & A & F & C & G & B & E \\
C & E & G & B & D & A & F \\
E & G & D & F & B & C & A \\
\end{array}
$$

In an animal experiment on piglets receiving 7 different diets *A–G*, columns could contain animals of the same sex from the same litter and rows could represent size order in the litter when the experiment began. Animals, like humans, vary considerably in their responses and all methods of controlling some of the variation, and removing systematic elements from it, are valuable.

The analysis is like that of the balanced incomplete block from which it is derived, with the addition of a term for rows which is orthogonal to treatments and to columns (blocks) and which therefore is computed just as in an orthogonal design.

When planning an experiment, care must be taken to have sufficient degrees of freedom in residual.

11.6   General block designs

The simplest orthogonal block design is randomized complete blocks, with every treatment appearing exactly once in each block. It is easy to carry out and analyse, and controls one source of systematic variation well. Sometimes more than one source can be controlled by the same blocking system. It is by far the most common design, partly because blocks can form convenient units for recording even if it is not certain that any heterogeneity does exist among the units.

So long as every block is made up in the same way, some treatments may be given more replication than others; for example a block containing *AAABBCDEF* can be the basis of an orthogonal design provided every block of 9 plots contains this same selection of treatments (suitably randomized afresh for each block). We have already mentioned simple cases of this in Chapter 4. In fact, as we shall see below, a design is still orthogonal if the *proportional* replication of treatments stays the same, as for example *ABCC* in block I, *AABBCCCC* in block II, *AAABBBCCCCCC* in block III where the ratio $r_A : r_B : r_C$ is always $1:1:2$. The treatment contrasts will be orthogonal to blocks in both a mathematical and a practical sense, even when block size does not remain the same, provided this proportionality holds.

Non-orthogonal designs

Balanced incomplete blocks, which we have studied earlier in this chapter, are not orthogonal; neither, strictly speaking, is a randomized complete block where a missing value has occurred – our methods of dealing with missing values do not in any sense recover the lost information. And certainly a design like I *AABCD*; II *ABCCD*; III *AACDD*; IV *BBBCD* is not orthogonal – still less is I *AABCC*; II *AAABBCC*. Non-orthogonal designs may arise

(1) when there are severe restrictions on block sizes,
(2) through errors in applying the treatments,
(3) because some observations are lost ('missing values'),
(4) when we deliberately wish to examine some contrasts with smaller variance (greater precision) than others.

General method of analysis

We will develop the matrix-based general method of analysis which can be used for any design when a statistical computing package is available. The notation we shall use is now reasonably standard, though there are still some variations in different texts. We define the following vectors:

r gives the replications of each of the treatments
k gives the sizes of each of the blocks
r^δ is r written as a diagonal matrix, with zeros off-diagonal
k^δ is k written as a diagonal matrix (see the example below)

A very important concept is that of the *incidence matrix*, which shows how often each treatment appears in each block; treatments are represented by the rows of the matrix and blocks by columns. We use the very simple design mentioned above, in two blocks illustrated by Fig. 11.1 randomized for use as a field trial. The plots are numbered for ease of reference later on.

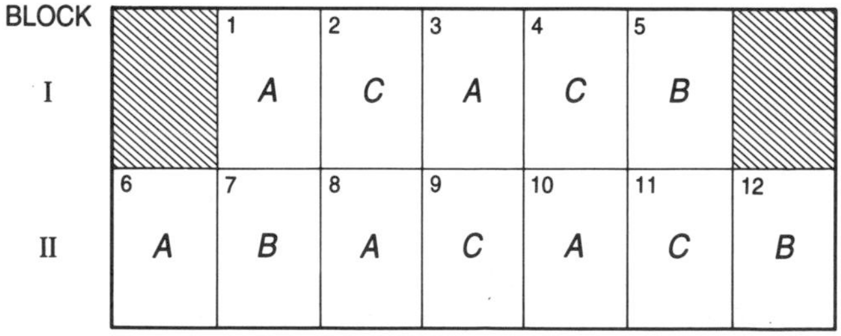

Fig. 11.1. Layout of a non-orthogonal block design

Unfortunately, N is commonly used for the incidence matrix, and so to avoid confusion we will change our notation for the total number of plots in the experiment to n, for the rest of this chapter.

There are 3 treatments and 2 blocks in our rather trivial example; the vectors r and k therefore are

$$r = \begin{pmatrix} 5 \\ 3 \\ 4 \end{pmatrix} \quad \text{and} \quad k = \begin{pmatrix} 5 \\ 7 \end{pmatrix}$$

To write r as a diagonal matrix, we place 5, 3, 4 in the leading diagonal and fill the rest of the matrix with 0's. Thus

$$r^\delta = \begin{pmatrix} 5 & 0 & 0 \\ 0 & 3 & 0 \\ 0 & 0 & 4 \end{pmatrix}$$

and

$$k^\delta = \begin{pmatrix} 5 & 0 \\ 0 & 7 \end{pmatrix}$$

By direct inspection, we see that

$$N = \begin{pmatrix} 2 & 3 \\ 1 & 2 \\ 2 & 2 \end{pmatrix}$$

Since there are 3 treatments, the vector of their means may be written

$$a = \begin{pmatrix} \alpha_1 \\ \alpha_2 \\ \alpha_3 \end{pmatrix}$$

(we will use a instead of τ in this notation), and the means in the two blocks are

$$\beta = \begin{pmatrix} \beta_1 \\ \beta_2 \end{pmatrix}$$

There is another way of developing these matrices, starting from *design matrices* D and Δ for blocks and treatments, respectively. Here we will need the plot numbering shown in Fig. 11.1. D has a row for each block, and as many columns as there are plots, numbered in order. The figure 1 is placed in the first row for those plot numbers that are in block I, and in the second row for those plot numbers in block II, and 0 is placed elsewhere. The design shown in Fig. 11.1 is therefore represented as

$$D = \begin{pmatrix} 1 & 1 & 1 & 1 & 1 & 0 & 0 & 0 & 0 & 0 & 0 & 0 \\ 0 & 0 & 0 & 0 & 0 & 1 & 1 & 1 & 1 & 1 & 1 & 1 \end{pmatrix}$$

Likewise, for treatments Δ has 3 rows; row 1 refers to A, row 2 to B and row 3 to C, and again we place 1's to show which treatment goes on which plot. For the design in Fig. 11.1 this gives

$$\Delta = \begin{pmatrix} 1 & 0 & 1 & 0 & 0 & 1 & 0 & 1 & 0 & 1 & 0 & 0 \\ 0 & 0 & 0 & 0 & 1 & 0 & 1 & 0 & 0 & 0 & 0 & 1 \\ 0 & 1 & 0 & 1 & 0 & 0 & 0 & 0 & 1 & 0 & 1 & 0 \end{pmatrix}$$

There are some important results, which can easily be checked for this example and are clear in general. Using ' to denote the transpose of a matrix, we have $\Delta D' = N$; also $DD' = k^\delta$, and finally $\Delta\Delta' = r^\delta$.

Also, defining **1** as a column vector with 1's in each row, and the right number of rows to be conformable for multiplication in each case, we find $N\mathbf{1} = r$ and $N'\mathbf{1} = k$.

11.7   Linear model and analysis

For any block design, complete or incomplete, as we have already remarked, the appropriate linear model is

$$y_{ij} = \mu + \beta_j + \alpha_i + \varepsilon_{ij} \qquad (i = 1 \text{ to } v, j = 1 \text{ to } b)$$

In the new notation this may be written

$$y = D'\beta + \Delta'a + \varepsilon$$

Some useful relationships are $D\mathbf{1} = k$; $\Delta\mathbf{1} = r$; $D'\mathbf{1} = \mathbf{1} = \Delta'\mathbf{1}$; $r^\delta\mathbf{1} = r$; $k^\delta\mathbf{1} = k$; $r'\mathbf{1} = n = k'\mathbf{1}$; and $y'\mathbf{1} = G$.

The analysis is based on the standard analysis of the general regression model (*see*, e.g. Draper and Smith, 1981), but there is a difficulty due to the singularity of the matrix which has to be inverted, C (see below). Previously we have imposed two extra constraints on the parameters being estimated in order to overcome this, namely $\sum_i \tau_i = 0 = \sum_j \beta_j$, but instead we will now use the idea of a *generalized inverse* matrix.

Any matrix C^- such that $CC^-C = C$ is a generalized inverse of C; there is no unique solution to the problem of finding C^- but we will mention the most useful ones.

Some further notation is required: let B be the vector of block totals, and T the vector of treatment totals. For non-orthogonal designs we also require a vector of *treatment effect totals*, which is called Q and is equal to $T - Nk^{-\delta}B$. The symbol $k^{-\delta}$ stands for the inverse of k^δ, whose elements are the reciprocals of the corresponding values in k.

Denoting the vector of observations as y, we have $Dy = B$ and $\Delta y = T$.

In the model $y = D'\beta + \Delta'a + \varepsilon$ we have to minimize $U = \varepsilon'\varepsilon$. Expanding this expression and using earlier results allows us to reduce U; then for least-squares estimation of the parameters we must set $\partial U/\partial\beta$ and $\partial U/\partial a$ equal to 0 to give the normal equations.

Note that in this general method of analysis we do not include a separate term μ, so there is no equation corresponding to $\partial U/\partial\mu$.

Now
$$U = (y - D'\beta - \Delta'a)'(y - D'\beta - \Delta'a)$$
$$= y'y + \beta'DD'\beta + a'\Delta\Delta'a + 2a'\Delta D'\beta - 2\beta'Dy - 2a'\Delta y$$
$$= y'y + \beta'k^\delta\beta + a'r^\delta a + 2a'N\beta - 2\beta'B - 2a'T$$

The identities mentioned above have been used in simplifying U.

$$\frac{\partial U}{\partial\beta} = 0 \text{ gives } 2k^\delta\hat{\beta} - 2B + 2N'\hat{a} = 0$$

and

$$\frac{\partial U}{\partial a} = 0 \quad \text{gives} \quad 2r^\delta\hat{a} - 2T + 2N\hat{\beta} = 0$$

These are the normal equations. Hence

$$k^\delta \hat\beta + N'\hat a = B$$

and

$$N\hat\beta + r^\delta \hat a = T$$

(For differentiation of matrices and vectors, see for example Graybill (1969, 1976).)
The first equation gives $\hat\beta = k^{-\delta}(B - N'\hat a)$, which we substitute in the second to find

$$Nk^{-\delta}(B - N'\hat a) + r^\delta \hat a = T$$

Writing $Q = T - Nk^{-\delta}B$, this becomes

$$(r^\delta - Nk^{-\delta}N')\hat a = Q$$

The matrix in the brackets has been studied by Chakrabati (1962), as well as other workers; it is generally called C. Therefore putting $C = (r^\delta - Nk^{-\delta}N')$ the solution is $C\hat a = Q$.

By using results noted above, we find $C\mathbf{1} = 0$, which shows that C is singular. Each row total and each column total in C is zero (and thus C will have at least one zero eigenvalue (page 238)).

Example 11.2. In the experiment shown in Fig. 11.1,

$$r = \begin{pmatrix} 5 \\ 3 \\ 4 \end{pmatrix}, \quad k = \begin{pmatrix} 5 \\ 7 \end{pmatrix}, \quad N = \begin{pmatrix} 2 & 3 \\ 1 & 2 \\ 2 & 2 \end{pmatrix}$$

Therefore

$$r^\delta = \begin{pmatrix} 5 & 0 & 0 \\ 0 & 3 & 0 \\ 0 & 0 & 4 \end{pmatrix}, \quad k^\delta = \begin{pmatrix} 5 & 0 \\ 0 & 7 \end{pmatrix}, \quad k^{-\delta} = \begin{pmatrix} 1/5 & 0 \\ 0 & 1/7 \end{pmatrix}$$

Hence

$$Nk^{-\delta}N' = \begin{pmatrix} 2 & 3 \\ 1 & 2 \\ 2 & 2 \end{pmatrix} \begin{pmatrix} 1/5 & 0 \\ 0 & 1/7 \end{pmatrix} \begin{pmatrix} 2 & 1 & 2 \\ 3 & 2 & 2 \end{pmatrix}$$

$$= \begin{pmatrix} 2 & 3 \\ 1 & 2 \\ 2 & 2 \end{pmatrix} \begin{pmatrix} 2/5 & 1/5 & 2/5 \\ 3/7 & 2/7 & 2/7 \end{pmatrix}$$

$$= \begin{pmatrix} 73/35 & 44/35 & 58/35 \\ 44/35 & 27/35 & 34/35 \\ 58/35 & 34/35 & 48/35 \end{pmatrix}$$

and therefore

$$C = \begin{pmatrix} 5 & 0 & 0 \\ 0 & 3 & 0 \\ 0 & 0 & 4 \end{pmatrix} - \begin{pmatrix} 73/35 & 44/35 & 58/35 \\ 44/35 & 27/35 & 34/35 \\ 58/35 & 34/35 & 48/35 \end{pmatrix}$$

$$= \frac{1}{35} \begin{pmatrix} 102 & -44 & -58 \\ -44 & 78 & -34 \\ -58 & -34 & 92 \end{pmatrix}$$

We note that, as expected, each row sum and each column sum is zero.
    Suppose that the observations y_{ij} from this experiment were:

Block								Total
I	–	A: 17	C: 24	A: 20	C: 22	B: 10	–	93
II	A: 14	B: 8	A: 15	C: 15	A: 12	C: 19	B: 11	94
								$G = 187$

Treatment totals are A: 78; B: 29; C: 80.
    Therefore we write

$$y = (17 \quad 24 \quad \cdots \quad 19 \quad 11)', \qquad B = \begin{pmatrix} 93 \\ 94 \end{pmatrix}, \qquad T = \begin{pmatrix} 78 \\ 29 \\ 80 \end{pmatrix}$$

We can check that $Dy = B$ and $\Delta y = T$.

11.8   Generalized inverses

We have already noted that there is no unique solution for C^-; furthermore, even if C is symmetrical, there is no need for C^- to be so. The original proposal by Tocher (1952) is commonly used: it is Ω defined by

$$\Omega^{-1} = C + \frac{1}{n} rr'$$

This is a generalized inverse, satisfying $C\Omega C = C$, because

$$C\Omega C = \left(\Omega^{-1} - \frac{1}{n} rr' \right) \Omega \left(\Omega^{-1} - \frac{1}{n} rr' \right)$$

$$= \left(\Omega^{-1} - \frac{1}{n} rr' \right) \left(I - \frac{1}{n} \Omega rr' \right)$$

$$= \Omega^{-1} - \frac{1}{n} rr' - \frac{1}{n} rr' + \frac{1}{n^2} rr' \Omega rr'$$

Using earlier results, $\Omega^{-1}1 = C1 + (1/n)rr'1 = 0 + r$, so $1 = \Omega r$.
Therefore

$$C\Omega C = \left(\Omega^{-1} - \frac{1}{n}rr'\right) - \frac{1}{n}rr' + \frac{1}{n^2}rr'1r'$$

$$= C - \frac{1}{n}rr' + \frac{1}{n^2}\cdot nrr' = C$$

Any method of adding a term to C which will remove the singularity, and so permit inversion, is acceptable.

In Example 11.2,

$$\frac{1}{n}rr' = \frac{1}{12}\begin{pmatrix}5\\3\\4\end{pmatrix}(5 \quad 3 \quad 4) = \frac{1}{12}\begin{pmatrix}25 & 15 & 20\\15 & 9 & 12\\20 & 12 & 16\end{pmatrix}$$

Adding this to C gives a symmetric matrix, so that Ω^{-1} and Ω will both be symmetric. It is clear from this very small example that the inversion will, in general, require a computer routine, especially in an experiment of normal size with only a moderately non-orthogonal pattern.

Sums of squares

For the general regression model $y = a\theta + \varepsilon$, the parameter estimates are $\hat{\theta} = (a'a)^{-1}a'y$, the 'fitted values' are $\hat{y} = a\hat{\theta}$ and the regression sum of squares is $\hat{\theta}a'y$. The residual sum of squares is then $y'y - \hat{\theta}a'y$. Using the convention that blocks are written first, before treatments,

$$a = (D' \ \Delta') \qquad \text{and} \qquad \theta = \begin{pmatrix}\beta\\a\end{pmatrix}$$

so that

$$\hat{y}'\hat{y} = (\hat{\beta}' \ \hat{a}')\begin{pmatrix}D\\\Delta\end{pmatrix}y = \hat{\beta}'Dy + \hat{a}'\Delta y = \hat{\beta}'B + \hat{a}'T$$

The uncorrected sum of squares for blocks, *not* adjusted for treatments, is $B'k^{-\delta}B$, and so 'treatments adjusted for blocks' is $\hat{\beta}'B + \hat{a}'T - B'k^{-\delta}B$, which reduces to $\hat{a}'Q$ because the first of the normal equations above gave $\hat{\beta} = k^{-\delta}(B - N'\hat{a})$. Thus, since $\hat{a} = \Omega Q$, the adjusted treatments sum of squares is $Q'\Omega Q$.

Example 11.3. Consider the following very simple design in 10 plots:

$$\text{Block I:} \quad A, 10; \quad A, 8; \quad B, 4; \quad B, 9; \quad B, 6$$

$$\text{II:} \quad A, \ 7; \quad A, 5; \quad A, 8; \quad A, 4; \quad B, 6$$

$$N = \begin{pmatrix}2 & 4\\3 & 1\end{pmatrix}, \quad r = \begin{pmatrix}6\\4\end{pmatrix}, \quad k = \begin{pmatrix}5\\5\end{pmatrix}, \quad B = \begin{pmatrix}37\\30\end{pmatrix}, \quad T = \begin{pmatrix}42\\25\end{pmatrix}, \quad G = 67, \quad n = 10.$$

$$y'y = 487, \qquad G^2/n = 448.9, \qquad \therefore S - S_0 = 38.1$$

$$C = r^\delta - Nk^{-\delta}N' = \begin{pmatrix} 6 & 0 \\ 0 & 4 \end{pmatrix} - \begin{pmatrix} 2 & 4 \\ 3 & 1 \end{pmatrix} \begin{pmatrix} 1/5 & 0 \\ 0 & 1/5 \end{pmatrix} \begin{pmatrix} 2 & 3 \\ 4 & 1 \end{pmatrix} = \begin{pmatrix} 2 & -2 \\ -2 & 2 \end{pmatrix}$$

$$Q = T - Nk^{-\delta}B = \begin{pmatrix} 42 \\ 25 \end{pmatrix} - \begin{pmatrix} 2 & 4 \\ 3 & 1 \end{pmatrix} \begin{pmatrix} 1/5 & 0 \\ 0 & 1/5 \end{pmatrix} \begin{pmatrix} 37 \\ 30 \end{pmatrix} = \frac{16}{5}\begin{pmatrix} 1 \\ -1 \end{pmatrix}$$

$$\Omega^{-1} = C + \frac{1}{n}rr' = \begin{pmatrix} 2 & -2 \\ -2 & 2 \end{pmatrix} + \frac{1}{10}\begin{pmatrix} 6 \\ 4 \end{pmatrix}(6 \quad 4) = \frac{1}{5}\begin{pmatrix} 28 & 2 \\ 2 & 18 \end{pmatrix}$$

Therefore

$$\Omega = \frac{5}{500}\begin{pmatrix} 18 & -2 \\ -2 & 28 \end{pmatrix} = \frac{1}{50}\begin{pmatrix} 9 & -1 \\ -1 & 14 \end{pmatrix}$$

$$\hat{a} = \Omega Q = \frac{1}{50}\begin{pmatrix} 9 & -1 \\ -1 & 14 \end{pmatrix}\frac{16}{5}\begin{pmatrix} 1 \\ -1 \end{pmatrix} = \frac{16}{250}\begin{pmatrix} 10 \\ -15 \end{pmatrix} = \frac{8}{25}\begin{pmatrix} 2 \\ -3 \end{pmatrix}$$

i.e. $\hat{a}_1 = 16/25 = 0.64$ and $\hat{a}_2 = -24/25 = -0.96$ are the adjusted treatment effects for the two treatments A, B. The unadjusted blocks sum of squares is

$$\frac{37^2 + 30^2}{5} - \frac{G^2}{n} = 4.9$$

This could have been found as

$$(37 \quad 30)\begin{pmatrix} 1/5 & 0 \\ 0 & 1/5 \end{pmatrix}\begin{pmatrix} 37 \\ 30 \end{pmatrix} - \frac{67^2}{10}$$

The adjusted treatments sum of squares is

$$Q'\Omega Q = \frac{1}{50} \times \frac{16}{5} \times \frac{16}{5} \times (1 \quad -1)\begin{pmatrix} 9 & -1 \\ -1 & 14 \end{pmatrix}\begin{pmatrix} 1 \\ -1 \end{pmatrix} = \frac{256}{1250} \times 25 = 5.12$$

Analysis of Variance

Source	d.f.	Sum of squares	Mean square	
Blocks ignoring treatments	1	4.90		
Treatments adjusted for blocks	1	5.12		$F_{(1,7)} = 1.28$ n.s.
Residual	7	28.08	4.011	
Total	9	38.10		

11.9 Application to designs with special patterns

1. A *completely randomized* design may be considered to have only one block. Thus $k = n$; $N = r$; $B = G$; β has only one element;

$$C = r^{\delta} - Nk^{-\delta}N' = r^{\delta} - \frac{1}{n}rr'$$

Hence $\Omega^{-1} = r^{\delta}$ and so $\Omega = r^{-\delta}$.

We find that the 'blocks' sum of squares $= B'k^{-\delta}B = G^2/n$, which is simply S_0, the correction term in the usual analysis by summation terms. The sum of squares for treatments, $Q'\Omega Q$, is likewise $S_T - S_0$; this follows because

$$Q = T - Nk^{-\delta}B = T - \frac{G}{n}r$$

so that

$$\left(T - \frac{G}{n}r\right)' r^{-\delta}\left(T - \frac{G}{n}r\right) = T'r^{-\delta}T - G^2/n$$

$$= S_T - S_0$$

2. In a *randomized complete block* we have $k = k1$; $r = r1$; and the incidence matrix $N = J$, where J denotes the matrix that contains 1 in every position; thus $N = 11'$. We may write

$$k^{-\delta} = \frac{1}{k}I$$

where in the usual notation I is the identity matrix.

$$Q = T - Nk^{-\delta}B = T - 11'IB/k = T - G1/k$$

Noting that $n = rv = bk$, and also $b = r$, $k = v$, we can develop the following results:

$$C = r^{\delta} - Nk^{-\delta}N' = rI - \frac{1}{k}11'I(11')^{-1} = rI - r11'/k$$

$$\Omega^{-1} = C + \frac{1}{n}rr' = rI - r11'/k + r^2 11'/n = rI$$

$$\Omega = \frac{1}{r}I$$

The sum of squares for blocks $= B'k^{-\delta}B - G^2/n = B'B/k - G^2/n = S_B - S_0$.

The sum of squares for treatments $= Q'\Omega Q$

$$= \left(T - \frac{G}{k}1\right)' \frac{1}{r}I\left(T - \frac{G}{k}1\right)$$

$$= \frac{1}{r}TT' - \frac{2G}{rk}1'T + \frac{G^2}{rk^2}11' \quad \text{(and here } 11' = v)$$

$$= \frac{1}{r} TT' - \frac{2G^2}{n} + \frac{G^2}{n}$$

$$= \frac{1}{r} TT' - \frac{G^2}{n} = S_T - S_0$$

3. *Balanced incomplete blocks* have $NN' = \lambda J + (r - \lambda)I$, with I and J defined as in (2) above, since there will be r's on the leading diagonal and λ's in every other place. Again, $k = k1$, $r = r1$, $rv = bk = n$; also for this design $\lambda(v-1) = r(k-1)$.

$$C = r^\delta - Nk^{-\delta}N' = rI - \frac{1}{k}(\lambda J + \{r - \lambda\}I)$$

$$= \frac{1}{k}\{[(k-1)r + \lambda]I - \lambda J\} = \frac{r(k-1)}{k(v-1)}(vI - J)$$

and by introducing $L = vI - J$ we may write this as

$$C = \frac{r(k-1)}{k(v-1)} L$$

Hence

$$\Omega^{-1} = \frac{r(k-1)}{k(v-1)}(vI - J) + \frac{1}{n} rr'$$

$$= \frac{r(k-1)}{k(v-1)}(vI - J) + \frac{r^2}{bk} 11'$$

$$= \frac{\lambda v}{k}\left(I + \frac{v-k}{v^2(k-1)} J\right)$$

using earlier results. Inversion gives

$$\Omega = \frac{k}{\lambda v}\left(I - \frac{(v-k)}{kv(v-1)} J\right)$$

and after some algebra the sum of squares for treatments reduces to the expression

$$\frac{(v-1)}{rvk(k-1)} \sum_{i=1}^{v} Q_i^2$$

given on page 217.

4. An interesting related design is the *supplemented balance* design in which every pair of main treatments $A, B, C, \ldots$ occurs together in a block the same number of times, λ, and there is also a *supplementing treatment*, sometimes a *control* (page 20) occurring equally often, λ_0 times, with each treatment. The main treatments are each replicated r times and the supplementing treatment is replicated r_0 times. In the simple application of this design, each block contains k plots (but see below).

An example is: I $OABC$; II $OABD$; III $OACD$; IV $OBCD$. Here $v=5$, $r=3$, $r_0=4$, $k=4$, $\lambda=2$, $\lambda_0=3$;

$$r=\begin{pmatrix}4\\3\\3\\3\\3\end{pmatrix}\qquad\text{and}\qquad k=\begin{pmatrix}4\\4\\4\\4\end{pmatrix}$$

$$N=\begin{pmatrix}1&1&1&1\\1&1&1&0\\1&1&0&1\\1&0&1&1\\0&1&1&1\end{pmatrix},\qquad k^{-\delta}=\begin{pmatrix}1/4&0&0&0\\0&1/4&0&0\\0&0&1/4&0\\0&0&0&1/4\end{pmatrix}$$

and these lead to

$$C=\begin{pmatrix}3&-3/4&-3/4&-3/4&-3/4\\-3/4&9/4&-1/2&-1/2&-1/2\\-3/4&-1/2&9/4&-1/2&-1/2\\-3/4&-1/2&-1/2&9/4&-1/2\\-3/4&-1/2&-1/2&-1/2&9/4\end{pmatrix}$$

This is of the form

$$\begin{pmatrix}p&q&q&q&q\\q&s&m&m&m\\q&m&s&m&m\\q&m&m&s&m\\q&m&m&m&s\end{pmatrix}$$

and has a generalized inverse

$$\begin{pmatrix}\dfrac{m}{q(s-m)}&0&0&\cdots&0\\[2ex]0&\dfrac{1}{(s-m)}&0&\cdots&0\\[2ex]0&0&\dfrac{1}{(s-m)}&&\vdots\\[2ex]\vdots&\vdots&&\ddots&0\\[2ex]0&0&\cdots&\cdots\quad0&\dfrac{1}{(s-m)}\end{pmatrix}$$

with entries only on the diagonal.

5. A design need not have all blocks the same size in order to achieve *total balance*. For example, I *AABC*; II *AABC*; III *BC* leads to

$$C = \begin{pmatrix} 2 & -1 & -1 \\ -1 & 2 & -1 \\ -1 & -1 & 2 \end{pmatrix}$$

which is the pattern of total balance. (Although this scheme would be too small for most experiments, another 3 blocks of the same type could be used with it, and the design would still have total balance.)

6. A very useful design in total balance has extra treatments added to blocks in a balanced way; e.g. I *ABCA*; II *ABCB*; III *ABCC* has each treatment in a special relationship with just one block, this time duplication instead of omission: *A* with I, *B* with II, *C* with III. It is easy to show that this has the correct pattern of the matrix *C* for a design in total balance.

7. A randomized complete block that suffers one missing value becomes a *supplemented balance* design, and could be analysed as such without resorting to the missing value techniques discussed earlier. The rôle of supplementing treatment is taken by the treatment that should have provided the lost observation.

8. The property of total balance can be extended to row-and-column designs which have additional treatments in rows rather than omitting some (to give a Youden square). The layout

$$\begin{array}{cccc} A & B & C & D \\ D & A & B & C \\ C & D & A & B \\ B & C & D & A \\ A & D & B & C \end{array}$$

has total balance because the first 4 rows form a Latin square while the final row adds one of the treatments to each column in such a way that the columns retain total balance. Useful extra replication can be added to a small design in this way.

Non-orthogonal designs are particularly useful in field trials, and further discussion may be found in Pearce (1983) and Pearce, Clarke, Dyke and Kempson (1988).

We have assumed that all treatments will be of equal importance when means are being compared, but often there will be contrasts of special interest, possibly involving some treatments more than others. Optimal designs (see, e.g. Jones (1976), Atkinson (1996)) may be constructed by computer searches of possible values of *r*, *k* and *b* when *v* treatments are used and the variances of some contrasts are required to be as small as possible, if necessary at the expense of others.

11.10 Exercises*

1. (i) In the standard notation for Balanced Incomplete Block (BIB) designs, show that the concurrence of treatments is given by

$$\lambda = \frac{r(k-1)}{v-1}$$

---

* Material from Appendices 11A and 11B on pages 238–40 is needed for some questions.

(ii) Obtain relationships between the standard parameters for a BIB design which are not generally true, but which necessarily apply when the design is

(a) unreduced,
(b) resolvable.

(iii) Prove that a Youden square always exists for $k = v - 1$ and show how such a design may be constructed. Express λ in terms of v for such a design.

(iv) A BIB design exists for 8 treatments in 14 blocks. Twelve of the blocks are as shown:

1	2	3	4		1	2	5	6
5	6	7	8		3	4	7	8
1	3	6	8		1	3	5	7
2	4	5	7		2	4	6	8
1	4	6	7		1	4	5	8
2	3	5	8		2	3	6	7

Obtain the other two blocks and comment on the special properties of this design.

(v) Suppose that the design in (iv) is rejected because an experimenter wishes to include one further treatment, but the block size must still remain at 4. It is known that a BIB design with less than 100 units can still be found for these conditions. Find the number of units required and explain how the result is deduced.

2. (i) Construct a balanced incomplete block design for 12 treatments in blocks of size 2. Write down the values of the parameters r, b, k, v, λ.

   (ii) Construct balanced incomplete block designs for 7 treatments in blocks of 3, 4 and 6 units. Write down the values of the parameters for each of these three designs.

   Arrange each design as a Youden square. Suggest an alternative way of constructing a 7×6 Youden square.

3. (i) Describe the properties of a balanced incomplete block (BIB) design and state its advantages and disadvantages when used instead of other common designs.

   (ii) Explain the meanings of the terms *resolvable* and *unreduced* in relation to BIB designs. Obtain an arrangement of a BIB design in 4 blocks which has 4 treatments each with 3 replicates. Explain whether this design is resolvable, unreduced or can be arranged as a Youden square.

   (iii) In a BIB design 10 treatments are arranged in 18 blocks of 5 units. Calculate the replication and the number of times treatments occur together in the same block.

   (iv) For the unreduced BIB designs with 6 treatments in blocks of size 3 calculate the level of replication and concurrence of treatments. Show that the unreduced design is resolvable but that it is not possible to obtain a resolvable design with 6 treatments in blocks of size 3 which is not unreduced.

(v) For any BIB design show that the Chakrabati coefficient matrix C is of the form

$$C=\frac{1}{k}[(rk-r+\lambda)I-\lambda J]$$

where I is the identity matrix and J is a matrix of 1's. Show that Tocher's Ω is also of the form $pI+qJ$ and calculate p and q for the design described in part (ii) above.

4. Discuss the meaning of the terms 'balanced incomplete blocks' and 'Youden squares', showing the connection between them. Show that it is possible to arrange a BIB design for 5 treatments in 10 blocks of size 3, and suggest how the treatments would be arranged between the blocks and describe how the design should be applied in practice.

With the usual notation, show that in general for BIB designs

$$\lambda=\frac{r(k-1)}{v-1}$$

and verify that the value of λ agrees for this design.

5. Explain fully the circumstances in which a *balanced incomplete block* design is appropriate. Derive the two equations connecting the parameters v, b, r, k and λ (in the usual notation), stating what each parameter stands for.

Six treatments are to be compared for their effect on the flavour of a food, and a trained consumer tasting panel will undertake the work. However, they can only examine a maximum of 4 treatments at each session. There are 30 people who can be called upon as members of this panel.

Three possible schemes for comparing the 6 treatments are:

(i) 15 panel members each examining 4 treatments;
(ii) 20 panel members each examining 3 treatments;
(iii) 30 panel members each examining 2 treatments.

Write down an experimental design for each of (i), (ii) and (iii), and state the values of the parameters.

The residual variance, σ^2, for scheme (ii) is thought to be 15% higher than that for scheme (iii); and σ^2 for scheme (i) is thought to be 30% higher than for scheme (iii). You may assume that the variance of the difference between 2 treatment means in a balanced incomplete block is $2k\sigma^2/(\lambda v)$.

Which design would you recommend? Give brief reasons.

6. In an industrial experiment 5 batches of metal ingots were selected at random from a production process where each batch comprised 4 ingots. Each ingot was melted and mixed with an amount of cadmium (Cd) and tin (Sn), then allowed to cool. When the treated ingot was reheated, its melting point (°C) was recorded, as shown.

Batch 1	A: 194	D: 205	E: 250	B: 214
Batch 2	B: 204	E: 243	D: 198	C: 238
Batch 3	D: 206	B: 205	C: 238	A: 186
Batch 4	A: 183	E: 247	B: 202	C: 229
Batch 5	E: 255	C: 244	D: 209	A: 198

The treatments were described as follows:

$$A: 10\% \text{ Cd, no Sn}$$
$$B: 20\% \text{ Cd, no Sn}$$
$$C: 30\% \text{ Cd, no Sn}$$
$$D: 10\% \text{ Cd, } 10\% \text{ Sn}$$
$$E: 30\% \text{ Cd, } 10\% \text{ Sn}$$

Obtain the Analysis of Variance for the data and use your analysis to comment on whether there is evidence that

(i) the melting point increases linearly with cadmium content,
(ii) the addition of tin makes a real difference to the melting point.

Estimate the mean and standard error of the increase in melting point through the addition of tin.

7. For the design:

Block 1	A	B	C	D
Block 2	A	B	C	E
Block 3	B	C	D	E
Block 4	A	C	D	E
Block 5	A	B	D	E

obtain Tocher's generalized inverse Ω and calculate the variance of the contrast between any 2 treatments as a multiple of the error variance σ^2. Obtain the effective replication of this contrast and the efficiency of the design.

Show how the given design is related to a Youden square and describe how such a design could be useful in practice. Discuss how a Youden square may be analysed.

8. Given that there are two resolvable balanced incomplete block designs in blocks of size 4 containing fewer than 100 units, obtain the values of their parameters λ.

9. Describe the properties of a balanced incomplete blocks (BIB) design and indicate why such a design may be useful in practice. Define the standard concurrence parameter λ and show that the coefficient matrix C may be written in the form

$$C = pI + qJ$$

where I is an identity matrix and J is a matrix with all elements unity.

Obtain Tocher's generalized inverse Ω for an unreduced BIB design with 4 treatments in 4 blocks of size 3, and calculate the effective replication of a contrast between any 2 treatments.

Analyse the data in the following BIB experiment:

Block 1	A: 20	B: 23	C: 16
Block 2	A: 22	C: 19	D: 17
Block 3	B: 25	C: 20	D: 21
Block 4	A: 16	B: 20	D: 14

and calculate the least significant difference between the treatments. Comment on the differences between adjusted treatment means.

10. A block design has the following arrangement:

Block 1	A	B	C	C	C	C
Block 2	A	A	A	A	B	B
Block 3	A	A	B	B	C	C
Block 4	A	B	B	B	C	C

The treatments X, Y, Z are to be allocated to the letters A, B, C in some order for the purposes of an experiment. X is the control and the contrast of principal interest is the comparison of the control with the mean of the other treatments. Deduce how the treatments should be allocated and find the variances of the resulting set of orthogonal contrasts if the error variance is σ^2. Compare these variances with those of the same contrasts for the corresponding randomized blocks design.

11. For the following design:

Block 1:	A	A	A	B	C
Block 2:	A	B	B	B	C
Block 3:	A	B	C	C	C

obtain the Chakrabati coefficient matrix C and its Moore–Penrose generalized inverse C^+. Hence or otherwise calculate the variance of simple contrasts between treatment means and comment on their efficiency. Comment also on the efficiency of the contrast between one treatment and the mean of the rest.

12. Three treatments were allocated at random to 20 units in a block design and the responses (y) were as shown below.

	A				B				C		
Block 1	20	26	21	24	16	18	21		23	25	26
Block 2	16	19	20		12	14	17	15	20	23	24

($\sum y^2 = 8320$)

(i) Obtain the Chakrabati coefficient matrix C for this design.

(ii) Show that Tocher's Ω is given by

$$\Omega = \frac{1}{966} \begin{bmatrix} 139 & -1 & 0 \\ -1 & 139 & 0 \\ 0 & 0 & 161 \end{bmatrix}$$

(iii) Obtain the Analysis of Variance and interpret the results.

(iv) Estimate the adjusted treatment effects.

(v) Comment on the differences between the means of treatment C and the other treatments, and the difference between A and B.

13. An experiment on rubber trees was supposed to use two replicates of each of the treatments A, B, C in two sites (blocks) in a randomized blocks design. Unfortunately one observation was lost in each block, but the number of units of latex

collected from the remaining trees was measured as follows:

	A		B		C	
Block 1	25		17	23	18	14
Block 2	26	29	25	23	26	

Show that for the realized design the value of Tocher's Ω is given by

$$\Omega = \frac{1}{84} \begin{bmatrix} 29 & 0 & -1 \\ 0 & 21 & 0 \\ -1 & 0 & 29 \end{bmatrix}$$

and hence or otherwise obtain the Analysis of Variance.

Comment on the evidence of explained variability in the experiment. Investigate also whether there is a difference between A and the mean of the other 2 treatments, and also its orthogonal contrast.

14. Three alloys A, B, C, were tested for wear by 4 operators. Each operator was given 5 samples to test and the allocation of alloys to operators and the amount of wear (y), in arbitrary units, of the samples were as indicated.

Operator 1	A: 20	A: 24	B: 16	B: 19	C: 20
Operator 2	A: 21	A: 20	B: 14	B: 16	C: 22
Operator 3	A: 18	B: 18	B: 13	C: 19	C: 17
Operator 4	A: 16	B: 12	B: 14	C: 17	C: 20

($\sum y^2 = 6522$)

(i) Obtain the Analysis of Variance, allowing for possible systematic differences between operators.
(ii) Show that the covariance between the adjusted treatment means for one particular treatment and that of either of the other two is zero.
(iii) Estimate the adjusted treatment effects and compare these values with the estimated effects where no allowance for operator difference is made.
(iv) Examine whether there is real evidence of a difference in wear between B and the mean of the other alloys.

15. An experiment for 3 treatments A, B, C is to be laid out in 3 blocks each of size 5. The following designs have been suggested for the experiment, in which one of the treatments is a control and the most important contrast is the control versus the mean of the other treatments.

Design 1						Design 2					
Block 1	A	A	C	C	C	Block 1	A	A	A	B	C
Block 2	A	A	B	B	B	Block 2	A	B	B	C	C
Block 3	A	B	B	C	C	Block 3	A	B	B	C	C

For each design compute Tocher's matrix, and discuss which is the more appropriate design and decide which letter should be assigned to the control. Assuming that the errors are distributed independently $\mathcal{N}(0, \sigma^2)$, obtain estimators for the variances of suitable treatment contrasts for the design you have chosen.

16. Describe the properties of a Balanced Incomplete Block (BIB) design.

Show that if the number of treatments v exceeds the block size by 1, then a BIB design is always possible, and indicate how it may be formed. For such a design calculate the concurrence of treatment pairs in terms of v, and discuss whether the design is resolvable, unreduced and may be arranged as a Youden square.

If I denotes the identity matrix and J is a matrix of 1's, show that the Chakrabati coefficient matrix C for any BIB design with v treatments can be expressed as a multiple of the matrix L, where

$$L = vI - J$$

Given that all the non-zero eigenvalues of L are equal to v, show that the Moore–Penrose generalized inverse of L is

$$L^+ = \frac{1}{v^2} L$$

and hence that C^+, the Moore–Penrose generalized inverse of C, is also a multiple of L.

Hence or otherwise evaluate C^+ for the unreduced BIB design with 5 treatments in blocks of size 3.

17. For the following design:

Block 1	A	A	B	B
Block 2	A	A	C	C
Block 3	B	B	C	C

obtain the Chakrabati coefficient matrix C and hence show that Tocher's Ω is given by

$$\Omega = \frac{1}{36} \begin{bmatrix} 11 & -1 & -1 \\ -1 & 11 & -1 \\ -1 & -1 & 11 \end{bmatrix}$$

For this design calculate the effective replication of the contrasts

(i) A versus B and C,
(ii) B versus C.

Discuss fully the nature of this design. Obtain an alternative design which uses 4 replications of each of 3 treatments in 3 blocks and is balanced with respect to treatments; comment briefly on this alternative design.

18. Define a *generalized inverse* of a real symmetric matrix and describe any generalized inverses you know that are useful for the analysis of general block designs.

A randomized blocks experiment with 3 treatments in 4 blocks suffered a missing value in one plot. The responses were as follows:

	Treatment		
	A	B	C
Block 1	20	16	22
Block 2	23	18	27
Block 3	25	20	30
Block 4	26	17	–

For the realized design show that the Chakrabati coefficient matrix is given by

$$C = \tfrac{1}{2} \begin{bmatrix} 5 & -3 & -2 \\ -3 & 5 & -2 \\ -2 & -2 & 4 \end{bmatrix}$$

and hence that Tocher's Ω is given by

$$\Omega = \frac{1}{968} \begin{bmatrix} 245 & 3 & -8 \\ 3 & 245 & -8 \\ -8 & -8 & 344 \end{bmatrix}$$

Obtain the correct Analysis of Variance for this experiment and comment on whether there is evidence of real treatment differences.

19. (i) Describe what is meant by the Moore–Penrose inverse C^+.
   (ii) The design and responses for a particular experiment were as follows:

Block 1	A	45	A	40	B	37		
Block 2	A	35	B	30	B	25		
Block 3	A	26	C	30	C	28	C	20

Given that the Moore–Penrose inverse for this design is

$$C^+ = \frac{1}{108} \begin{bmatrix} 25 & -2 & -23 \\ -2 & 52 & -50 \\ -23 & -50 & 73 \end{bmatrix}$$

obtain the Analysis of Variance for the data, using any valid method, and comment on the result.

Obtain the estimated variances of the contrasts between

(a) C and the mean of A and B,
(b) A and B,
(c) A and C,
(d) B and C.

By inspection of the design show that you would have expected the value obtained from (d) to be substantially greater than that for (b) and (c).

Estimate the adjusted treatment effects for this experiment and investigate the difference indicated by comparison (a).

20. (i) Show that post-multiplication of a matrix by the vector $\mathbf{1}$ gives the vector of its row sums.

   (ii) Establish the identities (pages 221–2) $\Delta D' = N$; $DD' = k^\delta$; $\Delta\Delta' = r^\delta$; $N\mathbf{1} = r$; $N'\mathbf{1} = k$; $D\mathbf{1} = k$; $\Delta\mathbf{1} = r$; $D'\mathbf{1} = 1$; $\Delta'\mathbf{1} = 1$; $r^\delta\mathbf{1} = r$; $k^\delta\mathbf{1} = k$; $r'\mathbf{1} = n$; $k'\mathbf{1} = n$; $y'\mathbf{1} = G$; $Dy = B$; $\Delta y = T$.

21. Show that every Youden 'square' (Youden rectangle) has a complementary 'square' (rectangle) such that when the two are combined a Latin square is formed.

Appendix 11A   Generalized inverse matrix by spectral decomposition

Another form of generalized inverse matrix, due to Moore and Penrose, is often denoted by C^+. It arises from the spectral decomposition theorem for a real symmetric matrix C, whose eigenvalues are λ_i and corresponding eigenvectors u_i; the theorem states that

$$C = \sum_{i=1}^{h} \lambda_i u_i u_i'$$

summed over the h non-zero eigenvalues. We may take

$$C^+ = \sum_{i=1}^{h} \frac{1}{\lambda_i} u_i u_i'$$

so that

$$CC^+C = \sum_{i=1}^{h} \frac{1}{\lambda_i} Cu_i u_i' C = \sum_{i=1}^{h} u_i u_i' C \quad \text{(because } Cu_i = \lambda_i u_i)$$

$$= \sum_{i=1}^{h} \lambda_i u_i u_i' \quad \text{(because } u_i' C = \lambda_i u_i')$$

$$= C$$

Example 11.4. For the balanced incomplete block design

$$\text{I}: ABC; \quad \text{II}: ABD; \quad \text{III}: ACD; \quad \text{IV}: BCD$$

we have

$$N = \begin{pmatrix} 1 & 1 & 1 & 0 \\ 1 & 1 & 0 & 1 \\ 1 & 0 & 1 & 1 \\ 0 & 1 & 1 & 1 \end{pmatrix}, \quad k = \begin{pmatrix} 3 \\ 3 \\ 3 \\ 3 \end{pmatrix}, \quad r = \begin{pmatrix} 3 \\ 3 \\ 3 \\ 3 \end{pmatrix}, \quad NN' = \begin{pmatrix} 3 & 2 & 2 & 2 \\ 2 & 3 & 2 & 2 \\ 2 & 2 & 3 & 2 \\ 2 & 2 & 2 & 3 \end{pmatrix} = I + 2J$$

$C = r^\delta - Nk^{-\delta}N' = 3I - \frac{1}{3}(I + 2J) = \frac{1}{3}(8I - 2J)$, with eigenvalues 8/3 (three times) and 0.
To find an eigenvector u, solve $(C - \lambda I)u = 0$ which gives

$$\left\{ \frac{1}{3}(8I - 2J) - \frac{8}{3}I \right\}u = 0 \quad \text{or} \quad Ju = 0$$

A set (not unique) of $(v - 1)$ independent contrasts is thus obtained by solving $1'u = 0$.
One such set is

$$u_1 = \frac{1}{\sqrt{2}}(1 \ -1 \ 0 \ 0)'; \qquad u_2 = \frac{1}{\sqrt{6}}(1 \ 1 \ -2 \ 0)'; \qquad u_3 = \frac{1}{\sqrt{12}}(1 \ 1 \ 1 \ -3)'$$

For these,

$$C^+ = \sum_{i=1}^{3} \frac{1}{\lambda_i} u_i u_i'$$

$$= \frac{1}{8} \left\{ \frac{1}{2} \begin{pmatrix} 1 & -1 & 0 & 0 \\ -1 & 1 & 0 & 0 \\ 0 & 0 & 0 & 0 \\ 0 & 0 & 0 & 0 \end{pmatrix} + \frac{1}{6} \begin{pmatrix} 1 & 1 & -2 & 0 \\ 1 & 1 & -2 & 0 \\ -2 & -2 & 4 & 0 \\ 0 & 0 & 0 & 0 \end{pmatrix} + \frac{1}{12} \begin{pmatrix} 1 & 1 & 1 & -3 \\ 1 & 1 & 1 & -3 \\ 1 & 1 & 1 & -3 \\ -3 & -3 & -3 & 9 \end{pmatrix} \right\}$$

$$= \frac{3}{8} \cdot \frac{1}{12} \left\{ \begin{pmatrix} 6 & -6 & 0 & 0 \\ -6 & 6 & 0 & 0 \\ 0 & 0 & 0 & 0 \\ 0 & 0 & 0 & 0 \end{pmatrix} + \begin{pmatrix} 2 & 2 & -4 & 0 \\ 2 & 2 & -4 & 0 \\ -4 & -4 & 8 & 0 \\ 0 & 0 & 0 & 0 \end{pmatrix} + \begin{pmatrix} 1 & 1 & 1 & -3 \\ 1 & 1 & 1 & -3 \\ 1 & 1 & 1 & -3 \\ -3 & -3 & -3 & 9 \end{pmatrix} \right\}$$

$$= \frac{1}{32} \begin{pmatrix} 9 & -3 & -3 & -3 \\ -3 & 9 & -3 & -3 \\ -3 & -3 & 9 & -3 \\ -3 & -3 & -3 & 9 \end{pmatrix} = \frac{3}{32}(4I - J)$$

Here

$$J = \begin{pmatrix} 1 & 1 & 1 & 1 \\ 1 & 1 & 1 & 1 \\ 1 & 1 & 1 & 1 \\ 1 & 1 & 1 & 1 \end{pmatrix}$$

so that

$$J^2 = \begin{pmatrix} 4 & 4 & 4 & 4 \\ 4 & 4 & 4 & 4 \\ 4 & 4 & 4 & 4 \\ 4 & 4 & 4 & 4 \end{pmatrix} = 4J$$

and

$$J^3 = J^2 J = 4JJ = 4J^2 = 16J$$

Thus

$$CC^+ = \tfrac{1}{32}(32I - 8J - 8J + 2J^2) = \tfrac{1}{32}(32I - 8J) = \tfrac{1}{4}(4I - J)$$

and so

$$CC^+C = \tfrac{1}{4}(4I - J)\tfrac{1}{3}(8I - 2J) = \tfrac{1}{12}(32I - 16J + 2J^2)$$
$$= \tfrac{1}{12}(32I - 8J)$$
$$= \tfrac{1}{3}(8I - 2J) = C$$

as required.

Appendix 11B   Natural contrasts and effective replication

Following on the use of eigenvectors in Appendix 11A, we may define the *natural contrasts* among a set of treatments as the *eigenvectors* of C. A (non-unique) set of eigenvectors for Example 11.4 is therefore given as the u_1, u_2, u_3 quoted there, but we could equally well take u_1 as the contrast between treatments A and D (instead of A and B) and u_2 as B versus A and D (instead of C versus A and B); then u_3 would have to be C versus A, B and D (instead of D versus A, B and C).

For the basic, simple versions of orthogonal designs, and also for balanced incomplete blocks, all treatments have equal replication r, and so contrasts or comparisons among them are all based on r replicates. From linear model theory, the variance of a contrast $c'a$, found from the treatment totals as $\sum_i c_i T_i$, is estimated as

$$\text{Var}(c'\hat{a}) = c' \, \text{Var}(\hat{a})c = c'C^+c\hat{\sigma}^2$$

where $\hat{\sigma}^2$ is the estimate of σ^2 found from the residual mean square in the Analysis of Variance. Therefore for unbalanced or non-orthogonal designs it is useful to define *effective replication*

$$R = \frac{c'c}{c'C^+c}$$

In a balanced orthogonal design, $R = r$ for any contrast being examined.

In a *general block design* the *effective replication of a contrast is its corresponding eigenvalue*.

To prove this we will use $C^+ = \sum_i (1/\lambda_i)u_i'\mu_i$. Therefore

$$c'C^+c = \sum_i \frac{1}{\lambda_i} c'u_i u_i' c$$

If we consider $c = \kappa u_i$, and use the fact that the set $\{u_i\}$ are mutually orthogonal, then

$$c'C^+c = \frac{\kappa^2}{\lambda_i} \quad \text{and} \quad c'c = \kappa^2$$

Thus $R = \lambda_i$.

Example 11.4: continued. The non-zero eigenvalues in this balanced incomplete block are each 8/3, which gives $R = 8/3$. The parameters of the design are $r = 3$, $k = 3$, $v = 4$, $b = 4$, $\lambda = 2$, giving $k/\lambda v = 3/8 = 1/R$ which is why we can call R the 'effective replication', since the variance of a difference between two means is $2k\sigma^2/\lambda v$.

12
More advanced designs

12.1 Introduction

In this chapter we consider extensions to our armoury of experimental designs. The new designs covered are *cross-overs*, *lattices*, *alpha designs* and *partially balanced incomplete blocks*. These topics will not be treated thoroughly here as this is an introductory text, but the practising statistician should be aware of the extensions to the designs considered so far in case the opportunity arises where such designs are appropriate. The crossover design has been in use for many years in agriculture, but its major importance at the present time is in medicine. Lattices and alpha designs are mostly of interest to experimenters who have large numbers of treatments, in particular those who conduct agricultural variety trials; lattices have been around for a long time but alpha designs are a more recent invention. The need for special designs such as partially balanced incomplete block designs, and the study of their properties, is not as great as it once was, the driving force in the past being ease of analysis which was attributable to the lack of computing power; nowadays we might be more concerned to plan experiments so that particular contrasts are well estimated, designs *optimal* in this sense being helpful.

12.2 Crossover designs

Crossover designs were commonly used in the 1950s for experiments on dairy cows receiving different diets, but are these days of considerable importance in medicine. The essential principle is that the same subject receives a sequence of different treatments so that a comparison may be made between the effects of the different treatments on the same subject. Both cows and patients are notoriously variable and the treatment differences sought may be considerably less than the differences between subjects. The basic design is the 2-period 2-treatment crossover:

Period		1	2
Patient group	1	*A*	*B*
	2	*B*	*A*

which indicates that some of the patients receive the sequence *A* followed by *B* while the rest receive treatment *B* followed by *A*. The rows of the design indicate the sequences applied to the two groups of patients. This means that every patient effectively acts as a 'block' of two 'plots', and so has received both treatments at some stage, whereas in an alternative 'parallel group' study each patient receives *either A or B*, but not both. This advantage of receiving both treatments may be important for the reason that variability between patients is often quite large, so many more patients may be required to detect the same treatment difference in the parallel group experiment than for the crossover design. Unfortunately there are also disadvantages, for example there may be a carry-over effect from one period to another so that in the sequence AB a patient may still be feeling the effects of A received in the first period during the second period when B is received. It may be possible to mitigate the carry-over with the help of a 'washout' stage between periods 1 and 2 when no treatment is given, or when a standard treatment is applied to every patient; this practice is regularly followed. Carry-over is one deficiency of the 2-period 2-treatment crossover but there are other problems too, notably the aliasing of the carry-over with other effects, so it is perhaps surprising that the design is so popular. Practical problems include systematic change in a patient's condition over time, or in the background conditions of the experiment.

The analysis of data with the basic design may be performed with an Analysis of Variance, but there are alternative and equivalent analyses that may be obtained with the help of t-tests.

If the patients are available for 3 periods then the basic design may be improved with the reversal to the first period treatment as in the following design:

Period		1	2	3
Patient group	1	A	B	A
	2	B	A	B

and this design has the additional benefit that every patient receives the sequence AB and the sequence BA at some stage.

If there are 3 treatments and 3 periods, then a Latin square is an obvious possibility for a suitable design: however there is a drawback with this design if carry-over effects are important. Suppose that the design is:

Period		1	2	3
Patient group	1	A	B	C
	2	B	C	A
	3	C	A	B

In two of the sequences A is followed by B, but B is never followed by A, and a similar feature is apparent for every other pair of treatments. The deficiency would be mitigated if A and B followed each other once, but that is not possible with the 3×3 Latin square. However, there is a better alternative to the design with 3 sequences and this design uses the complementary Latin square where the treatment sequences are reversed in addition. The subjects are then divided into 6 sets and each set receives one of the sequences shown in the scheme:

Period		1	2	3
Patient group	1	*A*	*B*	*C*
	2	*B*	*C*	*A*
	3	*C*	*A*	*B*
	4	*C*	*B*	*A*
	5	*A*	*C*	*B*
	6	*B*	*A*	*C*

made up from the rows of the original Latin square and its complementary square. Patients would have to be allocated to groups at random, and treatments to letters at random in order to construct a design for practical use. Note that each treatment is followed by each of the other treatments twice in this design.

For a detailed description of cross-over trials the reader is referred to Jones and Kenward (1989), which gives a thorough treatment of this topic, including experiments with many treatments continuing through many periods.

12.3 Lattices

Lattice designs are suitable for experiments that involve large numbers of treatments and they are particularly useful for variety trials in agriculture and horticulture. In such experiments it is not usual to have a strong treatment structure, as the main purpose of the experiment may be a screening test to see which varieties give the best responses so that the most promising ones may be subjected to further investigation.

The etymology of the word 'lattice' suggests that it is derived from a cross structure, and the reason for the name is evident from the layout of an example, such as the simple 3×3 lattice design:

Replicate 1					Replicate 2			
Block 1	1	2	3		Block 4	1	4	7
Block 2	4	5	6		Block 5	2	5	8
Block 3	7	8	9		Block 6	3	6	9

Observe that the 9 treatments are written out as rows in the first replicate and as columns in the second. This means that the blocks form two sets where each treatment occurs exactly once in each set, which means the design is resolvable. When there are two replications the design is called a simple lattice, but we shall see that this can easily be extended to more replicates. The simple 3×3 form of the design is not particularly useful as the design allows only 4 d.f. for the residual; this restriction would not be considered acceptable for experiments in agriculture but may be entertained in industry where unit costs can be considerable and residual variance may be less.

Further replicates may be obtained when orthogonal Latin squares are superimposed on the designs and further blocks are formed from the positions of the letters. In the present case suitable Latin squares are:

$$
\begin{array}{ccc}
A & B & C \\
B & C & A \\
C & A & B
\end{array}
\qquad
\begin{array}{ccc}
A & B & C \\
C & A & B \\
B & C & A
\end{array}
$$

These are mutually orthogonal because each occurrence of each letter in one square corresponds to all 3 letters once each in the same positions in the other square, as in a Graeco-Latin square.

When the first Latin square is superimposed on the first replicate square, then Replicate 3 is formed and the superimposition of the second Latin square on Replicate 1 gives Replicate 4, as shown below.

Replicate 3			
Block 7	1	6	8
Block 8	2	4	9
Block 9	3	5	7

Replicate 4			
Block 10	1	5	9
Block 11	2	6	7
Block 12	3	4	8

If all 4 replicates are used, then the design is balanced since each treatment concurs once in a block with every other treatment. However, if less replicates are used then some of the treatment pairs have a concurrence of 1 and others a concurrence of 0, which indicates that some treatments may be compared with greater precision than others.

Below are the concurrence matrices for (a) the *triple lattice* formed from Replicates 1, 2, 3 and (b) the *quadruple lattice* obtained when all four replicates are used.

$$
\begin{bmatrix}
3 & 1 & 1 & 1 & 0 & 1 & 1 & 1 & 0 \\
1 & 3 & 1 & 1 & 1 & 0 & 0 & 1 & 1 \\
1 & 1 & 3 & 0 & 1 & 1 & 1 & 0 & 1 \\
1 & 1 & 0 & 3 & 1 & 1 & 1 & 0 & 1 \\
0 & 1 & 1 & 1 & 3 & 1 & 1 & 1 & 0 \\
1 & 0 & 1 & 1 & 1 & 3 & 0 & 1 & 1 \\
1 & 0 & 1 & 1 & 1 & 0 & 3 & 1 & 1 \\
1 & 1 & 0 & 0 & 1 & 1 & 1 & 3 & 1 \\
0 & 1 & 1 & 1 & 0 & 1 & 1 & 1 & 3
\end{bmatrix}
\begin{bmatrix}
4 & 1 & 1 & 1 & 1 & 1 & 1 & 1 & 1 \\
1 & 4 & 1 & 1 & 1 & 1 & 1 & 1 & 1 \\
1 & 1 & 4 & 1 & 1 & 1 & 1 & 1 & 1 \\
1 & 1 & 1 & 4 & 1 & 1 & 1 & 1 & 1 \\
1 & 1 & 1 & 1 & 4 & 1 & 1 & 1 & 1 \\
1 & 1 & 1 & 1 & 1 & 4 & 1 & 1 & 1 \\
1 & 1 & 1 & 1 & 1 & 1 & 4 & 1 & 1 \\
1 & 1 & 1 & 1 & 1 & 1 & 1 & 4 & 1 \\
1 & 1 & 1 & 1 & 1 & 1 & 1 & 1 & 4
\end{bmatrix}
$$
$$\text{(a)} \qquad\qquad\qquad\qquad \text{(b)}$$

The diagonal elements of the concurrence matrices are merely the treatment replications, but it is interesting to note that the off-diagonal elements are either 1 or 0 in the case of the triple lattice while they are all 1 for the quadruple lattice. In fact the quadruple lattice satisfies the conditions for a Balanced Incomplete Block design (BIB) and the triple lattice is a Partially Balanced Incomplete Block design (PBIB, which will be discussed in more detail later). Simple lattice designs are clearly available for any number of treatments $v = k^2$ that is an exact square, but balanced lattices are usually available for the most important values of k except 6, which is explained by the absence of Graeco-Latin squares of this order.

As mentioned in the last paragraph, the complete design of four 3×3 lattice replicates provides a BIB design which has the important feature that all treatment contrasts have the same variance; this is useful if it is not known in advance which treatment contrasts are of greater importance. It is known that for blocks of size k the k^2 treatments in a lattice design may be arranged in a balanced design in $k+1$ replicates. That the number of replicates is $k+1$ is clear since every treatment needs to concur with k^2-1 other treatments in blocks of size k, so there are $k-1$ vacancies for them in each block. The number of replicates is therefore $(k^2-1)/(k-1)=k+1$. This explains why 4 replicates were required to provide a balanced lattice design for 9 treatments so that the complete design is a BIB. One of the important properties of the lattice design is the resolvability; it may not always be possible to perform the complete experiment and if one of the replicates needs to be dropped from, say, a quadruple lattice, then a considerable amount of useful information may still be obtained from the reduced experiment.

The quadruple lattice obtained from the 4 replicates possesses a further interesting property, as it may be adapted into the related but different design given below.

Replicate 1	Replicate 2
1 2 3	1 6 8
4 5 6	9 2 4
7 8 9	5 7 3

At first sight this appears to be a simple lattice design with 9 treatments in 6 blocks, but if concurrence is taken to indicate companionship in either rows or columns then it is evident that every treatment concurs with every other treatment exactly once. The whole design is therefore balanced over rows and columns and is called a *Balanced Lattice Square* design. With 9 treatments, 6 rows and 6 columns, the design itself is not particularly useful unless the replicates were repeated, in which case it would be better to use the quadruple lattice unless there were good reason to block by columns as well as rows. However, the principles of double blocking in lattice designs may be extended to higher-order designs without the necessity to repeat the design in order to be able to obtain an estimate of the residual variation.

There are further extensions of lattice designs to rectangular lattices, where the number of treatments is of the form $v=k(k-1)$, and cubic lattices where the number of treatments is very large and of the form $v=k^3$, but these will not be described here. However, Cochran and Cox (1992) and John (1987) discuss these designs more fully.

It may not always be possible to obtain a design with all the properties that are desired; however, the most useful designs and their properties are well known and the reader is referred to Cochran and Cox (1992) for a comprehensive set of plans which satisfy most requirements for block designs.

12.4 Alpha designs

Variety trials in agriculture and plant breeding are experiments that use a large number of treatments (varieties), the purpose being to decide which are the most promising for further investigation. Traditionally lattice designs have been one of the principal tools for these experiments, but these require the number of treatments to be a square number, which is often not true. There are various techniques whereby the lattice may be modified to remedy this problem, in particular treatments may be doubled; in a 3×3 triple lattice, for example, if only 7 treatments are available then treatments 1 and 2 may be repeated as treatments 8 and 9. This solution may be practically easy but it may not make the best use of resources since the variances of the contrasts between particular treatment pairs are then very different. In exploratory experiments it is often useful to try to equalize the variance of treatment contrasts as far as possible. Balanced incomplete blocks satisfy this requirement exactly, but a suitable BIB design is not always available.

A new class of designs was devised by Patterson and Williams (1976) which is of considerable importance for research workers in this context. These designs are available in many cases where the number of treatments v is an exact multiple of the block size, say $v = ks$. The construction of an α-design requires the following steps:

(i)   obtain a generating array α_1,
(ii) use the columns of α_1 to generate the design array α_2,
(iii) use the rows of α_2 to generate the design array α_3.

The basic arrays α_1 are listed by Patterson *et al.* (1978). All elements of the design consist of the residues modulo s. For step (ii) add the values $0, 1, 2, \ldots, (s-1)$ in turn to each of the columns of α_1 to give the columns of the array α_2. For step (iii) add the values $0, s, 2s, \ldots, (k-1)s$ to the rows of α_2 to obtain the array α_3. An example should help to clarify the procedure.

Example 12.1

Consider the basic array

$$\alpha_1 = \begin{bmatrix} 0 & 0 & 0 \\ 0 & 1 & 2 \\ 0 & 2 & 1 \end{bmatrix}$$

The array is formed from residues modulo 3, so $s = 3$. Now add the numbers 0, 1, 2 to column 1, then column 2, then column 3 in turn to give

$$\alpha_2 = \begin{bmatrix} 0 & 1 & 2 \\ 0 & 1 & 2 \\ 0 & 1 & 2 \end{bmatrix} \quad \begin{bmatrix} 0 & 1 & 2 \\ 1 & 2 & 0 \\ 2 & 0 & 1 \end{bmatrix} \quad \begin{bmatrix} 0 & 1 & 2 \\ 2 & 0 & 1 \\ 1 & 2 & 0 \end{bmatrix}$$

To obtain α_3 add 0 to the first row of α_2, 3 to the second row, 6 to the third:

$$\alpha_3 = \begin{bmatrix} 0 & 1 & 2 \\ 3 & 4 & 5 \\ 6 & 7 & 8 \end{bmatrix} \quad \begin{bmatrix} 0 & 1 & 2 \\ 4 & 5 & 3 \\ 8 & 6 & 7 \end{bmatrix} \quad \begin{bmatrix} 0 & 1 & 2 \\ 5 & 3 & 4 \\ 7 & 8 & 6 \end{bmatrix}$$

The columns of α_3 form the blocks of the final design, which may be given in this form or the alternative form where the levels are increased by one.

Block 1:	1	4	7	Block 4:	1	5	9	Block 7:	1	6	8
Block 2:	2	5	8	Block 5:	2	6	7	Block 8:	2	4	9
Block 3:	3	6	9	Block 6:	3	4	8	Block 9:	3	5	7

The final design has $v=9$, $k=3$, $r=3$, $b=9$, and of course $s=3$. The concurrence matrix for this design gives off-diagonal values of λ equal to 0 or 1, so contrasts between different treatment pairs form 2 sets accordingly. In fact this design is not only an alpha design but also a triple lattice and a PBIB, as discussed in the next section. However, although all alpha designs are formed in the manner described in this section, it does not follow that they will all form either of the other designs in general.

Patterson and Williams (1976) describe how generators may be formed from four different schemes. The designs are in s blocks of size k with r replications, where the condition $k\leqslant s$ is necessary but not sufficient. The generators are formed by the 4 series shown below.

Series 1		Series 2			Series 3			Series 4			
0	0	0	0	0	0	0	0	0	0	0	0
0	1	0	1	$s-1$	0	1	t	0	1	$s-1$	u
0	2	0	2	$s-2$	0	2	1	0	2	$s-2$	1
0	3	0	3	$s-3$	0	3	$t+1$	0	3	$s-3$	$u+1$
....		....			0	4	2	0	4	$s-4$	2
0	$s-1$	0	$s-1$	1	....			....			
					0	$s-1$	$s-2$	0	$s-2$	2	$s-1$
					0	$s-2$	$t-1$	0	$s-1$	1	$u-1$

$r=2$	$r=3$	$r=3$	$r=4$
$k\leqslant s$	$k\leqslant s$	$k\leqslant s-1$	$k\leqslant s$
	s odd	s even	s odd
		$t=s/2$	$u=(s+1)/2$

From this table it may be seen that Example 12.1 belongs to the class defined by Series 2 where $s=3$. The columns of the series are formed in a systematic way, the first column consisting of zeros and the second is an integer ascending sequence; other columns are either a descending sequence (almost) or else an alternate mixture of ascending and descending sequences.

For further study of this topic the reader should consult the references quoted earlier in the section.

12.5 Partially balanced incomplete blocks (PBIBs)

When all treatment comparisons are equally important, then it is desirable to make a suitable provision in the design to allow for comparisons to be made on the same basis in the analysis. It is a feature of the BIB design that all pairwise treatment contrasts have the same error variance, but unfortunately reduced BIBs are not always available for all values of v and k. This means that the experimenter may have to choose a large unreduced BIB with the number of blocks equal to the number of combinations of k

from v treatments. However, if the experimenter is willing to relax the condition that all treatment contrasts must have the same error variance, then a PBIB may provide a suitable alternative, particularly if he knows in advance which contrasts are of greater and which of lesser importance.

The definition of a PBIB design is not particularly easy to understand. Let the number of treatments be v; then for a PBIB

(i) the block size is constant over all blocks and equals k,
(ii) the replication is constant for all treatments and equals r,
(iii) the number of concurrences between any two treatments is one of m distinct values λ_a,
(iv) the association matrix p_{ij}^α is constant (see explanation later).

The first two conditions are clear, but the third needs a little amplification; it should be noted that in practice m is a small number, such as 1, 2 or perhaps 3. If $m=1$ then λ is constant and the design is a BIB; however, if $m=2$ then treatments with concurrence λ_1 are *first associates* and those with concurrence λ_2 are *second associates*, where $\lambda_1 > \lambda_2$. Higher values of m demand a similar ordering of associate classes so that the λ_i are in descending order. The most complicated condition of the PBIB design is the fourth. It is customary to use the symbol p_{ij}^α to denote the matrix of association where the ijth element is the number of treatments that concur λ_i times with one associate and λ_j times with the other. We shall consider an example that should help to clarify the point.

Example 12.2. Show that the triple 3×3 lattice satisfies the conditions of a PBIB design.

Recall the design:

Replicate 1				Replicate 2				Replicate 3			
Block 1	1	2	3	Block 4	1	4	7	Block 7	1	6	8
Block 2	4	5	6	Block 5	2	5	8	Block 8	2	4	9
Block 3	7	8	9	Block 6	3	6	9	Block 9	3	5	7

and the concurrence matrix

$$
\begin{bmatrix}
3 & 1 & 1 & 1 & 0 & 1 & 1 & 1 & 0 \\
1 & 3 & 1 & 1 & 1 & 0 & 0 & 1 & 1 \\
1 & 1 & 3 & 0 & 1 & 1 & 1 & 0 & 1 \\
1 & 1 & 0 & 3 & 1 & 1 & 1 & 0 & 1 \\
0 & 1 & 1 & 1 & 3 & 1 & 1 & 1 & 0 \\
1 & 0 & 1 & 1 & 1 & 3 & 0 & 1 & 1 \\
1 & 0 & 1 & 1 & 1 & 0 & 3 & 1 & 1 \\
1 & 1 & 0 & 0 & 1 & 1 & 1 & 3 & 1 \\
0 & 1 & 1 & 1 & 0 & 1 & 1 & 1 & 3
\end{bmatrix}
$$

There are two values of λ for the off-diagonal elements, $\lambda=1$ for first associates and $\lambda=0$ for second associates. We could now use the concurrence table to form lists of the concurrences with any two particular treatments:

Treatment 1	2	3	4	6	7	8
Treatment 2	1	3	4	5	8	9

We ignore Treatments 1 and 2 as these are the chosen treatments and observe that Treatments 3, 4, 8 concur with each of the remaining treatments once each, 6, 7 concur with Treatment 1 once, 5, 9 concur with Treatment 2 once but there are no other concurrences with the chosen treatments. These results may be summarized as follows:

		Treatment 2	
		$\lambda=1$	$\lambda=0$
Treatment 1	$\lambda=1$	3, 4, 8	6, 7
	$\lambda=0$	5, 9	–

Now Treatments 1, 2 are first associates and so $\alpha=1$ in this case. We have shown that

$$p_{12}^1 = \begin{bmatrix} 3 & 2 \\ 2 & 0 \end{bmatrix}$$

and it may be found that the same matrix is obtained for any two treatments that are first associates. Now consider 2 treatments that are second associates, such as Treatments 3 and 8. The concurrence table gives the associations

		Treatment 8	
		$\lambda=1$	$\lambda=0$
Treatment 3	$\lambda=1$	1, 2, 5, 6, 7, 9	–
	$\lambda=0$	–	–

so $p_{38}^2 = \begin{bmatrix} 6 & 0 \\ 0 & 0 \end{bmatrix}$

The associate matrices can be shown to be constant over all first associates and over all second associates, so the fourth condition for a PBIB design is satisfied.

For more details of the PBIB design consult Cochran and Cox (1992) and John (1987). The PBIB design has been extensively tabulated by Clatworthy (1973) but most investigation has been confined to the cases where $m=2$.

13
Random effects models: variance components and sampling schemes

13.1 Introduction

So far almost all the linear models used have contained only one random element, the residual term ε_{ij}: the only exception was in Chapter 7 where more than 1 level of variation existed and so 2 (or more) residual terms were needed for the different levels. Otherwise the block, row, column and treatment terms have all been assumed *fixed*, specific just to those treatments actually included in that experiment, and to that set of experimental units, machines, operatives or that piece of land available for the experiment.

Occasionally it is more appropriate to assume that the blocks of material were a random selection from a much larger population that could have been used; and obviously this idea would seem to give wider validity to the results of an experiment. We might also assume that the treatments actually selected were a random selection from a larger population of possible treatments, although this idea would only be applicable where there is available, for example, a large collection of similar varieties of a fruit crop or a corn crop, or where some variability is inevitable in preparing a mixture for an industrial experiment. Cochran and Cox (1992) and Dyke (1988) discuss a series of experiments carried out to study the same problem, at several different sites or on several different occasions, the whole series being considered together. These ideas of random selection are useful in that situation also.

Another related problem is when taking records in an experiment that involves sampling from a large amount of material and selecting relatively very small amounts upon which to make accurate measurements, such as chemical analyses. Each level of sampling or subsampling will carry its own component of variation or *variance component*. A linear model in which all the terms except the overall mean μ have their own characteristic variation is called a *random effects* model. If some terms are fixed and others random, we have a *mixed* model.

In this chapter we shall consider the estimation of variance components in their own right, and also their use in determining the most economical sampling schemes. We will limit our study to linear models like those for completely randomized and randomized complete block designs, including those that involve sub-sampling from larger units.

13.2 Two stages of sampling: between and within units

This is the same situation as in a completely randomized experiment, with between-unit variation being equivalent to treatments (which can also be called 'between groups'), and within-unit variation being equivalent to residual ('within groups'). The linear model contains the same terms as that on page 32, with slightly different properties.

$$y_{ij} = \mu + \tau_i + \varepsilon_{ij} \qquad (i = 1 \text{ to } v, j = 1 \text{ to } r_i)$$

where $\sum r_i = N$, $E[\tau_i] = 0$, $E[\varepsilon_{ij}] = 0$, $\text{Var}[\varepsilon_{ij}] = \sigma^2$ and now $\text{Var}[\tau_i] = \sigma_1^2$. All $\{\varepsilon_{ij}\}$ and all $\{\tau_i\}$ are uncorrelated.

The Analysis of Variance table is just as on page 35 as far as the calculation of mean squares; we now examine the expectations of these.

The 'within-groups', or residual, sum of squares is

$$\sum_{i=1}^{v} \sum_{j=1}^{r_i} (y_{ij} - \bar{y}_i) = \sum_i \sum_j \{(\mu + \tau_i + \varepsilon_{ij}) - (\mu + \tau_i + \bar{\varepsilon}_i)\}^2$$

$$= \sum_i \sum_j (\varepsilon_{ij} - \bar{\varepsilon}_i)^2$$

exactly as for the fixed model, so that it has the expected value $(N - v)\sigma^2$.

The 'between-groups', or treatment, sum of squares is calculated as $\sum_{i=1}^{v} T_i^2/r_i - G^2/N$, and its expectation is $\sum_{i=1}^{v} 1/r_i \, E[T_i^2] - 1/N \, E[G^2]$. In finding its expectation, we use the result that for any random variable X, $\text{Var}[x] = E[x^2] - (E[x])^2$ (see, e.g. Clarke and Cooke (1992)).

Now

$$T_i = r_i \mu + r_i \tau_i + \sum_{j=1}^{r_i} \varepsilon_{ij}$$

so that

$$E[T_i] = r_i \mu$$

using the properties of the terms in the linear model. Since T_i is a sum of independent random variables,

$$\text{Var}[T_i] = r_i^2 \sigma_1^2 + r_i \sigma^2$$

and hence

$$E[T_i^2] = \text{Var}[T_i] + (E[T_i])^2 = r_i^2 \sigma_1^2 + r_i \sigma^2 + r_i^2 \mu^2$$

Also

$$G = N\mu + \sum_i r_i \tau_i + \sum_i \sum_j \varepsilon_{ij}$$

and

$$E[G] = N\mu$$

so that

$$E[G^2] = \text{Var}[G] + (E[G])^2 = \sum_i r_i \sigma_1^2 + N\sigma^2 + N^2 \mu^2$$

Thus

$$E[\text{between-groups SS}] = \sum_i \frac{1}{r_i} E[T_i^2] - \frac{1}{N} E[G^2]$$

$$= \sum_i r_i \sigma_1^2 + \sum_i \sigma^2 + \sum_i r_i \mu^2 - \frac{1}{N} \sum_i r_i^2 \sigma_1^2 - \sigma^2 - N\mu^2$$

$$= \left(N - \frac{1}{N} \sum_i r_i^2\right) \sigma_1^2 + (v-1)\sigma^2$$

When each $r_i = r$, so that $N = rv$, this expectation reduces to

$$(N-r)\sigma_1^2 + (v-1)\sigma^2 \quad \text{or} \quad r(v-1)\sigma_1^2 + (v-1)\sigma^2$$

We summarize all this information in the following table:

		Expected value of mean square	
Source of variation	d.f.	General case	Equal replication
Between groups	$v-1$	$\sigma^2 + \lambda\sigma_1^2$	$\sigma^2 + r\sigma_1^2$
Within groups	$N-v$	σ^2	σ^2
Total	$N-1$		

The value of λ is

$$\frac{1}{v-1}\left(N - \frac{1}{N}\sum_{i=1}^{v} r_i^2\right)$$

Example 13.1. Four different laboratories, selected at random from all those available, are testing the quality of a food product. One of the measurements made is of the amount of a particular constituent present in a standard sample. In an experiment to examine sources of variation, samples from a well-mixed batch of the product are distributed to each laboratory *A, B, C, D* and they are asked to take 6 subsamples from each, carry out the test and report the measurement made on each subsample. However, for various operational reasons they do not all report on 6 subsamples; Fig. 13.1 illustrates the outcome.

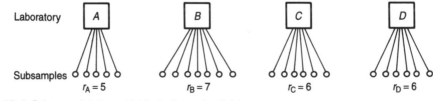

Fig. 13.1 Scheme of data available in Example 13.1

The person conducting the experiment considers that the samples distributed to the laboratories were as uniform as they could possibly be made, and the variation between the results from *A, B, C* and *D* can be ascribed to the laboratories. There are many more laboratories that could be involved in routine work of this nature, and the between-laboratory component of variance, σ_L^2, is therefore of interest. The within-laboratory component σ^2 is assumed to be the same in each laboratory. The experimental data are:

		Total
A	16.0, 17.1, 16.9, 17.2, 17.0	84.2
B	17.0, 17.3, 16.2, 17.1, 16.0, 17.2, 17.0	117.8
C	16.9, 16.1, 16.4, 16.1, 16.6, 16.3	98.4
D	15.0, 15.9, 16.0, 15.9, 16.2, 15.9	94.9
		$G = 395.3$

$$N = 24; \quad S_0 = 395.3^2/24 = 6510.9204; \quad S = 6518.95;$$

$$S_L = \frac{84.2^2}{5} + \frac{117.8^2}{7} + \frac{98.4^2 + 94.9^2}{6} = 6515.0954$$

Analysis of Variance

Source of variation	d.f.	Sum of squares	Mean square	E[M.S.]
Between laboratories	3	4.1750	1.3917	$\sigma^2 + \lambda \sigma_L^2$
Within laboratories	20	3.8546	0.1927	σ^2
Total	23	8.0296		

$$\lambda = \frac{1}{3}\left(24 - \frac{1}{24}\{5^2 + 7^2 + 6^2 + 6^2\}\right) = 5.972$$

$$\hat{\sigma}^2 = 0.1927, \quad \hat{\sigma}_L^2 = \frac{1}{5.972}(1.3917 - 0.1927) = 0.2008$$

13.3 Assessing alternative sampling schemes

If one laboratory took one subsample, the variance of the resulting measurement would be $\sigma_L^2 + \sigma^2$; if r subsamples were used in one laboratory the variance would be $\sigma_L^2 + \sigma^2/r$. When v laboratories are used, each taking r subsamples, the variance of the resulting mean measurement would be

$$\frac{1}{v}(\sigma_L^2 + \sigma^2/r) \quad \text{or} \quad \frac{\sigma_L^2}{v} + \frac{\sigma^2}{vr}$$

After finding estimates of σ^2 and σ_L^2, we may compare different possible sampling schemes and aim to find one which gives the minimum possible variance for the mean measurement using a given amount of sampling.

In the first place, let us ignore the relative costs of introducing more laboratories or asking for more subsamples; we return to this later. Still using 24 subsamples, we have:

Example 13.1. continued.

No. of laboratories, v	No. of samples per laboratory, r	Variance of mean
4	6	0.0582
2	12	0.1084
1	24	0.2088
3	8	0.0750
6	4	0.0415
8	3	0.0331
12	2	0.0248
24	1	0.0164

The first row is the scheme that was actually carried out, and the next two rows show that reducing the number of laboratories (even though keeping an unnecessarily large number of subsamples from each laboratory) increases the variance very greatly. The best plan appears to be to have as many laboratories as possible, each carrying out only a single sampling – although such an arrangement would be impractical in case there was an actual error in subsampling (at least 2 per laboratory are needed to give a check of this) and almost certainly on grounds of administrative complexity and cost. A balance with about 6 laboratories doing 4 samples each may be satisfactory.

Although (as with completely randomized experiments) the analysis is not seriously affected by unequal replication in a 2-stage scheme, it is much easier to assess the meaning of the results when r is the same for every unit, and we now look at the problem of relative costs in this case.

13.4 Using variance components in planning when sampling costs are given

Suppose we have estimates:

		Cost per unit	Number of units
Between units	σ_1^2	C_1	v units
Within units	σ^2	C_2	r measurements per unit

The variance of a mean or a contrast involving the units is proportional to

$$V = \left\{ \frac{\sigma_1^2}{v} + \frac{\sigma^2}{vr} \right\}$$

Total cost $C=(C_1+rC_2)v$, $CV=(C_1+rC_2)(\sigma_1^2+\sigma^2/r)$. To minimize C when V is given, or to minimize V when C is given,

$$\frac{d}{dr}(CV)=\frac{-C_1\sigma^2}{r^2}+C_2\sigma_1^2=0$$

giving

$$\hat{r}^2=\frac{\sigma^2C_1}{\sigma_1^2C_2}=\frac{R}{\theta}$$

where

$$R=\frac{C_1}{C_2}\quad\text{and}\quad\theta=\frac{\sigma_1^2}{\sigma^2}$$

Here $\hat{r}$ is the optimum value of r, and if $\theta>R$ then only one measurement should be taken on each unit.

The value of v if V is minimized is

$$C/(C_1+\sqrt{C_1C_2\sigma^2/\sigma_1^2})$$

and the value of v if C is minimized is

$$(\sigma_1^2+\sqrt{C_2\sigma_1^2\sigma^2/C_1})/V$$

A table of values of r for various R and θ is as follows (using $\hat{r}^2=R/\theta$):

$R=$	1/8	1/4	1/2	1	2	4
$\theta=1/8$	1	1.41	2	2.83	4	5.66
1/4	0.71	1	1.41	2	2.83	4
1/2	1/2	0.71	1	1.41	2	2.83
1	0.35	1/2	0.71	1	1.41	2
2	1/4	0.35	1/2	0.71	1	1.41
4	0.18	1/4	0.35	1/2	0.71	1
8	1/8	0.18	1/4	0.35	1/2	0.71

In Example 13.1, σ_1^2 (i.e. σ_L^2) and σ^2 were approximately equal, so we may take $\theta=1$. The cost per laboratory, C_1, is likely to be much larger than C_2 for subsampling: even the highest value given in the table, $R=C_1/C_2=4$, may be too low (it is easy to compute further columns of the table as required). Suppose that, using $R=4$, we take $r=2$. The number of laboratories will then depend on *either* the desired size of the variance *or* the total budget (C) available.

13.5   Three levels of variation

A common example is when material is available in large batches, from the same or a different source; samples are taken from it and subsamples from each sample are subjected to analysis (perhaps a chemical analysis, or a mechanical test after carrying out some operation on it). Figure 13.2 shows the process.

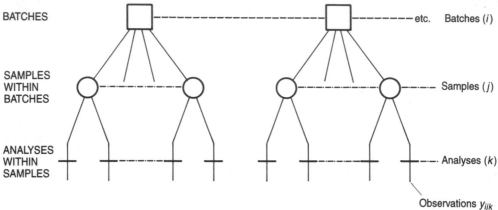

Fig. 13.2 Three-stage sampling scheme

The actual observations we can analyse statistically are those collected at the final stage of the sampling, y_{ijk}. They contain random variation from all 3 stages, and we wish to estimate the components of variance arising at each stage. Assume

$$y_{ijk} = \mu + \beta_i + \quad \gamma_{ij} + \quad \varepsilon_{ijk}$$
$$\text{(batch)} \quad \text{(sample)} \quad \text{(analysis)}$$

where μ is a grand mean and ε_{ijk} is, as usual, $\mathcal{N}(0, \sigma^2)$; the batches could have been selected from the whole large supply of material, and those actually used should be a random selection. Thus we now assume $\beta_i \sim \mathcal{N}(0, \sigma_B^2)$. In the same way the samples from a batch should be selected at random, and we assume $\gamma_{ij} \sim \mathcal{N}(0, \sigma_S^2)$.

As with experiments, we use sums of squares based on totals at various levels of variation in the data. But because we have made different assumptions about the distribution of each term in the model, the expected values of mean squares are different.

Suppose there are l batches ($i = 1$ to l); m samples taken within each batch ($j = 1$ to m); and n analyses per sample ($k = 1$ to n). (We do not use the same notation as for randomized complete blocks since we do not have exactly the same model.) The values of m and n should be kept the same within each batch and sample, respectively, to avoid very complicated calculations. There are lmn observations altogether, $\{y_{ijk}\}$. The total of all n analyses for batch i, sample j is Y_{ij}, and the total of all records from batch i is Y_i. We compute (i) the usual total sum of squares, $S - S_0$, for all observations; (ii) a sum of squares for samples, based on Y_{ij}; and (iii) a sum of squares for batches, based on Y_i.

As usual

$$S - S_0 = \sum_i \sum_j \sum_k y_{ijk}^2 - G^2/lmn$$

The corrected sum of squares for samples is

$$SS_S = \frac{1}{n} \sum_i \sum_j Y_{ij}{}^2 - G^2 / lmn$$

and that for batches is

$$SS_B = \frac{1}{mn} \sum_i Y_i{}^2 - G^2 / lmn$$

An Analysis of Variance table is constructed:

Source of variation	d.f.	Sum of squares	Expected value of mean square
Batches	$l-1$	SS_B	$\sigma^2 + n\sigma_S^2 + mn\sigma_B^2$
Samples within batches	$l(m-1)$	$SS_S - SS_B$	$\sigma^2 + n\sigma_S^2$
Samples	$lm-1$	SS_S	
Analyses within samples	$lm(n-1)$	$(S-S_o) - SS_S$	σ^2
Total	$lmn-1$	$S-S_o$	

The expected values of the mean squares are calculated on the assumptions stated above; the method is similar to that already shown in Section 13.2. The variance of a batch mean is

$$E \text{ [mean square within batches]}/\text{(no. of observations per batch)} = \frac{\sigma^2}{mn} + \frac{\sigma_S^2}{m}$$

The variance of the overall mean is

$$\frac{\sigma^2}{lmn} + \frac{\sigma_S^2}{lm} + \frac{\sigma_B^2}{l}$$

When estimates of σ^2, σ_S^2, σ_B^2 have been found, it is possible to study the effect of altering l, m, n so as to minimize the value of the overall mean's variance (or any other variance that is important in a multi-stage scheme). Often there is no need to do duplicate analyses in the laboratory except to check for mechanical errors, as σ^2 is usually smaller than the other components.

Any component that is estimated as less than 0 (which is impossible theoretically) is set equal to 0. A negative estimate can arise through sampling variation producing a rather unrepresentative sample in a particular experiment.

Example 13.2. Four batches of a chemical were sampled at random from a bulk supply; 5 samples were taken per batch; and 10 determinations were made per sample. A summary of results is

Batch totals	654	702	737	686	:	$G = 2779$
Sample totals	134	145	146	140		
	126	130	147	121		
	127	148	143	134		
	135	139	152	146		
	132	140	149	145		

Also the total sum of the squares of the 200 determinations was 39388. In the notation introduced above, we have $l = 4$, $m = 5$, $n = 10$, $lmn = 200$, $G^2/lmn = 38614.205$. Therefore $S - S_0 = 39388 - 38614.205 = 773.795$. The sum of squares between samples is

$$(134^2 + 126^2 + \cdots + 146^2 + 145^2)/10 - 38614.205 = 143.495$$

and that between batches is

$$(654^2 + 702^2 + 737^2 + 686^2)/50 - 38614.205 = 71.495$$

Analysis of Variance

Source of variation	d.f.	Sum of squares	Mean square	E[M.S.]
Between batches	3	71.495	23.832	$\sigma^2 + 10\sigma_S^2 + 50\sigma_B^2$
Between samples within batches	16	72.000	4.500	$\sigma^2 + 10\sigma_S^2$
Between samples	19	143.495		
Between determinations within samples	180	630.300	3.502	σ^2
Total	199	773.795		

$$\hat{\sigma}^2 = 3.502; \qquad \hat{\sigma}_S^2 = \frac{4.500 - 3.502}{10} = 0.0998; \qquad \hat{\sigma}_B^2 = \frac{23.832 - 4.500}{50} = 0.3866$$

Usually it is far more useful to estimate variance components than to test hypotheses about them, but since each of the mean squares follows a χ^2 distribution with the appropriate degrees of freedom, the usual comparisons can be made using F-tests.

To test the Null Hypothesis that $\sigma_S^2 = 0$, $F_{(16,180)} = 4.500/3.502 = 1.285$ n.s. We could assume $\sigma_S^2 = 0$. This would suggest that there was no need to take more than one sample per batch: resources could be diverted to other ways of reducing overall variability. For testing the Null Hypothesis '$\sigma_B^2 = 0$', we use $F_{(3,16)} = 23.832/4.500 = 5.30**$, and we should obviously reject that hypothesis. (If $\sigma_S^2 = 0$, we could instead consider $F_{(3,180)} = 23.83/3.50$, which gives an even stronger result for rejection of the Null Hypothesis about σ_B^2.)

Since mean squares follow χ^2 distributions, one way of providing standard errors of these estimated variance components is to use the fact that the variance

$\text{Var}\,[\hat\sigma^2]=2(\hat\sigma^2)^2/f$ when $\hat\sigma^2$ is based on f degrees of freedom. Therefore

$$\text{Var}\,[\hat\sigma^2]=\frac{2(3.502)^2}{180}=0.1363;\qquad \text{SE}[\hat\sigma^2]=0.369$$

Also

$$\text{Var}\,[\hat\sigma^2+10\hat\sigma_S^2]=\frac{2(4.500)^2}{16}=2.5313$$

$$\text{Var}\,[\hat\sigma^2+10\hat\sigma_S^2+50\hat\sigma_B^2]=\frac{2(23.832)^2}{3}=378.6428$$

There is an approximate method, which is quite simple, of giving standard errors to $\hat\sigma_S^2$ and $\hat\sigma_B^2$. (A better method, due to Healy (1963), is explained in Fisher and Yates' Statistical Tables, Table V_1 (1963)).

Approximately

$$\text{Var}\,[\hat\sigma_S^2]=\frac{1}{n^2}\left\{\text{Var}\,(\hat\sigma^2+n\hat\sigma_S^2)+\text{Var}\,(\hat\sigma^2)\right\}$$

$$=\frac{1}{100}(2.5313+0.1363)=0.02668$$

so that

$$\text{SE}[\hat\sigma_S^2]=0.163$$

Also

$$\text{Var}\,[\hat\sigma_B^2]=\frac{1}{2500}(378.6428+2.5313)=0.1524;\qquad \text{SE}[\hat\sigma_B^2]=0.390$$

We assume that the two variance estimates, of σ^2 and $\sigma^2+10\sigma_S^2$ for example, are independent in this method, so that there is no covariance term in the expressions we have to use. But clearly this is wrong, since the estimates come from a single analysis on a single set of data.

13.6   Costs in a three-stage scheme

If there are l first-stage units, each containing m second-stage units which are in turn split into n third-stage units, with sampling costs C_1, C_2, C_3 per unit for the 3 stages, it can be shown that the variance of an overall mean is

$$V=\frac{1}{l}\left(\sigma_1^2+\frac{1}{m}\{\sigma_2^2+\sigma_3^2/n\}\right)$$

and $C = l(C_1 + m\{C_2 + nC_3\})$ is the total cost. We have used σ_1^2, σ_2^2, σ_3^2 as the variance components at each stage to simplify notation. Eliminating l gives

$$\hat{m}^2 = \frac{C_1(\sigma_2^2 + \sigma_3^2/\hat{n})}{\sigma_1^2(C_2 + C_3\hat{n})}$$

and

$$\hat{n}^2 = \frac{\sigma_3^2(C_2 + C_1/\hat{m})}{C_3(\sigma_2^2 + \sigma_1^2\hat{m})}$$

In Example 13.2, $\sigma_1^2 = 0.3866$, $\sigma_2^2 = 0.0998$, $\sigma_3^2 = 3.502$.
   Suppose that the ratio $C_1 : C_2 : C_3$ is $5:1:3$. Then these equations are

$$\hat{m}^2 = \frac{5(0.0998 + 3.502/\hat{n})}{3.502(1 + 3\hat{n})}$$

and

$$\hat{n}^2 = \frac{3.502(1 + 5/\hat{m})}{5(0.0998 + 0.3866\hat{m})}$$

They have to be solved iteratively; let the value of m be guessed as 2. It will be found that after 3 cycles $\hat{m} = 1.805$ and $\hat{n} = 2.339$, and we can take $m = n = 2$ to the nearest integer. Then l is calculated according to design requirements. Suppose that this is part of an experiment in which 4 equally replicated treatments are to be compared; the standard error of the difference between two treatment means is to be not greater than $\frac{1}{2}$.
   Each treatment will be replicated $\frac{1}{4}l$ times. The variance of a difference between two means will be

$$\frac{8}{l}\left(\sigma_1^2 + \frac{1}{m}\left\{\sigma_2^2 + \frac{\sigma_3^2}{n}\right\}\right)$$

and this must be less than $(1/2)^2$. So, approximately,

$$\frac{1}{l}\left(0.39 + \frac{1}{2}\left\{0.1 + \frac{3.5}{2}\right\}\right) \leqslant \frac{1}{32}$$

which gives $\hat{l} \geqslant 32 \times 1.315 = 42.08$. Since we want to use 4 equally replicated treatments, we must round this up to the nearest multiple of 4, i.e. 44. Each treatment needs 11 replicates. This is a large number, and we ought to reconsider whether we are specifying too high a precision.

We may also assess alternative sampling schemes as in Section 13.3, by looking at

$$\frac{1}{lmn}[\sigma^2 + n\sigma_S^2 + mn\sigma_B^2] = \frac{\sigma^2}{lmn} + \frac{\sigma_S^2}{lm} + \frac{\sigma_B^2}{l}$$

The variance components have estimated values of $\sigma^2 = 3.502$, $\sigma_S^2 = 0.0998$, $\sigma_B^2 = 0.3866$; suppose that the maximum value possible for lmn is 24.

l	m	n	σ^2/lmn	σ_S^2/lm	σ_B^2/l	Total
24	1	1	0.1459	0.0042	0.0161	0.1662
12	2	1	0.1459	0.0042	0.0322	0.1823
12	1	2	0.1459	0.0083	0.0322	0.1864
8	3	1	0.1459	0.0042	0.0483	0.1984
6	2	2	0.1459	0.0083	0.0644	0.2186

Because σ^2 is the largest component, there is not very much to choose between these schemes. The only reason for having m or $n > 1$ would be to check for errors that were technical or mechanical, and not purely statistical, at either of these stages.

13.7   Example where one estimate is negative

Example 13.3. A food science company wished to study the effect of milling on the moisture content of cowpea. Five batches of cowpeas were selected at random from a large store, and from each batch 3 random samples of 100 g were taken and milled. From each milling, three 10 g samples were taken and moisture content measured.

Batch	Sample	Determination		
	1	9.3	9.2	8.8
1	2	8.6	8.7	9.9
	3	8.9	8.7	8.5
	1	8.0	8.2	9.2
2	2	9.7	9.4	8.2
	3	9.3	9.5	9.4
	1	11.0	10.7	9.9
3	2	9.3	13.9	9.2
	3	9.2	10.9	9.7
	1	10.1	10.2	9.9
4	2	8.6	9.4	8.3
	3	8.3	9.9	9.5
	1	12.0	9.3	10.8
5	2	12.2	9.6	11.7
	3	11.4	9.8	12.4

Necessary totals are:

Batch	Sample		Batch total	
	1	27.3		
1	2	27.2		
	3	26.1	80.6	
	1	25.4		
2	2	27.3		
	3	28.2	80.9	
	1	31.6		
3	2	32.4		
	3	29.8	93.8	
	1	30.2		
4	2	26.3		
	3	27.7	84.2	
	1	32.1		
5	2	33.5		Grand total
	3	33.6	99.2	438.7

The correction term for all sums of squares is $438.7^2/45 = 4276.8376$. The sum of squares for batches is

$$\tfrac{1}{9}(80.6^2 + \cdots + 99.2^2) - \text{correction term} = 30.9280$$

and that for samples is

$$\tfrac{1}{3}(27.3^2 + \cdots + 33.6^2) - \text{correction term} = 36.8391$$

Source of variation	d.f.	Sum of squares	Mean square	E[M.S.]
Batches	4	30.9280	7.732	$\sigma^2 + 3\sigma_S^2 + 9\sigma_B^2$
Samples within batches	10	5.9111	0.591	$\sigma^2 + 3\sigma_S^2$
Samples	14	36.8391		
Residual ($\equiv$ determinations within samples)	30	32.9533	1.098	σ^2
Total	44	69.7924		

$\hat{\sigma}^2 = 1.098$. The estimate of $\sigma^2 + 3\sigma_S^2$ is less than that of σ^2; hence $\hat{\sigma}_S^2 < 0$ and therefore we estimate σ_S^2 as 0.

$$\hat{\sigma}^2 + 3\hat{\sigma}_S^2 + 9\hat{\sigma}_B^2 = 7.732$$

so that

$$\hat{\sigma}_B^2 = \tfrac{1}{9}(7.732 - 1.098) = 0.737$$

(Because of the difficulty caused by the small value of the samples-within-batches mean square, we do not use that in estimating σ_B^2.) Batch variability is not very large, but the variance of individual determinations is quite high. Since $\hat{\sigma}_S^2 = 0$, there is no need to have any replication at this level in the scheme. The most important need is to make the milling technique as homogeneous as possible and to standardize the method of determination as carefully as possible. If σ^2 cannot be reduced very much, then more determinations will perhaps be needed per sample, but the number of samples per batch could be reduced correspondingly.

13.8 Exercises

1. An investigation was carried out to determine the sources of variation in the carbon content of metal ingots. It was decided to use metal from 3 different melts and to use 4 moulds from within each melt. Three test pieces were then selected at random from the ingots cast in the 4 moulds. The data are given below (in coded units).

Mould		Melt		
		1	2	3
1		22.0	17.6	26.2
		23.7	17.0	24.2
		21.2	17.8	25.5
	Total	66.9	52.4	75.9
2		21.6	17.2	28.3
		22.6	18.5	28.4
		21.9	17.3	30.6
	Total	66.1	53.0	87.3
3		20.7	12.3	24.2
		20.5	11.7	22.8
		20.2	11.6	26.3
	Total	61.4	35.6	73.3
4		23.8	22.3	30.2
		24.7	22.7	30.4
		23.0	21.2	30.6
	Total	71.5	66.2	91.2
Melt total		265.9	207.2	327.7

Analyse the data and determine estimates for the variances from the sources of variation present. You may use the fact that the total of the squares of all 36 observations is 18687.98.

2. (i) Explain the different objectives when analysing a 'fixed effects' linear model as opposed to a 'random effects' linear model, illustrating your remarks by appropriate simple practical examples. In particular discuss for each model the appropriate procedures following the derivation of an Analysis of Variance table.

(ii) The data given below are percentages of fibre in soya cotton cake and were obtained in the course of an investigation into the extent to which the results of routine analysis of samples of identical product may vary. The data were divided up into groups according to the laboratory where the cake was tested; each group is further divided into subgroups according to the technician who performed the analysis.

Using an Analysis of Variance table, estimate the extent of each of the sources of variability in these data. Discuss the difficulties that can arise when constructing confidence intervals for the parameters in the model you use.

Laboratory	A			B			C		
Technician	A_1	A_2	A_3	B_1	B_2	B_3	C_1	C_2	C_3
	12.44	12.38	12.90	12.80	12.88	12.95	12.94	12.83	12.77
	12.52	12.28	12.92	13.30	12.75	12.80	12.86	12.74	12.83
	12.72	12.48	12.83	13.00	12.98	12.65	12.72	13.08	12.81
	12.62	12.36	12.99	13.05	12.80	12.70	12.83	12.94	12.83

The sums of the observations made by each technician are given below, together with the corresponding sums of squares.

Technician	A_1	A_2	A_3
Sum	50.30	49.50	51.64
Sum of squares	632.5668	612.5828	666.6854

Technician	B_1	B_2	B_3
Sum	52.15	51.41	51.10
Sum of squares	680.0325	660.7773	652.8550

Technician	C_1	C_2	C_3
Sum	51.35	51.59	51.24
Sum of squares	659.2305	665.4465	656.3868

3. A tutorial college is testing 3 types of distance-learning material, A, B and C, which may be used in a course taken by a large number of students. The college considers that the difference between the scores achieved by individual students in two tests during the course, one early in the course and one late, will provide useful information about the types of material. A study is planned in which 3 groups of students are chosen at random, one group to receive A, one B and one C. Because of administrative difficulties in obtaining test scores from all the students, the number of results available for analysis is not the same in every group.

The college regards A, B and C as an effectively random sample from a larger population of types of material that could have been developed, and so intends to study the data using a random-effects model. Also, the tests contain two parts, 'short' and 'long' questions, and the data from the two parts will be analysed separately. The differences in scores are:

Short questions:

A: 22, 3, 16, 14, 8, 27, 11, 17;
B: 12, 17, 11, 10, 16, 18, 15, 13, 9, 20;
C: 4, 16, 32, 11, 9, 25, 27, 12, 26, 7, 14;

Long questions:

A: 12, 7, 19, 19, 11, 33, 20, 25;
B: 24, 6, 39, 14, 17, 10, 22, 35, 33, 21;
C: 15, 11, 17, 8, 2, 10, 16, 21, 9, 19, 23.

(i) Carry out the analysis proposed by the college, for each part of the tests.
(ii) The College has a programme of continuous change and development of its material, producing many more types of material that will require testing. Explain how the results of the present study may be used to help in planning and evaluating the programme.
(iii) What assumptions were made in the analysis in (i)?
(iv) Using any appropriate methods of Initial Data Analysis, describe the characteristics of the given data, and examine whether the assumptions (iii) seem to be reasonable for these data.
(v) Write a report advising the College whether their methods of analysis should be changed, or added to, in future studies.

14
Computer output using SAS

We illustrate several different types of experiment in the eight examples that follow, showing how the analysis would be produced using the SAS procedure PROC GLM.

The measurement y being analysed is called the *dependent* variable, and may be given a name. Residual is called *error*, total is *corrected total*, and the significance level is given in the last column using a '*P*-value' (see page 45).

By analogy with a regression analysis, *model* accounts for all the degrees of freedom for effects (and interactions) in the analysis. This is subdivided in a fuller analysis according to the type of design. *R-Square* is the proportion of the *total* sum of squares which is accounted for by the *model* sum of squares (and is not informative in analysing designed experiments). *RootMSE* is the square root of the residual ('error') mean square; *y mean* is the overall mean of the data, G/N, and CV (coefficient of variation) is the ratio of these expressed as a percentage,

$$CV = 100 \, (RootMSE)/(y \text{ mean})$$

The CV can help to show how precisely an experiment has been carried out (page 29).

Type I S.S. are sums of squares obtained when sums of squares are successively added to a model (e.g. blocks, then treatments; or rows, then columns, then treatments), and the resulting increase in the model-sum-of-squares is attributed to the new added source. *Type III S.S.* assume instead that terms are being excluded from the full model in turn, and the change in the sum of squares is due to the last one omitted. The default setting is to print Type I.

Since we assume, in the linear models used for analysis of simple designed experiments, that all observations have the same variance, the information *SD* given with each treatment mean is not useful. Nor will individual means be needed when the analysis is completed by studying contrasts, though they are necessary when graphs are being drawn.

Example 1

Question
The sugar content y (tonnes/ha) of four varieties of sugar beet roots was investigated in a randomized blocks experiment. The yields were as follows:

Variety	Block					
	1	2	3	4	5	6
Amono	38.2	37.6	39.3	40.3	38.4	39.1
Vytomo	42.1	40.3	40.1	44.1	43.4	40.8
Amber	39.6	38.2	39.7	40.7	38.7	40.1
Cora	40.8	41.2	42.1	39.5	39.6	41.6

Vytomo is susceptible to downy mildew, while the other three varieties are resistant. Amber has much smaller top-size than either Amono or Cora.

Analyse the data to see whether there are real differences in sugar content between the roots of the different varieties, and if so investigate where these differences occur.

$$\Sigma y = 965.5, \ \Sigma y^2 = 38901.61)$$

Answer

$$\text{Correction term} = S_0 = \frac{965.5^2}{24} = 38841.26$$

$$\text{Block S.S.} = S_B = \frac{160.7^2 + 157.3^2 + 161.2^2 + 164.6^2 + 160.1^2 + 161.6^2}{4} - S_0$$

$$= 6.977$$

$$\text{Varieties S.S.} = S_V = \frac{232.9^2 + 250.8^2 + 237.0^2 + 244.8^2}{6} - S_0$$

$$= 31.921$$

$$\text{Corrected total S.S.} = S - S_0 = \Sigma y^2 - S_0 = 60.350$$

Anova

Source	d.f.	S.S.	Mean square	F
Blocks	5	6.977	1.395	0.98
Residual	15	21.452	1.430	
Total	23	60.350		

There is strong evidence of real differences between varieties

$$\text{L.S.D.} = t_{(15;\,5\%)} \sqrt{s^2 \left(\frac{1}{r_1} + \frac{1}{r_2} \right)}$$

$$= 2.13 \sqrt{1.430 \left(\frac{1}{6} + \frac{1}{6} \right)}$$

$$= 1.47$$

Varietal means	Amono	Vytomo	Amber	Cora
	38.82	41.80	39.50	40.80

Amono's mean sugar beet content appears to be significantly different from that of Vytomo and Cora.

Appropriate contrasts are:

(1) susceptible variety (Vytomo) compared with the others;
(2) top-size comparison, Amber *versus* (Amono, Cora);
(3) Amono *versus* Cora, the remaining orthogonal contrast.

Replication $r = 6$.

Total	Amono 232.9	Vytomo 250.8	Amber 237.0	Cora 244.8	Value	Divisor	Sum of squares	$F_{(1,15)}$
(1)	−1	3	−1	−1	37.7	12 × 6	19.7401	13.80**
(2)	−1	0	2	−1	−3.7	6 × 6	0.3803	<1
(3)	1	0	0	−1	−11.9	2 × 6	11.8008	8.25*

The susceptible variety seems to have a different sugar content; top size appears unimportant; Amono appears to differ from Cora.

SAS programming commands

```
/*Randomized blocks with contrasts
*/

data d1;
   input variety $ block sugar @@;
   cards;
Amono    1  38.2      Amono    2  37.6      Amono    3  39.3
Amono    4  40.3      Amono    5  38.4      Amono    6  39.1
Vymoto   1  42.1      Vymoto   2  40.3      Vymoto   3  40.1
Vymoto   4  44.1      Vymoto   5  43.4      Vymoto   6  40.8
Amber    1  39.6      Amber    2  38.2      Amber    3  39.7
Amber    4  40.7      Amber    5  38.7      Amber    6  40.1
Cora     1  40.8      Cora     2  41.2      Cora     3  42.1
Cora     4  39.5      Cora     5  39.6      Cora     6  41.6
;
proc glm data=d1;
   class variety block;
   model sugar=block variety;
   contrast 'Mildew' variety 1 1 1 -3;
   contrast 'Top-size' variety -2 1 1 0;
   contrast 'Amono versus Cora' variety 0 -1 1 0;
   means variety;
   title 'Randomized blocks with contrasts';
run;
```

Example 1 output.   Randomized blocks with contrasts

```
        General Linear Models Procedure
           Class Level Information

   Class        Levels     Values

   VARIETY          4      Amber Amono Cora Vymoto

   BLOCK            6      1 2 3 4 5 6


     Number of observations in data set=24
```

General Linear Models Procedure

Dependent Variable: SUGAR

Source	DF	Sum of Squares	Mean Square	F Value	Pr>F
Model	8	38.89833333	4.86229167	3.40	0.0197
Error	15	21.45125000	1.43008333		
Corrected Total	23	60.34958333			

R-Square	C.V.	Root MSE	SUGAR Mean
0.644550	2.972622	1.19586092	40.22916667

Source	DF	Type I SS	Mean Square	F Value	Pr>F
BLOCK	5	6.97708333	1.39541667	0.98	0.4638
VARIETY	3	31.92125000	10.64041667	7.44	0.0028

Source	DF	Type III SS	Mean Square	F Value	Pr>F
BLOCK	5	6.97708333	1.39541667	0.98	0.4638
VARIETY	3	31.92125000	10.64041667	7.44	0.0028

General Linear Models Procedure

Dependent Variable: SUGAR

Contrast	DF	Contrast SS	Mean Square	F Value	Pr>F
Mildew	1	19.74013889	19.74013889	13.80	0.0021
Top-size	1	0.38027778	0.38027778	0.27	0.6136
Amono v. Cora	1	11.80083333	11.80083333	8.25	0.0116

General Linear Models Procedure

Level of VARIETY	N	------ SUGAR ------ Mean	SD
Amber	6	39.5000000	0.91433036
Amono	6	38.8166667	0.95376447
Cora	6	40.8000000	1.06018866
Vymoto	6	41.8000000	1.67809416

Example 2

Question
Explain what is meant by an interaction in the context of factorial experiments, and use illustrations to assist your explanation.

Twenty-four plots were used in a field trial in which three factors were applied in a completely randomized 2^3 arrangement with three replicates for each treatment. The factors were:

> *A*   Ammonium
> *B*   Straw
> *C*   Glucose

The soil amendment (y kg N/ha applied) was measured at a subsequent stage and the data were:

		*B*1							*B*2			
		*C*1			*C*2			*C*1			*C*2	
*A*1	29	33	30	40	40	44	63	59	56	60	68	58
*A*2	36	39	37	48	52	52	60	65	59	73	61	68

Analyse the data and comment in detail on the differences between treatment means.

Answer
Interaction is a combinatorial effect of treatments in addition to their additive effects. If there is no interaction then the interaction diagram indicates parallelism:

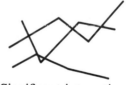

No interaction   Significant interaction

Treatment	(1)	*a*	*b*	*c*	*ab*	*ac*	*bc*	*abc*		
Total	92	112	178	124	184	152	186	202	Value	S.S.
A	−	+	−	−	+	+	−	+	70	204.167
B	−	−	+	−	+	−	+	+	270	3037.500
C	−	−	−	+	−	+	+	+	98	400.167
AB	+	−	−	+	+	−	−	+	−26	28.167
AC	+	−	+	−	−	+	−	+	18	13.500
BC	+	+	−	−	−	−	+	+	−46	88.167
ABC	−	+	+	+	−	−	−	+	2	0.167

$S = 67018$ $S_0 = 63037.5$

Anova

Source	d.f.	S.S.	Mean square	F
A	1	204.167		15.65***
B	1	3037.500		232.91***
AB	1	28.167		2.16
C	1	400.167		30.68***
AC	1	13.500		1.04
BC	1	88.167		6.76*
ABC	1	0.167		0.01
Residual	16	208.665	13.0416	
Total	23	3980.500		

There is strong evidence of real differences on account of main effects, but *BC* interaction is also significant.

Means	B^-	B^+
C^-	34.00	60.33
C^+	46.00	64.67

$$\text{LSD} = t^*_{16} \sqrt{\frac{2s^2}{6}}$$

$$= 2.120 \sqrt{\frac{13.0416}{3}}$$

$$= 4.42$$

The addition of *B* or *C* increases the soil amendment, but the effect of adding *C* as well as *B* is not so important.

SAS programming commands

```
/*Completely randomized design with 2^ 3 structure.
*/
data d2;
  input ammon straw gluc soil @@;
  cards;
1 1 1 29 1 1 1 33 1 1 1 30 1 1 2 40 1 1 2 40 1 1 2 44
1 2 1 63 1 2 1 59 1 2 1 56 1 2 2 60 1 2 2 68 1 2 2 58
2 1 1 36 2 1 1 39 2 1 1 37 2 1 2 48 2 1 2 52 2 1 2 52
2 2 1 60 2 2 1 65 2 2 1 59 2 2 2 73 2 2 2 61 2 2 2 68
;
proc glm data=d2;
  class ammon straw gluc;
  model soil=ammon|straw|gluc;
  means ammon straw gluc ammon*straw ammon*gluc straw*gluc;
  title 'Completely randomized design with 2 ^ 3 structure';
run;
```

Example 2 output. Completely randomized design with 2^3 structure

General Linear Models Procedure
Class Level Information

Class	Levels	Values
AMMON	2	1 2
STRAW	2	1 2
GLUC	2	1 2

Number of observations in data set=24

General Linear Models Procedure

Dependent Variable: SOIL

Source	DF	Sum of Squares	Mean Square	F Value	Pr>F
Model	7	3771.83333333	538.83333333	41.32	0.0001
Error	16	208.66666667	13.04166667		
Corrected Total	23	3980.50000000			

R-Square	C.V.	Root MSE	SOIL Mean
0.947578	7.046487	3.61132478	51.25000000

Source	DF	Type I SS	Mean Square	F Value	Pr>F
AMMON	1	204.16666667	204.16666667	15.65	0.0011
STRAW	1	3037.50000000	3037.50000000	232.91	0.0001
AMMON*STRAW	1	28.16666667	28.16666667	2.16	0.1611
GLUC	1	400.16666667	400.16666667	30.68	0.0001
AMMON*GLUC	1	13.50000000	13.50000000	1.04	0.3241
STRAW*GLUC	1	88.16666667	88.16666667	6.76	0.0193
AMMON*STRAW*GLUC	1	0.16666667	0.16666667	0.01	0.9114

General Linear Models Procedure

Dependent Variable: SOIL

Source	DF	Type III SS	Mean Square	F Value	Pr>F
AMMON	1	204.16666667	204.16666667	15.65	0.0011
STRAW	1	3037.50000000	3037.50000000	232.91	0.0001
AMMON*STRAW	1	28.16666667	28.16666667	2.16	0.1611
GLUC	1	400.16666667	400.16666667	30.68	0.0001
AMMON*GLUC	1	13.50000000	13.50000000	1.04	0.3241
STRAW*GLUC	1	88.16666667	88.16666667	6.76	0.0193
AMMON*STRAW*GLUC	1	0.16666667	0.16666667	0.01	0.9114

General Linear Models Procedure

Level of		------ SOIL ------	
AMMON	N	Mean	SD
1	12	48.3333333	13.8388999
2	12	54.1666667	12.3202224

Level of		------ SOIL ------	
STRAW	N	Mean	SD
1	12	40.0000000	7.76940386
2	12	62.5000000	5.03623236

Level of		------ SOIL ------	
GLUC	N	Mean	SD
1	12	47.1666667	14.1795841
2	12	55.3333333	11.1545615

Level of	Level of		------ SOIL ------	
AMMON	STRAW	N	Mean	SD
1	1	6	36.0000000	6.16441400
1	2	6	60.6666667	4.27395211
2	1	6	44.0000000	7.50999334
2	2	6	64.3333333	5.42832080

General Linear Models Procedure

Level of AMMON	Level of GLUC	N	------ SOIL ------ Mean	SD
1	1	6	45.0000000	15.9122594
1	2	6	51.6666667	11.8939761
2	1	6	49.3333333	13.3366663
2	2	6	59.0000000	10.0000000

Level of STRAW	Level of GLUC	N	------ SOIL ------ Mean	SD
1	1	6	34.0000000	4.00000000
1	2	6	46.0000000	5.51361950
2	1	6	60.3333333	3.20416396
2	2	6	64.6666667	5.85377371

Example 3

Question
In an experiment designed to test the resistance of 5 varieties of second early potatoes to foliage blight, 25 plots are selected for the trial which takes place in a sloping field with a hedge on one side. During the trial the crop is exposed to the disease and the percentage y of plants showing signs of the disease is recorded for each plot. The key to the variety names and their date of development, the design of the experiment and the percentage of plants infected in each plot are as shown below.

A	Craigs Alliance	(1948)	B 21	A 20	D 14	E 19	C 18
B	Maris Peer	(1962)	C 16	D 13	E 19	B 24	A 16
C	Estima	(1973)	D 14	B 17	C 17	A 19	E 13
D	Wilja	(1974)	A 18	E 15	B 19	C 9	D 15
E	Maris Page	(1962)	E 19	C 15	A 22	D 15	B 25

///////////////////////////////

Hedge
----------> Slope downwards

$(\sum y = 432 \qquad \sum y^2 = 7780)$

(a) Say what you think the design is, and describe how it relates to the hedge and slope. Perform a suitable analysis.
(b) The two Maris varieties were produced at Cambridge, Craigs Alliance in Scotland and the other two are Dutch. Discuss the interesting features of differences between varieties in view of this information.
(c) Comment on the blocking scheme in this experiment.

Answer
(a) Latin square design:

$$S_0 = \frac{432^2}{25} = 7464.96$$

Anova

Source	d.f.	S.S.	Mean square	F
Rows	4	55.04	13.76	2.03
Columns	4	13.04	3.26	0.48
Treatments	4	165.44	41.36	6.088**
Residual	12	81.52	6.7933	
Total	24	315.04		

$$\text{Rows S.S.} = \frac{92^2 + 88^2 + 80^2 + 76^2 + 96^2}{5} - S_0 = 55.04$$

$$\text{Columns S.S.} = \frac{88^2 + 80^2 + 91^2 + 86^2 + 87^2}{5} - S_0 = 13.04$$

$$\text{Treatments S.S.} = \frac{95^2 + 106^2 + 75^2 + 71^2 + 85^2}{5} - S_0 = 165.44.$$

Some evidence of real differences between treatments.

(b) Orthogonal contrasts or LSD:

$$\text{LSD} = \sqrt{s^2\left(\frac{1}{r_1} + \frac{1}{r_2}\right)} \times t_{12} = 2.18 \sqrt{6.7933\left(\frac{1}{5} + \frac{1}{5}\right)} = 3.59(4)$$

Means	A	B	C	D	E
	19.0	21.2	15.0	14.2	17.0

C, D, E give no significant differences, nor do A and B, or A and E; otherwise significant differences throughout.

The two Maris varieties and Craigs appear to be generally different from the rest, i.e. the British varieties seem to have higher yields than the Dutch.

Contents/Total	A	B	C	D	E	Value	Divisor	S.S.	$F_{(1,12)}$
	95	106	75	71	85				
ABE versus *CD*	2	2	−3	−3	2	134	150	119.7067	17.62**
A versus BE	2	−1	0	0	−1	−1	30	0.0333	0.005
B versus E	0	1	0	0	−1	21	10	44.1	6.49*
C versus D	0	0	1	−1	0	4	10	1.6	0.24

The two Maris varieties appear to be different also.

(c) Blocking with respect to columns has been disadvantageous, but blocking with respect to rows has been helpful in reducing residual variation substantially.

SAS programming commands

```
/*Latin Square with contrasts
*/
data d3;
  input row col variety $ infect @@;
  cards;
1  1  B  21  1  2  A  20  1  3  D  14  1  4  E  19  1  5  C  18
2  1  C  16  2  2  D  13  2  3  E  19  2  4  B  24  2  5  A  16
3  1  D  14  3  2  B  17  3  3  C  17  3  4  A  19  3  5  E  13
4  1  A  18  4  2  E  15  4  3  B  19  4  4  C   9  4  5  D  15
5  1  E  19  5  2  C  15  5  3  A  22  5  4  D  15  5  5  B  25
;
proc glm data=d3;
  class row col variety;
  model infect=row col variety;
  contrast 'British v. Dutch' variety -2 -2 3 3 -2;
  contrast 'Scot v. English' variety -2 1 0 0 1;
  contrast 'Within Dutch' variety 0 0 -1 1 0;
  contrast 'Within English' variety 0 -1 0 0 1;
  means variety;
  title  'Latin Square with contrasts';
run;
```

Example 3 output. Latin Square with contrasts

```
                General Linear Models Procedure
                   Class Level Information
        Class                 Levels      Values

        ROW                      5        1 2 3 4 5

        COL                      5        1 2 3 4 5

        VARIETY                  5        A B C D E


    Number of observations in data set=25
```

General Linear Models Procedure

Dependent Variable: INFECT

Source	DF	Sum of Squares	Mean Square	F Value	Pr>F
Model	12	233.52000000	19.46000000	2.86	0.0403
Error	12	81.52000000	6.79333333		
Corrected Total	24	315.04000000			

R-Square	C.V.	Root MSE	INFECT Mean
0.741239	15.08335	2.60640237	17.28000000

Source	DF	Type I SS	Mean Square	F Value	Pr>F
ROW	4	55.04000000	13.76000000	2.03	0.1547
COL	4	13.04000000	3.26000000	0.48	0.7502
VARIETY	4	165.44000000	41.36000000	6.09	0.0065

Source	DF	Type III SS	Mean Square	F Value	Pr>F
ROW	4	55.04000000	13.76000000	2.03	0.1547
COL	4	13.04000000	3.26000000	0.48	0.7502
VARIETY	4	165.44000000	41.36000000	6.09	0.0065

General Linear Models Procedure

Dependent Variable: INFECT

Contrast	DF	Contrast SS	Mean Square	F Value	Pr>F
British v. Dutch	1	119.70666667	119.70666667	17.62	0.0012
Scot v. English	1	0.03333333	0.03333333	0.00	0.9453
Within Dutch	1	1.60000000	1.60000000	0.24	0.6362
Within English	1	44.10000000	44.10000000	6.49	0.0256

General Linear Models Procedure

Level of VARIETY	N	------INFECT------ Mean	SD
A	5	19.0000000	2.23606798
B	5	21.2000000	3.34664011
C	5	15.0000000	3.53553391
D	5	14.2000000	0.83666003
E	5	17.0000000	2.82842712

Example 4

Question
Four apple trees are used in an experiment in which eight treatments are to be tested. The treatments involve three factors, each at two levels, and the factors A, B, C denote pruning, level of spray, and time of spraying, according to the following table.

Treatment	Pruning	Spray	Time
(1)	no	once	late
a	yes	once	late
b	no	twice	late
ab	yes	twice	late
c	no	once	early
ac	yes	once	early
bc	no	twice	early
abc	yes	twice	early

Unfortunately the treatments could only be applied to branches of the trees used in the experiment, and there were only four usable branches per tree. Since there were eight treatments and only four branches per tree, the ABC interaction was confounded with trees, but four trees were available altogether so the treatments could be replicated. Analyse the data given that the layout and results were as shown below.

	Replicate 1				Replicate 2		
Tree 1		Tree 2		Tree 3		Tree 4	
(1)	20	c	22	b	26	ac	36
ac	32	b	27	a	26	(1)	22
bc	30	abc	35	abc	36	bc	31
ab	33	a	26	c	24	ab	34

Answer
For the moment we shall consider the design to be that for randomized blocks where the eight treatments are allocated to two blocks (replicates). We shall deal with the confounding problem later. The tree totals are 115, 110, 112, 123 on 4 observations each, so

$$S_B = \frac{115^2 + 110^2 + 112^2 + 123^2}{4} = 13249.5$$

$$S_0 = \frac{460^2}{16} = 13225$$

$$S = 20^2 + 33^2 + \cdots + 31^2 = 13648$$

and the treatment totals are

	(1)	a	b	ab	c	ac	bc	abc
	42	52	53	67	46	68	61	71

To find the treatments S.S. we first need to calculate

$$S_T = \frac{42^2 + 52^2 + \cdots + 71^2}{2} = 13634$$

It may appear that the analysis of variance table may now be formed as

Source	d.f.	S.S.	Mean square	*F*
Trees	3	24.5	8.1667	–
Treatments	7	409.00	58.4286	
Residual	Diff.	Diff.	Ratio	
Total	15	423.00		

but this is not the final analysis table since we have not dealt with the confounding problem, nor is the treatments S.S. partitioned according to the factorial treatment structure. To achieve this let us now construct a table for the contrasts.

Treatment	(1)	a	b	c	ab	ac	bc	abc		
Total	42	52	53	46	67	68	61	71	Value	Divisor
A	−	+	−	−	+	+	−	+	56	16
B	−	−	+	−	+	−	+	+	44	16
C	−	−	−	+	−	+	+	+	32	16
AB	+	−	−	+	+	−	−	+	−8	16
AC	+	−	+	−	−	+	−	+	8	16
BC	+	+	−	−	−	−	+	+	−8	16
[ABC	−	+	+	+	−	−	−	+	−16	16]

Now the variation accounted for by the *ABC* interaction cannot be attributed to that source since we cannot tell whether this variation is variability between trees or due to the treatments. So we must assign this variability to tree differences, which are already accounted for in the analysis, so the final table is as shown below.

Source	d.f.	S.S.	Mean square	F
Trees	3	24.50	8.1667	–
A	1	196.00	196.00	213.81***
B	1	121.00	121.00	132.00***
AB	1	4.00	4.00	4.36
C	1	64.00	64.00	69.82***
AC	1	4.00	4.00	4.36
BC	1	4.00	4.00	4.36
Residual	6	5.50	0.9167	
Total	15	423.00		

It is clear that there are important differences between the main effects for all three
factors, but there is no evidence of any interaction.

SAS programming commands

```
/*2 ^ 3 factorial with full confounding.
*/
data d5;
   input tree a b c yield @@;
   cards;
1 1 1 1 20 1 2 1 2 32 1 1 2 2 30 1 2 2 1 33
2 1 1 2 22 2 1 2 1 27 2 2 2 2 35 2 2 1 1 26
3 1 2 1 26 3 2 1 1 26 3 2 2 2 36 3 1 1 2 24
4 2 1 2 36 4 1 1 1 22 4 1 2 2 31 4 2 2 1 34
   ;
proc glm data=d5;
   class tree a b c;
   model yield = tree a|b|c@2;
   lsmeans a|b|c @2;
   title '2 ^ 3 factorial with full confounding';
run;
```

Example 4 output. 2^3 factorial with full confounding

General Linear Models Procedure

Class Level Information

Class	Levels	Values
TREE	4	1 2 3 4
A	2	1 2
B	2	1 2
C	2	1 2

Number of observations in data set=16

General Linear Models Procedure

Dependent Variable: YIELD

Source	DF	Sum of Squares	Mean Square	F Value	Pr>F
Model	9	417.50000000	46.38888889	50.61	0.0001
Error	6	5.50000000	0.91666667		
Corrected Total	15	423.00000000			

R-Square	C.V.	Root MSE	YIELD Mean
0.986998	3.330181	0.95742711	28.75000000

Source	DF	Type I SS	Mean Square	F Value	Pr>F
TREE	3	24.50000000	8.16666667	8.91	0.0125
A	1	196.00000000	196.00000000	213.82	0.0001
B	1	121.00000000	121.00000000	132.00	0.0001
A*B	1	4.00000000	4.00000000	4.36	0.0817
C	1	64.00000000	64.00000000	69.82	0.0002
A*C	1	4.00000000	4.00000000	4.36	0.0817
B*C	1	4.00000000	4.00000000	4.36	0.0817

General Linear Models Procedure

Dependent Variable: YIELD

Source	DF	Type III SS	Mean Square	F Value	Pr>F
TREE	3	24.50000000	8.16666667	8.91	0.0125
A	1	196.00000000	196.00000000	213.82	0.0001
B	1	121.00000000	121.00000000	132.00	0.0001
A*B	1	4.00000000	4.00000000	4.36	0.0817
C	1	64.00000000	64.00000000	69.82	0.0002
A*C	1	4.00000000	4.00000000	4.36	0.0817
B*C	1	4.00000000	4.00000000	4.36	0.0817

General Linear Models Procedure
Least Squares Means

A	YIELD LSMEAN
1	25.2500000
2	32.2500000

B	YIELD LSMEAN
1	26.0000000
2	31.5000000

A	B	YIELD LSMEAN
1	1	22.0000000
1	2	28.5000000
2	1	30.0000000
2	2	34.5000000

General Linear Models Procedure
Least Squares Means

C	YIELD LSMEAN
1	26.7500000
2	30.7500000

A	C	YIELD LSMEAN
1	1	23.7500000
1	2	26.7500000
2	1	29.7500000
2	2	34.7500000

```
General Linear Models Procedure
    Least Squares Means

 B    C         YIELD
                LSMEAN

 1    1      23.5000000
 1    2      28.5000000
 2    1      30.0000000
 2    2      33.0000000
```

Example 5

Question: Non-orthogonal design

The following experiment was conducted several years ago at the East Malling Research Station (Pearce, 1953) and its data have been used for purposes of demonstration many times since.

It was intended to find out if the application of herbicides would harm the growth of strawberry plants. The treatments comprised four kinds of herbicide, *A–D*, and an untreated control, *O*. The data represent the total spread in inches of twelve sample plants from each plot:

Block I	Block II	Block III	Block IV
D. 107	A. 136	C. 118	O. 173
A. 166	O. 146	A. 117	D. 95
B. 133	D. 104	O. 176	D. 109
C. 166	C. 152	B. 132	A. 130
O. 177	B. 119	C. 139	B. 103
A. 163	O. 164	O. 186	O. 185
O. 190	B. 132	D. 103	C. 147

$$N = \begin{bmatrix} 2 & 1 & 1 & 1 \\ 1 & 2 & 1 & 1 \\ 1 & 1 & 2 & 1 \\ 1 & 1 & 1 & 2 \\ 2 & 2 & 2 & 2 \end{bmatrix}, \quad Nk^{-\delta}N' = \frac{1}{7}\begin{bmatrix} 7 & 6 & 6 & 6 & 10 \\ 6 & 7 & 6 & 6 & 10 \\ 6 & 6 & 7 & 6 & 10 \\ 6 & 6 & 6 & 7 & 10 \\ 10 & 10 & 10 & 10 & 16 \end{bmatrix}$$

$$C = \frac{1}{7}\begin{bmatrix} 28 & -6 & -6 & -6 & -10 \\ -6 & 28 & -6 & -6 & -10 \\ -6 & -6 & 28 & -6 & -10 \\ -6 & -6 & -6 & 28 & -10 \\ -10 & -10 & -10 & -10 & 40 \end{bmatrix}, \quad \Omega^{-1} = \frac{1}{28}\begin{bmatrix} 137 & 1 & 1 & 1 & 0 \\ 1 & 137 & 1 & 1 & 0 \\ 1 & 1 & 137 & 1 & 0 \\ 1 & 1 & 1 & 137 & 0 \\ 0 & 0 & 0 & 0 & 224 \end{bmatrix}$$

SAS programming commands

```
/*   Balanced non-orthogonal design
*/

data d8;
  input block herb $ spread @@;
  cards;
1  D 107  1  A 166  1  B 133  1  C 166  1  O 177  1  A 163  1  O 190
2  A 136  2  O 146  2  D 104  2  C 152  2  B 119  2  O 164  2  B 132
3  C 118  3  A 117  3  O 176  3  B 132  3  C 139  3  O 186  3  D 103
4  O 173  4  D  95  4  D 109  4  A 130  4  B 103  4  O 185  4  C 147
;
proc glm data = d8;
  class block herb;
  model spread = block herb/ssl;
  lsmeans herb;
  title 'Balanced non-orthogonal design';
run;
```

Example 5 output. Balanced non-orthogonal design

```
        General Linear Models Procedure
            Class Level Information

        Class       Levels    Values

        BLOCK          4      1 2 3 4

        HERB           5      A B C D O


    Number of observations in data set=28
```

General Linear Models Procedure

Dependent Variable: SPREAD

Source	DF	Sum of Squares	Mean Square	F Value	Pr>F
Model	7	19395.45546218	2770.77935174	16.01	0.0001
Error	20	3460.25882353	173.01294118		
Corrected Total	27	22855.71428571			

R-Square	C.V.	Root MSE	SPREAD Mean
0.848604	9.281660	13.15343838	141.71428571

Source	DF	Type I SS	Mean Square	F Value	Pr>F
BLOCK	3	2366.00000000	788.66666667	4.56	0.0137
HERB	4	17029.45546218	4257.36386555	24.61	0.0001

General Linear Models Procedure

Least Squares Means

HERB	SPREAD LSMEAN
A	139.572059
B	124.807353
C	145.483824
D	104.336765
O	174.625000

OBS	BATCH	TREAT	MPOINT
1	1	A	194
2	1	D	205
3	1	E	250
4	1	B	214
5	2	B	204
6	2	E	243
7	2	D	198
8	2	C	238
9	3	D	206
10	3	B	205
11	3	C	238
12	3	A	186
13	4	A	183
14	4	E	247
15	4	B	202
16	4	C	229
17	5	E	255
18	5	C	244
19	5	D	209
20	5	A	198

Example 6

Question
Four varieties of autumn kale were grown in a randomized blocks experiment on 24 units arranged in split plots, the sub-plot treatments being applications of 40 and 80 kg/ha of nitrogen. The design and the amount (tonnes/ha) of digestible organic matter were as shown.

  (i) Compute the Analysis of Variance table and report on the significance of the results.
 (ii) Obtain the standard errors of the differences between nitrogen levels and between varieties and their interactions.
(iii) Criticize the suitability of the design.

$N1$: 40 kg/ha nitrogen
$N2$: 80 kg/ha nitrogen

Varieties A: Maris Kestrel
          B: Proteor
          C: Midas
          D: Vulcan

	Block 1			Block 2			Block 3	
A	$N1$	$N2$	B	$N2$	$N1$	C	$N2$	$N1$
	5.3	5.6		5.6	5.3		5.0	4.8
B	$N1$	$N2$	D	$N1$	$N2$	A	$N2$	$N1$
	5.8	5.8		5.0	5.4		5.3	5.5
C	$N2$	$N1$	A	$N2$	$N1$	D	$N1$	$N2$
	5.3	4.7		5.2	5.3		4.8	5.1
D	$N2$	$N1$	C	$N1$	$N2$	B	$N2$	$N1$
	5.2	5.0		4.6	4.7		5.4	5.3

Nitrogen × Varieties total table

	A	B	C	D	Total
$N1$	16.1	16.4	14.1	14.8	61.4
$N2$	16.1	16.8	15.0	15.7	63.6
Total	32.2	33.2	29.1	30.5	125.0

Answer

$$S_0 = 125.0^2/24 = 651.0417$$

(i) Total Sum of Squares $= \sum y_{ijk}^2 - S_0 = 653.62 - \dfrac{125.0^2}{24} = 2.57833$

$$NSS = \dfrac{61.4^2 + 63.6^2}{12} - S_0 = 651.24333 - S_0 = 0.20167$$

$$VSS = \dfrac{32.2^2 + 33.2^2 + 29.1^2 + 30.5^2}{6} - S_0 = 652.69 - S_0 = 1.64833$$

$$N \times V = \dfrac{16.1^2 + \cdots + 15.7^2}{3} - S_0 - NSS - VSS$$

$$= 652.98667 - S_0 - 0.20167 - 1.6483$$

$$= 0.095$$

$$BSS = \dfrac{42.7^2 + 41.1^2 + 41.2^2}{8} - S_0 = 0.20083$$

$$B \times V = \dfrac{10.9^2 + \cdots + 9.9^2}{2} - S_0 - VSS - BSS$$

$$= 653.15 - S_0 - 1.64833 - 0.20083$$

$$= 0.25917$$

Totals	B1	B2	B3
Varieties *A*	10.9	10.5	10.8
B	11.6	10.9	10.7
C	10.0	9.3	9.8
D	10.2	10.4	9.9
	42.7	41.1	41.2

Analysis of Variance

Source of variation	d.f.	Sum of squares	Mean square	F
Blocks	2	0.20083	0.1004	2.32
Varieties (V)	3	1.64833	0.5494	12.72**
Main-plot residual	6	0.25917	0.0432	
Main-plot total	11	2.10833		
Nitrogen (N)	1	0.20167	0.2017	9.30*
N × V	3	0.09500	0.0317	1.46
Sub-plot residual	8	0.17333	0.0217	
Total	23	2.57833		

(ii) $\sigma_m^2 = $ MPE variance, $\sigma^2 = $ SPE variance, $b = $ block no., $m = $ MP treatment no., $t = $ SPT no. We have

$$\text{MPE variance } \sigma^2 + t\sigma_m^2 \qquad \text{estimated by } 0.04320$$
$$\text{SPE variance } \sigma^2 \qquad \text{estimated by } 0.02167$$

Variances of differences:

$$\text{Varieties } \frac{2}{bt}(\sigma^2 + t\sigma_m^2) \qquad \text{estimated by } 0.01441, \text{ SED} = 0.1200$$

$$\text{Nitrogen } \frac{2}{mb}\sigma^2 \qquad \text{estimated by } 0.003612, \text{ SED} = 0.0601$$

$V \times N$

$$\frac{2}{b}\sigma^2 \rightarrow 0.01445, \qquad \text{SED} = 0.1202 \text{ same variety}$$

$$\frac{2}{b}(\sigma^2 + \sigma_m^2) \rightarrow 0.02162, \qquad \text{SED} = 0.1470 \text{ different varieties}$$

(iii) The design has detected a difference between N means, but the SPE has only a small number of degrees of freedom. The experiment is a little on the small side, perhaps. Some evidence of block variation and SPE suggests that the reduction in error variance justified the use of split plots. There is strong evidence of varietal differences.

SAS programming commands

```
/*Randomized blocks with split plots.
*/
data d11;
   input block variety $ nitrogen $ orgmat @@;
   cards;
1 A N1 5.3 1 A N2 5.6 2 B N2 5.6 2 B N1 5.3 3 C N2 5.0 3 C N1 4.8
1 B N1 5.8 1 B N2 5.8 2 D N1 5.0 2 D N2 5.4 3 A N2 5.3 3 A N1 5.5
1 C N2 5.3 1 C N1 4.7 2 A N2 5.2 2 A N1 5.3 3 D N1 4.8 3 D N2 5.1
1 D N2 5.2 1 D N1 5.0 2 C N1 4.6 2 C N2 4.7 3 B N2 5.4 3 B N1 5.3
;
proc glm data=d11;
   class block variety nitrogen;
   model orgmat=block variety block*variety nitrogen variety*nitrogen;
   means variety|nitrogen;
   test h=variety e=variety*block;
   title 'Randomized blocks with split plots';
run;
```

Example 6 output. Randomized blocks with split plots

```
General Linear Models Procedure
   Class Level Information

Class       Levels    Values

BLOCK          3       1 2 3

VARIETY        4       A B C D

NITROGEN       2       N1 N2

Number of observations in data set=24
```

General Linear Models Procedure

Dependent Variable: ORGMAT

Source	DF	Sum of Squares	Mean Square	F Value	Pr>F
Model	15	2.40500000	0.16033333	7.40	0.0038
Error	8	0.17333333	0.02166667		
Corrected Total	23	2.57833333			

R-Square	C.V.	Root MSE	ORGMAT Mean
0.932773	2.826163	0.14719601	5.20833333

Source	DF	Type I SS	Mean Square	F Value	Pr>F
BLOCK	2	0.20083333	0.10041667	4.63	0.0461
VARIETY	3	1.64833333	0.54944444	25.36	0.0002
BLOCK*VARIETY	6	0.25916667	0.04319444	1.99	0.1803
NITROGEN	1	0.20166667	0.20166667	9.31	0.0158
VARIETY*NITROGEN	3	0.09500000	0.03166667	1.46	0.2962

Source	DF	Type III SS	Mean Square	F Value	Pr>F
BLOCK	2	0.20083333	0.10041667	4.63	0.0461

General Linear Models Procedure

Dependent Variable: ORGMAT

Source	DF	Type III SS	Mean Square	F Value	Pr>F
VARIETY	3	1.64833333	0.54944444	25.36	0.0002
BLOCK*VARIETY	6	0.25916667	0.04319444	1.99	0.1803
NITROGEN	1	0.20166667	0.20166667	9.31	0.0158
VARIETY*NITROGEN	3	0.09500000	0.03166667	1.46	0.2962

General Linear Models Procedure

| Level of VARIETY | N | ------- ORGMAT ------- | |
		Mean	SD
A	6	5.36666667	0.15055453
B	6	5.53333333	0.23380904
C	6	4.85000000	0.25884358
D	6	5.08333333	0.20412415

| Level of NITROGEN | N | ------- ORGMAT ------- | |
		Mean	SD
N1	12	5.11666667	0.35887028
N2	12	5.30000000	0.29541958

| Level of VARIETY | Level of NITROGEN | N | ------- ORGMAT ------- | |
			Mean	SD
A	N1	3	5.36666667	0.11547005
A	N2	3	5.36666667	0.20816660
B	N1	3	5.46666667	0.28867513
B	N2	3	5.60000000	0.20000000
C	N1	3	4.70000000	0.10000000
C	N2	3	5.00000000	0.30000000
D	N1	3	4.93333333	0.11547005
D	N2	3	5.23333333	0.15275252

General Linear Models Procedure

Dependent Variable: ORGMAT

Tests of Hypotheses using the Type III MS for BLOCK*VARIETY as an error term

Source	DF	Type III SS	Mean Square	F Value	Pr>F
VARIETY	3	1.64833333	0.54944444	12.72	0.0052

Example 7

Question
Describe the uses of *partial confounding* and show how a 2^3 design with four replicates may be arranged in eight blocks of size 4 where all main effects are estimable but all interactions are partially confounded.

A 2^3 experiment with two replicates measured the yield of Brussels sprouts (kg/plot) as follows, where the factors were *A*, *B*, *C*.

	Replicate 1				Replicate 2		
ab	28	*c*	26	*a*	17	(1)	21
ac	27	*a*	22	*bc*	26	*abc*	32
(1)	20	*abc*	35	*ab*	26	*b*	24
bc	26	*b*	21	*c*	24	*ac*	29
(Block 1)		(Block 2)		(Block 3)		(Block 4)	

Obtain the Analysis of Variance for the above data, ignoring the block structure but accounting for replicates.

Given that the four columns in the table denote the four blocks as indicated in brackets, describe the confounding scheme and adjust your analysis accordingly. In this case discuss the importance of the treatment contrasts identified in your analysis.

Answer
Partial confounding. Confounding is the deliberate sacrifice of information on particular contrasts in order that required contrasts are estimable. Partial confounding allows different contrasts to be confounded in some replicates so they are estimable in the others where they are not confounded.

$$ABC \quad (1) \quad ab \quad ac \quad bc \,|\, a \quad b \quad c \quad abc \,\|\, AC \,(1) \quad ac \quad b \quad abc \,|\, a \quad c \quad ab \quad bc$$
$$AB \quad (1) \quad ab \quad c \quad abc \,|\, a \quad b \quad ac \quad bc \,\|\, BC \,(1) \quad bc \quad a \quad abc \,|\, b \quad c \quad ab \quad ac$$

gives the partial confounding scheme.

Treatment	(1)	*a*	*b*	*c*	*ab*	*ac*	*bc*	*abc*		Sum of
Total	41	39	45	50	54	56	52	67	Value	squares
A	−	+	−	−	+	+	−	+	28	49.00
B	−	−	+	−	+	−	+	+	32	64.00
C	−	−	−	+	−	+	+	+	46	132.25
AB	+	−	−	+	+	−	−	+	20	25.00
AC	+	−	+	−	−	+	−	+	14	(12.25)
BC	+	+	−	−	−	−	+	+	−6	(2.25)
ABC	−	+	+	+	−	−	−	+	−2	0.25

$S = 10514$ $S_0 = 10201$ $S_T = 10486$ $S_B = 10225.5$

The design is partially confounded with *BC* confounded in the first replicate, *AC* confounded in the second replicate. The S.S.s for these must be estimated separately.

$$BC: \text{use } abc + a + bc + (1) - ac - ab - a - b$$

In replicate 2 this total is 18, hence

$$\text{S.S.} = (18)^2/8 = 2.25$$

$$AC: \text{use } abc + ac + b + (1) - bc - c - a - ab$$

In replicate 1 this total is 1, hence

$$\text{S.S.} = 1^2/8 = 0.125$$

Source of variation	Degrees of freedom	Sum of squares	Mean square	*F*
Blocks	3	24.500	8.167	
A	1	49.000	49.000	19.21***
B	1	64.000	64.000	25.10***
C	1	132.250	132.250	51.86***
AB	1	25.000	25.000	9.80*
AC	1	0.125	0.125	< 1
BC	1	2.250	2.250	< 1
ABC	1	3.125	3.125	1.23
Residual	5	12.750	2.550	
Total	15	313.000		

It appears that all main effects are important, but there is also evidence of an **AB** interaction.

```
/*2 ^ 3 design with partial confounding.
*/
data d14;
   input rep block a b c yield @@;
   cards;
1 1 2 2 1 28 1 1 2 1 2 27 1 1 1 1 1 20 1 1 1 2 2 26
1 2 1 1 2 26 1 2 2 1 1 22 1 2 2 2 2 35 1 2 1 2 1 21
2 3 2 1 1 17 2 3 1 2 2 26 2 3 2 2 1 26 2 3 1 1 2 24
2 4 1 1 1 21 2 4 2 2 2 32 2 4 1 2 1 24 2 4 2 1 2 29
;
proc glm data = d14;
   class rep a b c;
   mode yield = rep a|b|c/ssl;
   lsmeans a|b|c;
   title '2 ^ 3 design with no confounding.';
run;
proc glm data = d14;
   class block a b c;
   model yield = block a|b|c/ss1;
   lsmeans a|b|c;
   title '2 ^ 3 design with partial confounding.';
run;
```

Example 7 output. 2^3 Design with no confounding

```
General Linear Models Procedure
      Class Level Information
```

Class	Levels	Values
REP	2	1 2
A	2	1 2
B	2	1 2
C	2	1 2

```
Number of observations in data set=16
```

General Linear Models Procedure

Dependent Variable: YIELD

Source	DF	Sum of Squares	Mean Square	F Value	Pr>F
Model	8	287.25000000	35.90625000	9.76	0.0035
Error	7	25.75000000	3.67857143		
Corrected Total	15	313.00000000			

R-Square	C.V.	Root MSE	YIELD Mean
0.917732	7.595882	1.91796023	25.25000000

Source	DF	Type I SS	Mean Square	F Value	Pr>F
REP	1	2.25000000	2.25000000	0.61	0.4598
A	1	49.00000000	49.00000000	13.32	0.0082
B	1	64.00000000	64.00000000	17.40	0.0042
A*B	1	25.00000000	25.00000000	6.80	0.0351
C	1	132.25000000	132.25000000	35.95	0.0005
A*C	1	12.25000000	12.25000000	3.33	0.1108
B*C	1	2.25000000	2.25000000	0.61	0.4598
A*B*C	1	0.25000000	0.25000000	0.07	0.8018

General Linear Models Procedure
Least Squares Means

A	YIELD LSMEAN
1	23.5000000
2	27.0000000

B	YIELD LSMEAN
1	23.2500000
2	27.2500000

A	B	YIELD LSMEAN
1	1	22.7500000
1	2	24.2500000
2	1	23.7500000
2	2	30.2500000

General Linear Models Procedure
Least Squares Means

C	YIELD LSMEAN
1	22.3750000
2	28.1250000

A	C	YIELD LSMEAN
1	1	21.5000000
1	2	25.5000000
2	1	23.2500000
2	2	30.7500000

General Linear Models Procedure
Class Level Information

Class	Levels	Values
BLOCK	4	1 2 3 4
A	2	1 2
B	2	1 2
C	2	1 2

Number of observations in data set=16

General Linear Models Procedure
Least Squares Means

B	C	YIELD LSMEAN
1	1	20.0000000
1	2	26.5000000
2	1	24.7500000
2	2	29.7500000

A	B	C	YIELD LSMEAN
1	1	1	20.5000000
1	1	2	25.0000000
1	2	1	22.5000000
1	2	2	26.0000000
2	1	1	19.5000000
2	1	2	28.0000000
2	2	1	27.0000000
2	2	2	33.5000000

General Linear Models Procedure

Dependent Variable: YIELD

Source	DF	Sum of Squares	Mean Square	F Value	Pr>F
Model	10	300.25000000	30.02500000	11.17	0.0070
Error	5	12.75000000	2.55000000		
Corrected Total	15	313.00000000			

R-Square	C.V.	Root MSE	YIELD Mean
0.959265	6.324245	1.59687194	25.25000000

Source	DF	Type I SS	Mean Square	F Value	Pr>F
BLOCK	3	24.50000000	8.16666667	3.20	0.1212
A	1	49.00000000	49.00000000	19.22	0.0071
B	1	64.00000000	64.00000000	25.10	0.0041
A*B	1	25.00000000	25.00000000	9.80	0.0259
C	1	132.25000000	132.25000000	51.86	0.0008
A*C	1	0.12500000	0.12500000	0.05	0.8335
B*C	1	2.25000000	2.25000000	0.88	0.3907
A*B*C	1	3.12500000	3.12500000	1.23	0.3187

General Linear Models Procedure
Least Squares Means

A	YIELD LSMEAN
1	23.5000000
2	27.0000000

B	YIELD LSMEAN
1	23.2500000
2	27.2500000

A	B	YIELD LSMEAN
1	1	22.7500000
1	2	24.2500000
2	1	23.7500000
2	2	30.2500000

General Linear Models Procedure
Least Squares Means

C	YIELD LSMEAN
1	22.3750000
2	28.1250000

A	C	YIELD LSMEAN
1	1	20.7500000
1	2	26.2500000
2	1	24.0000000
2	2	30.0000000

General Linear Models Procedure
Least Squares Means

B	C	YIELD LSMEAN
1	1	20.0000000
1	2	26.5000000
2	1	24.7500000
2	2	29.7500000

A	B	C	YIELD LSMEAN
1	1	1	20.2500000
1	1	2	25.2500000
1	2	1	21.2500000
1	2	2	27.2500000
2	1	1	19.7500000
2	1	2	27.7500000
2	2	1	28.2500000
2	2	2	32.2500000

Example 8

The final example for Chapter 14 uses data from Richert *et al.* (1974).

Variables are: x_1, heating temperature

x_2, pH

x_3, redox potential

x_4, sodium oxalate

x_5, sodium lauryl sulphate.

Coding:

	Units	-2	-1	0	1	2
x_1	°C/30 min	65	70	75	80	85
x_2	–	4	5	6	7	8
x_3	volt	-0.025	0.075	0.175	0.275	0.375
x_4	molar	0.00	0.0125	0.025	0.0375	0.05
x_5	% of solids	0.00	0.05	0.10	0.15	0.20

GLM analysis in preparation for a Contour plot

```
data d1;
input x1 x2 x3 x4 x5 y;
x1x1=x1*x1;
x2x2=x2*x2;
x3x3=x3*x3;
x4x4=x4*x4;
x5x5=x5*x5;
x1x2=x1*x2;
x1x3=x1*x3;
x1x4=x1*x4;
x1x5=x1*x5;
x2x3=x2*x3;
x2x4=x2*x4;
x2x5=x2*x5;
x3x4=x3*x4;
x3x5=x3*x5;
x4x5=x4*x5;
cards;
```

	C1	C2	C3	C4	C5
80.6	-1	-1	-1	-1	-1
67.9	-1	-1	-1	-1	-1
83.1	-1	-1	-1	-1	-1
38.1	-1	-1	-1	-1	-1
79.7	-1	-1	-1	-1	-1
74.7	-1	-1	-1	-1	-1
71.2	-1	-1	-1	-1	-1
36.8	-1	-1	-1	-1	-1
81.7	-1	-1	-1	-1	-1
66.8	-1	-1	-1	-1	-1
73.0	-1	-1	-1	-1	-1
40.5	-1	-1	-1	-1	-1
74.9	-1	-1	-1	-1	-1
74.2	-1	-1	-1	-1	-1
63.5	-1	-1	-1	-1	-1
42.8	-1	-1	-1	-1	-1
80.9	-1	-1	-1	-1	-2
42.4	0	0	0	0	2
73.4	0	0	0	-2	0
45.0	0	0	0	2	0
66.0	0	0	-2	0	0
71.7	0	0	2	0	0
77.5	0	-2	0	0	0
76.3	-2	2	0	0	0
67.4	2	0	0	0	0
86.5	0	0	0	0	0
77.4	0	0	0	0	0
74.6	0	0	0	0	0
79.8	0	0	0	0	0
78.3	0	0	0	0	0
74.8	0	0	0	0	0
80.9	0	0	0	0	0

```
proc glm data=dl;
  model y=xl x2 x3 x4 x5 xlxl x2x2 x3x3 x4x4 x5x5
         xlx2 xlx3 xlx4 xlx5 x2x3 x2x4 x2x5 x3x4 x3x5 x4x5;
  title 'GLM analysis for full quadratic model with all variables';

proc glm data=dl;
  model y=xl x2 x3 x4 x5 xlxl x2x2 x3x3 x4x4 x5x5
         xlx2 xlx3 xlx4 xlx5 x2x3 x2x4 x2x5 x3x4 x3x5;
  title 'GLM analysis for full quadratic model with all variables except the last term';

proc glm data=dl;
  model y=xl x2 x3 xlxl x2x2 x3x3 xlx2 xlx3 x2x3;
  title 'GLM analysis for full quadratic model with variables xl, x2, x3';

proc glm data=dl;
  model y=xl x2 xlxl x2x2 xlx2;
  title 'GLM analysis for full quadratic model with variables xl, x2';

quit;
run;
```

Example 8 output. GLM analysis for full quadratic model with all variables

General Linear Models Procedure
Number of observations in data set=32

Dependent Variable: Y

Source	DF	Sum of Squares	Mean Square	F Value	Pr>F
Model	20	6386.90803030	319.34540152	14.73	0.0001
Error	11	238.55196970	21.68654270		
Corrected Total	31	6625.46000000			

R-Square	C.V.	Root MSE	Y Mean
0.963995	6.766264	4.65688122	68.82500000

Source	DF	Type I SS	Mean Square	F Value	Pr>F
X1	1	2458.35041667	2458.35041667	113.36	0.0001
X2	1	1807.87041667	1807.87041667	83.36	0.0001
X3	1	0.26041667	0.26041667	0.01	0.9147
X4	1	12.18375000	12.18375000	0.56	0.4693
X5	1	45.10041667	45.10041667	2.08	0.1771
X1X1	1	396.76033333	396.76033333	18.30	0.0013
X2X2	1	619.04288095	619.04288095	28.55	0.0002
X3X3	1	157.84640110	157.84640110	7.28	0.0207
X4X4	1	2.96205128	2.96205128	0.14	0.7187
X5X5	1	3.22969697	3.22969697	0.15	0.7069
X1X2	1	616.27806250	616.28062500	28.42	0.0002
X1X3	1	122.65562500	122.65562500	5.66	0.0366
X1X4	1	50.05562500	50.05562500	2.31	0.1569
X1X5	1	7.98062500	7.98062500	0.37	0.5564
X2X3	1	45.22562500	45.22562500	2.09	0.1766
X2X4	1	1.05062500	1.05062500	0.05	0.8298
X2X5	1	3.70562500	3.70562500	0.17	0.6873
X3X4	1	0.03062500	0.03062500	0.00	0.9707
X3X5	1	36.30062500	36.30062500	1.67	0.2222
X4X5	1	0.01562500	0.01562500	0.00	0.9791

Source	DF	Type III SS	Mean Square	F Value	Pr>F
X1	1	2458.35041667	2458.35041667	113.36	0.0001
X2	1	1807.87041667	1807.87041667	83.36	0.0001
X3	1	0.26041667	0.26041667	0.01	0.9147
X4	1	12.18375000	12.18375000	0.56	0.4693
X5	1	45.10041667	45.10041667	2.08	0.1771
X1X1	1	506.85469697	506.85469697	23.37	0.0005
X2X2	1	667.22761364	667.22761364	30.77	0.0002
X3X3	1	162.93469697	162.93469697	7.51	0.0192
X4X4	1	3.47761364	3.47761364	0.16	0.6965
X5X5	1	3.22969697	3.22969697	0.15	0.7069
x1X2	1	616.28062500	616.28062500	28.42	0.0002
X1X3	1	122.65562500	122.65562500	5.66	0.0366
X1X4	1	50.05562500	50.05562500	2.31	0.1569
X1X5	1	7.98062500	7.98062500	0.37	0.5564
X2X3	1	45.22562500	45.22562500	2.09	0.1766
X2X4	1	1.05062500	1.05062500	0.05	0.8298
X2X5	1	3.70562500	3.70562500	0.17	0.6873
X3X4	1	0.03062500	0.03062500	0.00	0.9707
X3X5	1	36.30062500	36.30062500	1.67	0.2222

General Linear Models Procedure

Dependent Variable: Y

Source	DF	Type III SS	Mean Square	F Value	Pr>F
X4X5	1	0.01562500	0.01562500	0.00	0.9791

Parameter	Estimate	T for H0: Parameter=0	Pr>|T|	Std Error of Estimate
INTERCEPT	77.79431818	41.88	0.0001	1.85745304
X1	-10.12083333	-10.65	0.0001	0.95058190
X2	-8.67916667	-9.13	0.0001	0.95058190
X3	-0.10416667	-0.11	0.9147	0.95058190
X4	-0.71250000	-0.75	0.4693	0.95058190
X5	1.37083333	1.44	0.1771	0.95058190
X1X1	-4.15681818	-4.83	0.0005	0.85983368
X2X2	-4.76931818	-5.55	0.0002	0.85983368
X3X3	-2.35681818	-2.74	0.0192	0.85983368
X4X4	-0.34431818	-0.40	0.6965	0.85983368
X5X5	-0.33181818	-0.39	0.7069	0.85983368
X1X2	-6.20625000	-5.33	0.0002	1.16422031
X1X3	2.76875000	2.38	0.0366	1.16422031
X1X4	1.76875000	1.52	0.1569	1.16422031
X1X5	0.70625000	0.61	0.5564	1.16422031
X2X3	-1.68125000	-1.44	0.1766	1.16422031
X2X4	-0.25625000	-0.22	0.8298	1.16422031
X2X5	0.48125000	0.41	0.6873	1.16422031
X3X4	0.04375000	0.04	0.9707	1.16422031
X3X5	1.50625000	1.29	0.2222	1.16422031
X4X5	0.03125000	0.03	0.9791	1.16422031

GLM analysis for full quadratic model with all variables except the last term

General Linear Models Procedure
Number of observations in data set=32

Dependent Variable: Y

Source	DF	Sum of Squares	Mean Square	F Value	Pr>F
Model	19	6386.89240530	336.15223186	16.91	0.0001
Error	12	238.56759470	19.88063289		
Corrected Total	31	6625.46000000			

R-Square	C.V.	Root MSE	Y Mean
0.963992	6.478417	4.45877033	68.8250000000

Source	DF	Type I SS	Mean Square	F Value	Pr>F
X1	1	2458.35041667	2458.35041667	123.66	0.0001
X2	1	1807.87041667	1807.87041667	90.94	0.0001
X3	1	0.26041667	0.26041667	0.01	0.9108
X4	1	12.18375000	12.18375000	0.61	0.4489
X5	1	45.10041667	45.10041667	2.27	0.1579
X1X1	1	396.76033333	396.76033333	19.96	0.0008
X2X2	1	619.04288095	619.04288095	31.14	0.0001
X3X3	1	157.84640110	157.84640110	7.94	0.0155
X4X4	1	2.96205128	2.96205128	0.15	0.7063
X5X5	1	3.22969697	3.22969697	0.16	0.6940
X1X2	1	616.28062500	616.28062500	31.00	0.0001
X1X3	1	122.65562500	122.65562500	6.17	0.0288
X1X4	1	50.05562500	50.05562500	2.52	0.1386
X1X5	1	7.98062500	7.98062500	0.40	0.5382
X2X3	1	45.22562500	45.22562500	2.27	0.1574
X2X4	1	1.05062500	1.05062500	0.05	0.8221
X2X5	1	3.70562500	3.70562500	0.19	0.6736
X3X4	1	0.03062500	0.03062500	0.00	0.9693
X3X5	1	36.30062500	36.30062500	1.83	0.2015

General Linear Models Procedure

Dependent Variable: Y

Parameter	Estimate	T for HO: Parameter=0	Pr>|T|	Std Error of Estimate
INTERCEPT	77.79431818	43.74	0.0001	1.77843413
X1	-10.12083333	-11.12	0.0001	0.91014268
X2	-8.67916667	-9.54	0.0001	0.91014268
X3	-0.10416667	-0.11	0.9108	0.91014268
X4	-0.71250000	-0.78	0.4489	0.91014268
X5	1.37083333	1.51	0.1579	0.91014268
X1X1	-4.15681818	-5.05	0.0003	0.82325503
X2X2	-4.76931818	-5.79	0.0001	0.82325503
X3X3	-2.35681818	-2.86	0.0143	0.82325503
X4X4	-0.34431818	-0.42	0.6832	0.82325503
X5X5	-0.33181818	-0.40	0.6940	0.82325503
X1X2	-6.20625000	-5.57	0.0001	1.11469258
X1X3	2.76875000	2.48	0.0288	1.11469258
X1X4	1.76875000	1.59	0.1386	1.11469258
X1X5	0.70625000	0.63	0.5382	1.11469258
X2X3	-1.68125000	-1.51	0.1574	1.11469258
X2X4	-0.25625000	-0.23	0.8221	1.11469258
X2X5	0.48125000	0.43	0.6736	1.11469258
X3X4	0.04375000	0.04	0.9693	1.11469258
X3X5	1.50625000	1.35	0.2015	1.11469258

GLM analysis for full quadratic model with variables x1, x2, x3

```
                    General Linear Models Procedure
                    Number of observations in data set=32
```

Dependent Variable: Y

Source	DF	Sum of Squares	Mean Square	F Value	Pr>F
Model	9	6224.29274038	691.58808226	37.93	0.0001
Error	22	401.16725962	18.23487544		
Corrected Total	31	6625.46000000			

R-Square	C.V.	Root MSE	Y Mean
0.939451	6.204477	4.27023131	68.82500000

Source	DF	Type I SS	Mean Square	F Value	Pr>F
X1	1	2458.35041667	2458.35041667	134.82	0.0001
X2	1	1807.87041667	1807.87041667	99.14	0.0001
X3	1	0.26041667	0.26041667	0.01	0.9060
X1X1	1	396.76033333	396.76033333	21.76	0.0001
X2X2	1	619.04288095	619.04288095	33.95	0.0001
X3X3	1	157.84640110	157.84640110	8.66	0.0075
X1X2	1	616.28062500	616.28062500	33.80	0.0001
X1X3	1	122.65562500	122.65562500	6.73	0.0166
X2X3	1	45.22562500	45.22562500	2.48	0.1296

Source	DF	Type III SS	Mean Square	F Value	Pr>F
X1	1	2458.35041667	2458.35041667	134.82	0.0001
X2	1	1807.87041667	1807.87041667	99.14	0.0001
X3	1	0.26041667	0.26041667	0.01	0.9060
X1X1	1	500.66925824	500.66925824	27.46	0.0001
X2X2	1	661.23175824	661.23175824	36.26	0.0001
X3X3	1	157.84640110	157.84640110	8.66	0.0075
X1X2	1	616.28062500	616.28062500	33.80	0.0001
X1X3	1	122.65562500	122.65562500	6.73	0.0166
X2X3	1	45.22562500	45.22562500	2.48	0.1296

Parameter	Estimate	T for H0: Parameter=0	Pr>|T|	Std Error of Estimate
INTERCEPT	77.17019231	58.28	0.0001	1.32414252
X1	-10.12083333	-11.61	0.0001	0.87165732
X2	-8.67916667	-9.96	0.0001	0.87165732
X3	-0.10416667	-0.12	0.9060	0.87165732
X1X1	-4.10480769	-5.24	0.0001	0.78337328
X2X2	-4.71730769	-6.02	0.001	0.78337328
X3X3	-2.30480769	-2.94	0.0075	0.78337328
X1X2	-6.20625000	-5.81	0.0001	1.06755783
X1X3	2.76875000	2.59	0.0166	1.06766783
X2X3	-1.68125000	-1.57	0.1296	1.06755783

GLM analysis for full quadratic model with variables x1, x2

General Linear Models Procedure
Number of observations in data set=32

Dependent Variable: Y

Source	DF	Sum of Squares	Mean Square	F Value	Pr>F
Model	5	5898.30467262	1179.66093452	42.18	0.0001
Error	26	727.15532738	27.96751259		
Corrected Total	31	6625.46000000			

R-Square	C.V.	Root MSE	Y Mean
0.890248	7.683882	5.28843196	68.82500000

Source	DF	Type I SS	Mean Square	F Value	Pr>F
X1	1	2458.35041667	2458.35041667	87.90	0.0001
X2	1	1807.87041667	1807.87041667	64.64	0.0001
X1X1	1	396.76033333	396.76033333	14.19	0.0009
X2X2	1	619.04288095	619.04288095	22.13	0.0001
X1X2	1	616.28062500	616.28062500	22.04	0.0001

Source	DF	Type III SS	Mean Square	F Value	Pr>F
X1	1	2458.35041667	2458.35041667	87.90	0.0001
X2	1	1807.87041667	1807.87041667	64.64	0.0001
X1X1	1	463.68021429	463.68021429	16.58	0.0004
X2X2	1	619.04288095	619.04288095	22.13	0.0001
X1X2	1	616.28062500	616.28062500	22.04	0.0001

Parameter	Estimate	T for H0: Parameter=0	Pr>|T|	Std Error of Estimate
INTERCEPT	75.19464286	53.20	0.0001	1.41339289
X1	-10.12083333	-9.38	0.0001	1.07949665
X2	-8.67916667	-8.04	0.0001	1.07949665
X1X1	-3.94017857	-4.07	0.0004	0.96768396
X2X2	-4.55267857	-4.70	0.0001	0.96768396
X1X2	-6.20625000	-4.69	0.0001	1.32210799

```
data d2;
  do x1=-2 to 2 by 0.1;
    do x2=-2 to 2 by 0.1;
      z=75.19464286-10.12083333*x1-8.67916667*x2
        -3.94017857*x1*x1-4.55267857*x2*x2-6.20625000*x1*x2;
      output;
    end;
end;

title 'Response surface for Foaming against Temp and pH';
filename plot1 'c:\ck\fig1.plt';
goptions device=hp7550a gsfname=plot1 gsfmode=replace
noprompt
  autofeed;
proc g3d data=d2;
options ps=64 nonumber nodate;
  plot x1*x2=z;
run;
```

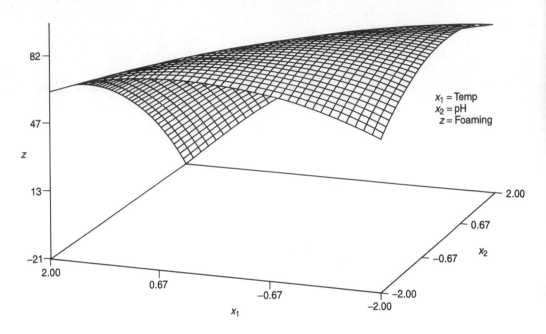

Response surface for example 8

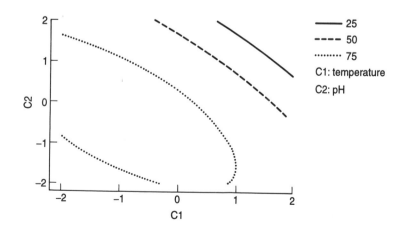

Contour plot for example 8

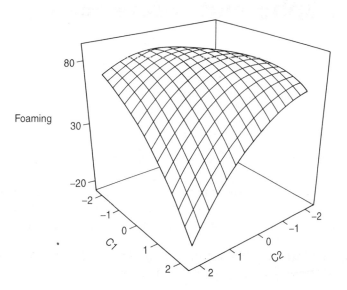

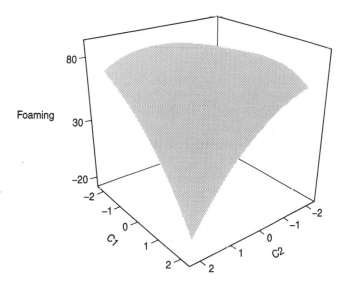

Alternative forms of the response surface for example 8

Bibliography and references

Ali, A. (1983). Interpretation of multivariate data from fruit nutrition experiments. Unpublished D.Phil. thesis, University of Sussex.

Ali, A., G. M. Clarke and K. Trustrum (1986). Log-linear response functions and their use to model data from plant nutrition experiments. J. Sci. Food Agric., 37, 1165.

Altman, D. G. (1991). Practical statistics for medical research. Wiley, New York & Chichester.

American Supplier Institute (1987). Fifth Symposium on Taguchi Methods.

Anscombe, F. J. and J. W. Tukey (1963). The examination and analysis of residuals. Technometrics, 5, 141.

Atkinson, A. C. (1996). The usefulness of Optimum Experimental Designs. Journal of the Royal Statistical Society, B, 58, 59.

Barnett, V. (1991). Sample Survey Principles and Methods, Arnold, London.

Bennett, C. A. and N. L. Franklin (1954). Statistical analysis in chemistry and the chemical industry. Wiley, New York.

Bissell, A. F. (1989). Interpreting mean squares in saturated factorial designs. Journal of Applied Statistics, 16, 7.

Bissell, A. F. (1994). Statistical methods in SPC and TQM, Chapman and Hall, London.

Box, G. E. P. and D. R. Cox (1964). An analysis of transformations, Journal of the Royal Statistical Society, B, 26, 211.

Box, G. E. P. and J. S. Hunter (1957). Multifactor Experimental Designs for exploring Response Surfaces. Ann. Math. Statistics, 28, 195.

Box, G. E. P., W. G. Hunter and J. S. Hunter (1978). Statistics for Experimenters. Wiley, New York & Chichester.

Chakrabati, M. C. (1962). Mathematics of design and analysis of experiments. Asia Publishing House, Bombay.

Clarke, G. M. (1994). Statistics and experimental design, 3rd edn. Arnold, London.

Clarke, G. M. and D. Cooke (1992). A basic course in statistics, 3rd edn. Arnold, London.

Clatworthy, W. H. (1973). Tables of two-associate partially balanced designs. National Bureau of Standards, US Dept. of Commerce.

Cochran, W. G. (1934). The distribution of quadratic forms in a normal system, with applications to the analysis of covariance. Proc. Camb. Phil. Soc., 30, 178.

Cochran, W. G. (1992). Sampling techniques, 3rd edn. Wiley, New York & Chichester.

Cochran, W. G. and G. M. Cox (1992). Experimental designs. Wiley, New York & Chichester.

Cornell, J. A. (1990). Experiments with mixtures. Wiley, New York.

Cox, D. R. (1992). Planning of experiments. Wiley, New York & Chichester.

Daniel, C. (1959). Use of half-normal plots in interpreting factorial two-level experiments. Technometrics, *1*, 311.

Daniel, C. (1976). Applications of Statistics to Industrial Experimentation. Wiley, New York & Chichester.

Davies, O. L. (ed). (1954). The Design & Analysis of Industrial Experiments. Oliver and Boyd, London.

Draper, N. R. and H. Smith (1981). Applied regression analysis. Wiley, New York.

Duncan, D. B. (1955). Multiple range and multiple F-tests. Biometrics, *11*, 1.

Dyke, G. V. (1988). Comparative experiments with field crops. 2nd edn. Arnold, London.

Fisher, R. A. (1958). Statistical methods for research workers. Oliver and Boyd, Edinburgh.

Fisher, R. A. and F. Yates (1963). Statistical tables for biological, agricultural and medical research. Oliver and Boyd, Edinburgh.

Fleiss, J. L. (1985). The design and analysis of clinical experiments. Wiley, New York & Chichester.

Freund, J. E. and R. E. Walpole (1987). Mathematical Statistics, 3rd edn. Prentice-Hall, Englewood Cliffs, N.J., U.S.A.

Graybill, F. A. (1969). Introduction to Matrices with Applications in Statistics. Wadsworth, Belmont, NC.

Graybill, F. A. (1976). Theory and application of the linear model. Duxbury, Massachusetts.

Healy, M. J. R. (1963). Fiducial limits for a variance component. Journal of the Royal Statistical Society, B, *25*, 128.

Hogg, R. V. and A. T. Craig (1994). Introduction to mathematical statistics. Collier-Macmillan, London.

John, J. A. (1987). Cyclic designs. Chapman and Hall, London.

John, P. W. M. (1973). Statistical design and analysis of experiments. Macmillan, New York.

John, J. A. and M. H. Quenouille (1977). Experiments: Design and Analysis, 2nd edn., Griffin, London.

Jones, B. (1976). An algorithm for deriving optimal block designs. Technometrics, *18*(4), 451.

Jones, B. and M. G. Kenward (1989). Design and analysis of crossover trials. Chapman and Hall, London.

Kempthorne, O. (1952). Design and analysis of experiments. Krieger, New York.

Kempthorne, O. (1977). Why randomise? Jour. Statist. Plan. Infer., *1*, 1.

Kendall, M. G., A. Stuart and J. K. Ord (1991). The advanced theory of statistics, Vol. 2. Griffin, London.

Kendall, M. G., A Stuart and J. K. Ord (1994). The Advanced Theory of Statistics. Vol. 1. Griffin, London.

Khuri, A. I. and J. A. Cornell (1987). Response surfaces: designs and analyses. Marcel Dekker, New York.

Lancaster, H. O. (1954). Traces and cumulants of quadratic forms in normal variables. Journal of the Royal Statistical Society, B, *16*, 247.

McCullagh, P. and J. A. Nelder (1989). Generalized linear models. Chapman and Hall, London.

Mead, R. (1988). The design of experiments. Cambridge.

Mead, R. and R. N. Curnow (1983). Statistical methods in agriculture and experimental biology. Chapman and Hall, London.

Montgomery, D. C. (1984). Design and analysis of experiments. Wiley, New York.

Moser, C. A. and G. Kalton (1985). Survey methods in social investigation. Heinemann, London.

Nelder, J. A. (1966). Inverse Polynomials, a useful group of multifactor Response Functions. Biometrics, *22*, 128.

Nelder, J. A. (1968). Weighted regression, quantal response data and Inverse Polynomials. Biometrics, *24*, 979.

O'Neill, R. and G. B. Wetherill (1971). The present state of multiple comparison methods. Journal of the Royal Statistical Society, B, *33*, 218.

Patterson, H. D. and E. R. Williams (1976). A new class of resolvable incomplete block designs. Biometrika, *63*(1), 83.

Patterson, H. D., E. R. Williams and E. A. Hunter (1978). Block designs for variety trials. J. Agric. Sci., *90*, 395.

Peace, K. E. (ed.) (1988). Biopharmaceutical statistics for drug development. Marcel Dekker, New York.

Pearce, S. C. (1953). Field Experimentation with fruit trees and other perennial plants. Commwth. Bureau Hort. East Malling, Maidstone.

Pearce, S. C. (1983). The agricultural field experiment. Wiley, Chichester and New York.

Pearce, S. C. and J. N. R. Jeffers (1971). Block designs and missing data. Journal of the Royal Statistical Society, B, *33*, 131.

Pearce, S. C., G. M. Clarke, G. V. Dyke and R. E. Kempson (1988). Manual of crop experimentation. Griffin, London.

Plackett, R. L. and J. P. Burman (1946). The design of optimum multifactorial experiments. Biometrika, *33*, 305.

Preece, D. A., R. A. Bailey and H. D. Patterson (1978). A randomisation problem in forming designs with superimposed treatments. Aust. J. Statist., *20*, 111.

Rao, C. R. (1946). Confounded factorial designs in quasi-Latin squares. Sankhyā, *7*, 295.

Richert, S. H., C. V. Morr and C. M. Cooney (1974). Effect of heat and other factors upon foaming properties of whey protein concentrates. Journal of Food Science, *39*, 42.

Satterthwaite, F. E. (1946). An approximate distribution of estimates of variance components. Biometrics Bulletin, *2*, 110.

Searle, S. R. (1971). Linear models. Wiley, New York & Chichester.

Scheffé, H. W. (1957). The analysis of variance. Wiley, New York & Chichester.

Scheffé, H. W. (1958). Experiments with mixtures. Journal of the Royal Statistical Society, B, *20*, 344.

Scheffé, H. W. (1963). The simplex-centroid design for experiments with mixtures. Journal of the Royal Statistical Society, B, *25*, 235.

Snedecor, G. W. and W. G. Cochran (1967). Statistical methods, 6th edn. Iowa State University Press, Iowa.

Steel, R. G. D. and J. H. Torrie (1980). Principles and procedures of statistics. McGraw-Hill, New York.

Taguchi, G. and S. Konishi (1987). Orthogonal arrays and linear graphs, American Supplier Institute.

Tobias, R. D. (1989). Factex and Optex, SAS/QC. SAS Institute, Cary, NC.

Tocher, K. D. (1952). The design and analysis of block experiments. Journal of the Royal Statistical Society, B, *14*, 45.

Yates, F. (1937). Design and analysis of factorial experiments, Technical Communication No. 35. Imperial Bureau of Soil Sciences, London.

Yates, F. (1940). The recovery of inter-block information in balanced incomplete block designs. Ann. Eugen. *10*, 317.

Youden, W. J. (1937). Use of incomplete block replications in estimating tobacco-mosaic virus. Contr. Boyce Thompson Inst. *9*, 41.

Youden, W. J. (1940). Experimental designs to increase accuracy of greenhouse studies. Contr. Boyce Thompson Inst. *11*, 219.

Youden, W. J. (1972). Randomisation and experimentation. Technometrics, *14*, 13.

Wetherill, G. B. (1986). Regression analysis. Chapman and Hall, London.

Tables

Table I. Student's t-distribution.
Values exceeded in two-tail test with probability P.

d.f.	$P = 0.1$	0.05	0.02	0.01	0.002	0.001
1	6·314	12·706	31·821	63·657	318·31	636·62
2	2·920	4·303	6·965	9·925	22·327	31·598
3	2·353	3·182	4·541	5·841	10·214	12·924
4	2·132	2·776	3·747	4·604	7·173	8·610
5	2·015	2·571	3·365	4·032	5·893	6·869
6	1·943	2·447	3·143	3·707	5·208	5·959
7	1·895	2·365	2·998	3·499	4·785	5·408
8	1·860	2·306	2·896	3·355	4·501	5·041
9	1·833	2·262	2·821	3·250	4·297	4·781
10	1·812	2·228	2·764	3·169	4·144	4·587
11	1·796	2·201	2·718	3·106	4·025	4·437
12	1·782	2·179	2·681	3·055	3·930	4·318
13	1·771	2·160	2·650	3·012	3·852	4·221
14	1·761	2·145	2·624	2·977	3·787	4·140
15	1·753	2·131	2·602	2·947	3·733	4·073
16	1·746	2·120	2·583	2·921	3·686	4·015
17	1·740	2·110	2·567	2·898	3·646	3·965
18	1·734	2·101	2·552	2·878	3·610	3·922
19	1·729	2·093	2·539	2·861	3·579	3·883
20	1·725	2·086	2·528	2·845	3·552	3·850
21	1·721	2·080	2·518	2·831	3·527	3·819
22	1·717	2·074	2·508	2·819	3·505	3·792
23	1·714	2·069	2·500	2·807	3·485	3·767
24	1·711	2·064	2·492	2·797	3·467	3·745
25	1·708	2·060	2·485	2·787	3·450	3·725
26	1·706	2·056	2·479	2·779	3·435	3·707
27	1·703	2·052	2·473	2·771	3·421	3·690
28	1·701	2·048	2·467	2·763	3·408	3·674
29	1·699	2·045	2·462	2·756	3·396	3·659
30	1·697	2·042	2·457	2·750	3·385	3·646
40	1·684	2·021	2·423	2·704	3·307	3·551
60	1·671	2·000	2·390	2·660	3·232	3·460
120	1·658	1·980	2·358	2·617	3·160	3·373
∞	1·645	1·960	2·326	2·576	3·090	3·291

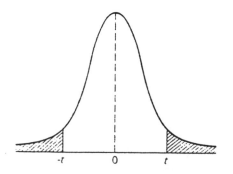

Table II. Table of F-distribution. Upper 5% points.

v_2 \ v_1	1	2	3	4	5	6	7	8	9	10	12	15	20	24	30	40	60	120	∞
1	161·4	199·5	215·7	224·6	230·2	234·0	236·8	238·9	240·5	241·9	243·9	245·9	248·0	249·1	250·1	251·1	252·2	253·3	254·3
2	18·51	19·00	19·16	19·25	19·30	19·33	19·35	19·37	19·38	19·40	19·41	19·43	19·45	19·45	19·46	19·47	19·48	19·49	19·50
3	10·13	9·55	9·28	9·12	9·01	8·94	8·89	8·85	8·81	8·79	8·74	8·70	8·66	8·64	8·62	8·59	8·57	8·55	8·53
4	7·71	6·94	6·59	6·39	6·26	6·16	6·09	6·04	6·00	5·96	5·91	5·86	5·80	5·77	5·75	5·72	5·69	5·66	5·63
5	6·61	5·79	5·41	5·19	5·05	4·95	4·88	4·82	4·77	4·74	4·68	4·62	4·56	4·53	4·50	4·46	4·43	4·40	4·36
6	5·99	5·14	4·76	4·53	4·39	4·28	4·21	4·15	4·10	4·06	4·00	3·94	3·87	3·84	3·81	3·77	3·74	3·70	3·67
7	5·59	4·74	4·35	4·12	3·97	3·87	3·79	3·73	3·68	3·64	3·57	3·51	3·44	3·41	3·38	3·34	3·30	3·27	3·23
8	5·32	4·46	4·07	3·84	3·69	3·58	3·50	3·44	3·39	3·35	3·28	3·22	3·15	3·12	3·08	3·04	3·01	2·97	2·93
9	5·12	4·26	3·86	3·63	3·48	3·37	3·29	3·23	3·18	3·14	3·07	3·01	2·94	2·90	2·86	2·83	2·79	2·75	2·71
10	4·96	4·10	3·71	3·48	3·33	3·22	3·14	3·07	3·02	2·98	2·91	2·85	2·77	2·74	2·70	2·66	2·62	2·58	2·54
11	4·84	3·98	3·59	3·36	3·20	3·09	3·01	2·95	2·90	2·85	2·79	2·72	2·65	2·61	2·57	2·53	2·49	2·45	2·40
12	4·75	3·89	3·49	3·26	3·11	3·00	2·91	2·85	2·80	2·75	2·69	2·62	2·54	2·51	2·47	2·43	2·38	2·34	2·30
13	4·67	3·81	3·41	3·18	3·03	2·92	2·83	2·77	2·71	2·67	2·60	2·53	2·46	2·42	2·38	2·34	2·30	2·25	2·21
14	4·60	3·74	3·34	3·11	2·96	2·85	2·76	2·70	2·65	2·60	2·53	2·46	2·39	2·35	2·31	2·27	2·22	2·18	2·13
15	4·54	3·68	3·29	3·06	2·90	2·79	2·71	2·64	2·59	2·54	2·48	2·40	2·33	2·29	2·25	2·20	2·16	2·11	2·07
16	4·49	3·63	3·24	3·01	2·85	2·74	2·66	2·59	2·54	2·49	2·42	2·35	2·28	2·24	2·19	2·15	2·11	2·06	2·01
17	4·45	3·59	3·20	2·96	2·81	2·70	2·61	2·55	2·49	2·45	2·38	2·31	2·23	2·19	2·15	2·10	2·06	2·01	1·96
18	4·41	3·55	3·16	2·93	2·77	2·66	2·58	2·51	2·46	2·41	2·34	2·27	2·19	2·15	2·11	2·06	2·02	1·97	1·92
19	4·38	3·52	3·13	2·90	2·74	2·63	2·54	2·48	2·42	2·38	2·31	2·23	2·16	2·11	2·07	2·03	1·98	1·93	1·88
20	4·35	3·49	3·10	2·87	2·71	2·60	2·51	2·45	2·39	2·35	2·28	2·20	2·12	2·08	2·04	1·99	1·95	1·90	1·84
21	4·32	3·47	3·07	2·84	2·68	2·57	2·49	2·42	2·37	2·32	2·25	2·18	2·10	2·05	2·01	1·96	1·92	1·87	1·81
22	4·30	3·44	3·05	2·82	2·66	2·55	2·46	2·40	2·34	2·30	2·23	2·15	2·07	2·03	1·98	1·94	1·89	1·84	1·78
23	4·28	3·42	3·03	2·80	2·64	2·53	2·44	2·37	2·32	2·27	2·20	2·13	2·05	2·01	1·96	1·91	1·86	1·81	1·76
24	4·26	3·40	3·01	2·78	2·62	2·51	2·42	2·36	2·30	2·25	2·18	2·11	2·03	1·98	1·94	1·89	1·84	1·79	1·73
25	4·24	3·39	2·99	2·76	2·60	2·49	2·40	2·34	2·28	2·24	2·16	2·09	2·01	1·96	1·92	1·87	1·82	1·77	1·71
26	4·23	3·37	2·98	2·74	2·59	2·47	2·39	2·32	2·27	2·22	2·15	2·07	1·99	1·95	1·90	1·85	1·80	1·75	1·69
27	4·21	3·35	2·96	2·73	2·57	2·46	2·37	2·31	2·25	2·20	2·13	2·06	1·97	1·93	1·88	1·84	1·79	1·73	1·67
28	4·20	3·34	2·95	2·71	2·56	2·45	2·36	2·29	2·24	2·19	2·12	2·04	1·96	1·91	1·87	1·82	1·77	1·71	1·65
29	4·18	3·33	2·93	2·70	2·55	2·43	2·35	2·28	2·22	2·18	2·10	2·03	1·94	1·90	1·85	1·81	1·75	1·70	1·64
30	4·17	3·32	2·92	2·69	2·53	2·42	2·33	2·27	2·21	2·16	2·09	2·01	1·93	1·89	1·84	1·79	1·74	1·68	1·62
40	4·08	3·23	2·84	2·61	2·45	2·34	2·25	2·18	2·12	2·08	2·00	1·92	1·84	1·79	1·74	1·69	1·64	1·58	1·51
60	4·00	3·15	2·76	2·53	2·37	2·25	2·17	2·10	2·04	1·99	1·92	1·84	1·75	1·70	1·65	1·59	1·53	1·47	1·39
120	3·92	3·07	2·68	2·45	2·29	2·17	2·09	2·02	1·96	1·91	1·83	1·75	1·66	1·61	1·55	1·50	1·43	1·35	1·25
∞	3·84	3·00	2·60	2·37	2·21	2·10	2·01	1·94	1·88	1·83	1·75	1·67	1·57	1·52	1·46	1·39	1·32	1·22	1·00

v_1, v_2 are upper, lower d.f. respectively.
Tabulated values are those exceeded with probability ·05 in a one-tail test.

Table II. (cont.) Upper 2.5% points of F-distribution.

v_2 \ v_1	1	2	3	4	5	6	7	8	9	10	12	15	20	24	30	40	60	120	∞
1	647.8	799.5	864.2	899.6	921.8	937.1	948.2	956.7	963.3	968.6	976.7	984.9	993.1	997.2	1001	1006	1010	1014	1018
2	38.51	39.00	39.17	39.25	39.30	39.33	39.36	39.37	39.39	39.40	39.41	39.43	39.45	39.46	39.46	39.47	39.48	39.49	39.50
3	17.44	16.04	15.44	15.10	14.88	14.73	14.62	14.54	14.47	14.42	14.34	14.25	14.17	14.12	14.08	14.04	13.99	13.95	13.90
4	12.22	10.65	9.98	9.60	9.36	9.20	9.07	8.98	8.90	8.84	8.75	8.66	8.56	8.51	8.46	8.41	8.36	8.31	8.26
5	10.01	8.43	7.76	7.39	7.15	6.98	6.85	6.76	6.68	6.62	6.52	6.43	6.33	6.28	6.23	6.18	6.12	6.07	6.02
6	8.81	7.26	6.60	6.23	5.99	5.82	5.70	5.60	5.52	5.46	5.37	5.27	5.17	5.12	5.07	5.01	4.96	4.90	4.85
7	8.07	6.54	5.89	5.52	5.29	5.12	4.99	4.90	4.82	4.76	4.67	4.57	4.47	4.42	4.36	4.31	4.25	4.20	4.14
8	7.57	6.06	5.42	5.05	4.82	4.65	4.53	4.43	4.36	4.30	4.20	4.10	4.00	3.95	3.89	3.84	3.78	3.73	3.67
9	7.21	5.71	5.08	4.72	4.48	4.32	4.20	4.10	4.03	3.96	3.87	3.77	3.67	3.61	3.56	3.51	3.45	3.39	3.33
10	6.94	5.46	4.83	4.47	4.24	4.07	3.95	3.85	3.78	3.72	3.62	3.52	3.42	3.37	3.31	3.26	3.20	3.14	3.08
11	6.72	5.26	4.63	4.28	4.04	3.88	3.76	3.66	3.59	3.53	3.43	3.33	3.23	3.17	3.12	3.06	3.00	2.94	2.88
12	6.55	5.10	4.47	4.12	3.89	3.73	3.61	3.51	3.44	3.37	3.28	3.18	3.07	3.02	2.96	2.91	2.85	2.79	2.72
13	6.41	4.97	4.35	4.00	3.77	3.60	3.48	3.39	3.31	3.25	3.15	3.05	2.95	2.89	2.84	2.78	2.72	2.66	2.60
14	6.30	4.86	4.24	3.89	3.66	3.50	3.38	3.29	3.21	3.15	3.05	2.95	2.84	2.79	2.73	2.67	2.61	2.55	2.48
15	6.20	4.77	4.15	3.80	3.58	3.41	3.29	3.20	3.12	3.06	2.96	2.86	2.76	2.70	2.64	2.59	2.52	2.46	2.40
16	6.12	4.69	4.08	3.73	3.50	3.34	3.22	3.12	3.05	2.99	2.89	2.79	2.68	2.63	2.57	2.51	2.45	2.38	2.32
17	6.04	4.62	4.01	3.66	3.44	3.28	3.16	3.06	2.98	2.92	2.82	2.72	2.62	2.56	2.50	2.44	2.38	2.32	2.25
18	5.98	4.56	3.95	3.61	3.38	3.22	3.10	3.01	2.93	2.87	2.77	2.67	2.56	2.50	2.44	2.38	2.32	2.26	2.19
19	5.92	4.51	3.90	3.56	3.33	3.17	3.05	2.96	2.88	2.82	2.72	2.62	2.51	2.45	2.39	2.33	2.27	2.20	2.13
20	5.87	4.46	3.86	3.51	3.29	3.13	3.01	2.91	2.84	2.77	2.68	2.57	2.46	2.41	2.35	2.29	2.22	2.16	2.09
21	5.83	4.42	3.82	3.48	3.25	3.09	2.97	2.87	2.80	2.73	2.64	2.53	2.42	2.37	2.31	2.25	2.18	2.11	2.04
22	5.79	4.38	3.78	3.44	3.22	3.05	2.93	2.84	2.76	2.70	2.60	2.50	2.39	2.33	2.27	2.21	2.14	2.08	2.00
23	5.75	4.35	3.75	3.41	3.18	3.02	2.90	2.81	2.73	2.67	2.57	2.47	2.36	2.30	2.24	2.18	2.11	2.04	1.97
24	5.72	4.32	3.72	3.38	3.15	2.99	2.87	2.78	2.70	2.64	2.54	2.44	2.33	2.27	2.21	2.15	2.08	2.01	1.94
25	5.69	4.29	3.69	3.35	3.13	2.97	2.85	2.75	2.68	2.61	2.51	2.41	2.30	2.24	2.18	2.12	2.05	1.98	1.91
26	5.66	4.27	3.67	3.33	3.10	2.94	2.82	2.73	2.65	2.59	2.49	2.39	2.28	2.22	2.16	2.09	2.03	1.95	1.88
27	5.63	4.24	3.65	3.31	3.08	2.92	2.80	2.71	2.63	2.57	2.47	2.36	2.25	2.19	2.13	2.07	2.00	1.93	1.85
28	5.61	4.22	3.63	3.29	3.06	2.90	2.78	2.69	2.61	2.55	2.45	2.34	2.23	2.17	2.11	2.05	1.98	1.91	1.83
29	5.59	4.20	3.61	3.27	3.04	2.88	2.76	2.67	2.59	2.53	2.43	2.32	2.21	2.15	2.09	2.03	1.96	1.89	1.81
30	5.57	4.18	3.59	3.25	3.03	2.87	2.75	2.65	2.57	2.51	2.41	2.31	2.20	2.14	2.07	2.01	1.94	1.87	1.79
40	5.42	4.05	3.46	3.13	2.90	2.74	2.62	2.53	2.45	2.39	2.29	2.18	2.07	2.01	1.94	1.88	1.80	1.72	1.64
60	5.29	3.93	3.34	3.01	2.79	2.63	2.51	2.41	2.33	2.27	2.17	2.06	1.94	1.88	1.82	1.74	1.67	1.58	1.48
120	5.15	3.80	3.23	2.89	2.67	2.52	2.39	2.30	2.22	2.16	2.05	1.94	1.82	1.76	1.69	1.61	1.53	1.43	1.31
∞	5.02	3.69	3.12	2.79	2.57	2.41	2.29	2.19	2.11	2.05	1.94	1.83	1.71	1.64	1.57	1.48	1.39	1.27	1.00

v_1, v_2 are upper, lower d.f. respectively.

Table II. (cont.) Upper 1% points of F-distribution.

$v_2 \backslash v_1$	1	2	3	4	5	6	7	8	9	10	12	15	20	24	30	40	60	120	∞
1	4052	4999·5	5403	5625	5764	5859	5928	5981	6022	6056	6106	6157	6209	6235	6261	6287	6313	6339	6366
2	98·50	99·00	99·17	99·25	99·30	99·33	99·36	99·37	99·39	99·40	99·42	99·43	99·45	99·46	99·47	99·47	99·48	99·49	99·50
3	34·12	30·82	29·46	28·71	28·24	27·91	27·67	27·49	27·35	27·23	27·05	26·87	26·69	26·60	26·50	26·41	26·32	26·22	26·13
4	21·20	18·00	16·69	15·98	15·52	15·21	14·98	14·80	14·66	14·55	14·37	14·20	14·02	13·93	13·84	13·75	13·65	13·56	13·46
5	16·26	13·27	12·06	11·39	10·97	10·67	10·46	10·29	10·16	10·05	9·89	9·72	9·55	9·47	9·38	9·29	9·20	9·11	9·02
6	13·75	10·92	9·78	9·15	8·75	8·47	8·26	8·10	7·98	7·87	7·72	7·56	7·40	7·31	7·23	7·14	7·06	6·97	6·88
7	12·25	9·55	8·45	7·85	7·46	7·19	6·99	6·84	6·72	6·62	6·47	6·31	6·16	6·07	5·99	5·91	5·82	5·74	5·65
8	11·26	8·65	7·59	7·01	6·63	6·37	6·18	6·03	5·91	5·81	5·67	5·52	5·36	5·28	5·20	5·12	5·03	4·95	4·86
9	10·56	8·02	6·99	6·42	6·06	5·80	5·61	5·47	5·35	5·26	5·11	4·96	4·81	4·73	4·65	4·57	4·48	4·40	4·31
10	10·04	7·56	6·55	5·99	5·64	5·39	5·20	5·06	4·94	4·85	4·71	4·56	4·41	4·33	4·25	4·17	4·08	4·00	3·91
11	9·65	7·21	6·22	5·67	5·32	5·07	4·89	4·74	4·63	4·54	4·40	4·25	4·10	4·02	3·94	3·86	3·78	3·69	3·60
12	9·33	6·93	5·95	5·41	5·06	4·82	4·64	4·50	4·39	4·30	4·16	4·01	3·86	3·78	3·70	3·62	3·54	3·45	3·36
13	9·07	6·70	5·74	5·21	4·86	4·62	4·44	4·30	4·19	4·10	3·96	3·82	3·66	3·59	3·51	3·43	3·34	3·25	3·17
14	8·86	6·51	5·56	5·04	4·69	4·46	4·28	4·14	4·03	3·94	3·80	3·66	3·51	3·43	3·35	3·27	3·18	3·09	3·00
15	8·68	6·36	5·42	4·89	4·56	4·32	4·14	4·00	3·89	3·80	3·67	3·52	3·37	3·29	3·21	3·13	3·05	2·96	2·87
16	8·53	6·23	5·29	4·77	4·44	4·20	4·03	3·89	3·78	3·69	3·55	3·41	3·26	3·18	3·10	3·02	2·93	2·84	2·75
17	8·40	6·11	5·18	4·67	4·34	4·10	3·93	3·79	3·68	3·59	3·46	3·31	3·16	3·08	3·00	2·92	2·83	2·75	2·65
18	8·29	6·01	5·09	4·58	4·25	4·01	3·84	3·71	3·60	3·51	3·37	3·23	3·08	3·00	2·92	2·84	2·75	2·66	2·57
19	8·18	5·93	5·01	4·50	4·17	3·94	3·77	3·63	3·52	3·43	3·30	3·15	3·00	2·92	2·84	2·76	2·67	2·58	2·49
20	8·10	5·85	4·94	4·43	4·10	3·87	3·70	3·56	3·46	3·37	3·23	3·09	2·94	2·86	2·78	2·69	2·61	2·52	2·42
21	8·02	5·78	4·87	4·37	4·04	3·81	3·64	3·51	3·40	3·31	3·17	3·03	2·88	2·80	2·72	2·64	2·55	2·46	2·36
22	7·95	5·72	4·82	4·31	3·99	3·76	3·59	3·45	3·35	3·26	3·12	2·98	2·83	2·75	2·67	2·58	2·50	2·40	2·31
23	7·88	5·66	4·76	4·26	3·94	3·71	3·54	3·41	3·30	3·21	3·07	2·93	2·78	2·70	2·62	2·54	2·45	2·35	2·26
24	7·82	5·61	4·72	4·22	3·90	3·67	3·50	3·36	3·26	3·17	3·03	2·89	2·74	2·66	2·58	2·49	2·40	2·31	2·21
25	7·77	5·57	4·68	4·18	3·85	3·63	3·46	3·32	3·22	3·13	2·99	2·85	2·70	2·62	2·54	2·45	2·36	2·27	2·17
26	7·72	5·53	4·64	4·14	3·82	3·59	3·42	3·29	3·18	3·09	2·96	2·81	2·66	2·58	2·50	2·42	2·33	2·23	2·13
27	7·68	5·49	4·60	4·11	3·78	3·56	3·39	3·26	3·15	3·06	2·93	2·78	2·63	2·55	2·47	2·38	2·29	2·20	2·10
28	7·64	5·45	4·57	4·07	3·75	3·53	3·36	3·23	3·12	3·03	2·90	2·75	2·60	2·52	2·44	2·35	2·26	2·17	2·06
29	7·60	5·42	4·54	4·04	3·73	3·50	3·33	3·20	3·09	3·00	2·87	2·73	2·57	2·49	2·41	2·33	2·23	2·14	2·03
30	7·56	5·39	4·51	4·02	3·70	3·47	3·30	3·17	3·07	2·98	2·84	2·70	2·55	2·47	2·39	2·30	2·21	2·11	2·01
40	7·31	5·18	4·31	3·83	3·51	3·29	3·12	2·99	2·89	2·80	2·66	2·52	2·37	2·29	2·20	2·11	2·02	1·92	1·80
60	7·08	4·98	4·13	3·65	3·34	3·12	2·95	2·82	2·72	2·63	2·50	2·35	2·20	2·12	2·03	1·94	1·84	1·73	1·60
120	6·85	4·79	3·95	3·48	3·17	2·96	2·79	2·66	2·56	2·47	2·34	2·19	2·03	1·95	1·86	1·76	1·66	1·53	1·38
∞	6·63	4·61	3·78	3·32	3·02	2·80	2·64	2·51	2·41	2·32	2·18	2·04	1·88	1·79	1·70	1·59	1·47	1·32	1·00

Table II. (cont.) Upper 0·1% points of F-distribution.

v_2 \ v_1	1	2	3	4	5	6	7	8	9	10	12	15	20	24	30	40	60	120	∞
1	4053*	5000*	5404*	5625*	5764*	5859*	5929*	5981*	6023*	6056*	6107*	6158*	6209*	6235*	6261*	6287*	6313*	6340*	6366*
2	998·5	999·0	999·2	999·2	999·3	999·3	999·4	999·4	999·4	999·4	999·4	999·4	999·4	999·5	999·5	999·5	999·5	999·5	999·5
3	167·0	148·5	141·1	137·1	134·6	132·8	131·6	130·6	129·9	129·2	128·3	127·4	126·4	125·9	125·4	125·0	124·5	124·0	123·5
4	74·14	61·25	58·18	53·44	51·71	50·53	49·66	49·00	48·47	48·05	47·41	46·76	46·10	45·77	45·43	45·09	44·75	44·40	44·05
5	47·18	37·12	33·20	31·09	29·75	28·84	28·16	27·64	27·24	26·92	26·42	25·91	25·39	25·14	24·87	24·60	24·33	24·06	23·79
6	35·61	27·00	23·70	21·92	20·81	20·03	19·48	19·03	18·69	18·41	17·99	17·58	17·12	16·89	16·67	16·44	16·21	15·99	15·75
7	29·25	21·69	18·77	17·19	16·21	15·52	15·02	14·63	14·33	14·08	13·71	13·32	12·93	12·73	12·53	12·33	12·12	11·91	11·70
8	25·42	18·49	15·83	14·39	13·49	12·86	12·40	12·04	11·77	11·54	11·19	10·84	10·48	10·30	10·11	9·92	9·73	9·53	9·33
9	22·86	16·39	13·90	12·56	11·71	11·13	10·70	10·37	10·11	9·89	9·57	9·24	8·90	8·72	8·55	8·37	8·19	8·00	7·81
10	21·04	14·91	12·55	11·28	10·48	9·92	9·52	9·20	8·96	8·75	8·45	8·13	7·80	7·64	7·47	7·30	7·12	6·94	6·74
11	19·69	13·81	11·56	10·36	9·58	9·05	8·66	8·35	8·12	7·92	7·63	7·32	7·01	6·85	6·68	6·52	6·35	6·17	6·00
12	18·64	12·97	10·80	9·63	8·89	8·38	8·00	7·71	7·48	7·29	7·00	6·71	6·40	6·25	6·09	5·93	5·76	5·59	5·42
13	17·81	12·31	10·21	9·07	8·35	7·86	7·49	7·21	6·98	6·80	6·52	6·23	5·93	5·78	5·63	5·47	5·30	5·14	4·97
14	17·14	11·78	9·73	8·62	7·92	7·43	7·08	6·80	6·58	6·40	6·13	5·85	5·56	5·41	5·25	5·10	4·94	4·77	4·60
15	16·59	11·34	9·34	8·25	7·57	7·09	6·74	6·47	6·26	6·08	5·81	5·54	5·25	5·10	4·95	4·80	4·64	4·47	4·31
16	16·12	10·97	9·00	7·94	7·27	6·81	6·46	6·19	5·98	5·81	5·55	5·27	4·99	4·85	4·70	4·54	4·39	4·23	4·06
17	15·72	10·66	8·73	7·68	7·02	6·56	6·22	5·96	5·75	5·58	5·32	5·05	4·78	4·63	4·48	4·33	4·18	4·02	3·85
18	15·38	10·39	8·49	7·46	6·81	6·35	6·02	5·78	5·56	5·39	5·13	4·87	4·59	4·45	4·30	4·15	4·00	3·84	3·67
19	15·08	10·16	8·28	7·26	6·62	6·18	5·85	5·59	5·39	5·22	4·97	4·70	4·43	4·29	4·14	3·99	3·84	3·68	3·51
20	14·82	9·95	8·10	7·10	6·46	6·02	5·69	5·44	5·24	5·08	4·82	4·56	4·29	4·15	4·00	3·86	3·70	3·54	3·38
21	14·59	9·77	7·94	6·95	6·32	5·88	5·56	5·31	5·11	4·95	4·70	4·44	4·17	4·03	3·88	3·74	3·58	3·42	3·26
22	14·38	9·61	7·80	6·81	6·19	5·76	5·44	5·19	4·99	4·83	4·58	4·33	4·08	3·92	3·78	3·63	3·48	3·32	3·15
23	14·19	9·47	7·67	6·69	6·08	5·65	5·33	5·09	4·89	4·73	4·48	4·23	3·96	3·82	3·68	3·53	3·38	3·22	3·05
24	14·03	9·34	7·55	6·59	5·98	5·55	5·23	4·99	4·80	4·64	4·39	4·14	3·87	3·74	3·59	3·45	3·29	3·14	2·97
25	13·88	9·22	7·45	6·49	5·88	5·46	5·15	4·91	4·71	4·56	4·31	4·06	3·79	3·66	3·52	3·37	3·22	3·06	2·89
26	13·74	9·12	7·36	6·41	5·80	5·38	5·07	4·83	4·64	4·48	4·24	3·99	3·72	3·59	3·44	3·30	3·15	2·99	2·82
27	13·61	9·02	7·27	6·33	5·73	5·31	5·00	4·76	4·57	4·41	4·17	3·92	3·68	3·52	3·38	3·23	3·08	2·92	2·75
28	13·50	8·93	7·19	6·25	5·66	5·24	4·93	4·69	4·50	4·35	4·11	3·86	3·60	3·46	3·32	3·18	3·02	2·86	2·69
29	13·39	8·85	7·12	6·19	5·59	5·18	4·87	4·64	4·45	4·29	4·05	3·80	3·54	3·41	3·27	3·12	2·97	2·81	2·64
30	13·29	8·77	7·05	6·12	5·53	5·12	4·82	4·58	4·39	4·24	4·00	3·75	3·40	3·36	3·22	3·07	2·92	2·76	2·59
40	12·61	8·25	6·60	5·70	5·13	4·73	4·44	4·21	4·02	3·87	3·64	3·40	3·15	3·01	2·87	2·73	2·67	2·41	2·23
60	11·97	7·76	6·17	5·31	4·76	4·37	4·09	3·87	3·69	3·54	3·31	3·08	2·83	2·69	2·65	2·41	2·25	2·08	1·89
120	11·38	7·32	5·79	4·95	4·42	4·01	3·77	3·55	3·38	3·24	3·02	2·78	2·63	2·40	2·26	2·11	1·95	1·76	1·54
∞	10·83	6·91	5·42	4·62	4·10	3·74	3·47	3·27	3·10	2·96	2·74	2·51	2·27	2·13	1·99	1·84	1·66	1·45	1·00

v_1, v_2 are upper, lower d.f. respectively. *Multiply these entries by 100.

Table III. Values of the χ^2 distribution exceeded with probability P.

d.f.	P 0·995	0·975	0·050	0·025	0·010	0·005	0·001
1	3.9×10^{-5}	9.8×10^{-4}	3·84	5·02	6·63	7·88	10·83
2	0·010	0·051	5·99	7·38	9·21	10·60	13·81
3	0·071	0·22	7·81	9·35	11·34	12·84	16·27
4	0·21	0·48	9·49	11·14	13·28	14·86	18·47
5	0·41	0·83	11·07	12·83	15·09	16·75	20·52
6	0·68	1·24	12·59	14·45	16·81	18·55	22·46
7	0·99	1·69	14·07	16·01	18·48	20·28	24·32
8	1·34	2·18	15·51	17·53	20·09	21·96	26·13
9	1·73	2·70	16·92	19·02	21·67	23·59	27·88
10	2·16	3·25	18·31	20·48	23·21	25·19	29·59
11	2·60	3·82	19·68	21·92	24·73	26·76	31·26
12	3·07	4·40	21·03	23·34	26·22	28·30	32·91
13	3·57	5·01	22·36	24·74	27·69	29·82	34·53
14	4·07	5·63	23·68	26·12	29·14	31·32	36·12
15	4·60	6·26	25·00	27·49	30·58	32·80	37·70
16	5·14	6·91	26·30	28·85	32·00	34·27	39·25
17	5·70	7·56	27·59	30·19	33·41	35·72	40·79
18	6·26	8·23	28·87	31·53	34·81	37·16	42·31
19	6·84	8·91	30·14	32·85	36·19	38·58	43·82
20	7·43	9·59	31·41	34·17	37·57	40·00	45·32
21	8·03	10·28	32·67	35·48	38·93	41·40	46·80
22	8·64	10·98	33·92	36·78	40·29	42·80	48·27
23	9·26	11·69	35·17	38·08	41·64	44·18	49·73
24	9·89	12·40	36·42	39·36	42·98	45·56	51·18
25	10·52	13·12	37·65	40·65	44·31	46·93	52·62
26	11·16	13·84	38·89	41·92	45·64	48·29	54·05
27	11·81	14·57	40·11	43·19	46·96	49·64	55·48
28	12·46	15·31	41·34	44·46	48·28	50·99	56·89
29	13·12	16·05	42·56	45·72	49·59	52·34	58·30
30	13·79	16·79	43·77	46·98	50·89	53·67	59·70
40	20·71	24·43	55·76	59·34	63·69	66·77	73·40
50	27·99	32·36	67·50	71·42	76·16	79·49	86·66
60	35·53	40·48	79·08	83·30	88·38	91·95	99·61
70	43·28	48·76	90·53	95·02	100·43	104·22	112·32
80	51·17	57·15	101·88	106·63	112·33	116·32	124·84
90	59·20	65·65	113·15	118·14	124·12	128·30	137·21
100	67·33	74·22	124·34	129·56	135·81	140·17	149·44

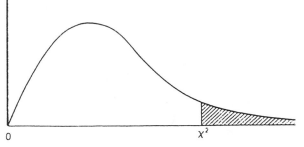

P is the shaded area

Table IV. Random digits.

19211	73336	80586	08681	28012	48881	34321	40156	03776	45150
94520	44451	07032	36561	41311	28421	95908	91280	74627	86359
70986	03817	40251	61310	25940	92411	34796	85416	00993	99487
65249	79677	03155	09232	96784	17126	50350	86469	41300	62715
82102	03098	01785	00653	39438	43660	02406	08404	24540	80000
91600	94635	35392	81737	01505	04967	91097	02011	26642	38540
20559	85361	20093	46000	83304	96624	62541	41722	79676	98970
53305	79544	99937	87727	32210	19438	58250	77265	02998	02973
57108	86498	14158	60697	41673	18087	46088	11238	82135	79035
08270	11929	92040	37390	71190	58952	98702	41638	95725	22798
90119	23206	75634	60053	90724	29080	69423	66815	11896	18607
45124	69607	17078	61747	15891	69904	79589	68137	19006	19045
83084	02589	37660	63882	99025	34831	92048	23671	68895	73795
06485	31035	93828	16159	05015	54800	76534	22974	13589	01801
61349	04538	89318	27693	02674	34368	24720	40682	20940	37392
14082	65020	49956	01336	41685	01758	49242	52122	01030	60378
82615	53477	58014	62229	72640	32042	73521	14166	45850	02372
50942	78633	16588	19275	62258	20773	67601	93065	69002	03985
76381	77455	81218	02520	22900	80130	61554	98901	26939	78732
05845	35063	85932	22410	31357	54790	39707	94348	11969	89755
78591	83750	46137	74989	39931	33068	35155	49486	28156	04556
31945	87960	04852	41411	63105	44116	95250	04046	59211	07270
08648	89822	04170	38365	23842	61917	57453	03495	61430	20154
32511	07999	18920	77045	44299	85057	51395	17457	24207	02730
79348	56194	58145	88645	84867	41594	28148	84985	89949	26689
61973	03660	32988	70689	17794	61340	58311	32569	23949	85626
92032	60127	34066	28149	22352	12907	53788	86648	57649	07887
74609	71072	63958	58336	67814	40598	12626	30754	75895	42194
98668	76074	25634	56913	88254	41647	05398	69463	49778	31382
85248	72078	58634	88678	21764	67940	45666	84664	35714	43081

Index